TENSOR METHODS FOR ENGINEERS

MATHEMATICS AND ITS APPLICATIONS
Series Editor: G. M. BELL,
Professor of Mathematics, King's College London, University of London

STATISTICS, OPERATIONAL RESEARCH AND COMPUTATIONAL MATHEMATICS Section
Editor: B. W. CONOLLY,
Emeritus Professor of Mathematics (Operational Research), Queen Mary College, University of London

Mathematics and its applications are now awe-inspiring in their scope, variety and depth. Not only is there rapid growth in pure mathematics and its applications to the traditional fields of the physical sciences, engineering and statistics, but new fields of application are emerging in biology, ecology and social organization. The user of mathematics must assimilate subtle new techniques and also learn to handle the great power of the computer efficiently and economically.

The need for clear, concise and authoritative texts is thus greater than ever and our series endeavours to supply this need. It aims to be comprehensive and yet flexible. Works surveying recent research will introduce new areas and up-to-date mathematical methods. Undergraduate texts on established topics will stimulate student interest by including applications relevant at the present day. The series will also include selected volumes of lecture notes which will enable certain important topics to be presented earlier than would otherwise be possible.

In all these ways it is hoped to render a valuable service to those who learn, teach, develop and use mathematics.

Mathematics and its Applications
Series Editor: G. M. BELL,
Professor of Mathematics, King's College London, University of London

Anderson, I.	**Combinatorial Designs: Construction Methods**
Artmann, B.	**Concept of Number: From Quaternions to Monads and Topological Fields**
Arczewski, K. & Pietrucha, J.	**Mathematical Modelling in Discrete Mechanical Systems**
Arczewski, K. and Pietrucha, J.	**Mathematical Modelling in Continuous Mechanical Systems**
Bainov, D.D. & Konstantinov, M.	**The Averaging Method and its Applications**
Baker, A.C. & Porteous, H.L.	**Linear Algebra and Differential Equations**
Balcerzyk, S. & Jösefiak, T.	**Commutative Rings**
Balcerzyk, S. & Jösefiak, T.	**Commutative Noetherian and Krull Rings**
Baldock, G.R. & Bridgeman, T.	**Mathematical Theory of Wave Motion**

Series continued at back of book

TENSOR METHODS FOR ENGINEERS

M. FARRASHKHALVAT B.Sc., M.Sc., Ph.D.
Department of Engineering
University of Bath

J. P. MILES B.A., M.A., Ph.D.
Department of Mathematics
University of Manchester Institute of Science and Technology

ELLIS HORWOOD
NEW YORK LONDON TORONTO SYDNEY TOKYO SINGAPORE

First published in 1990 by
ELLIS HORWOOD LIMITED
Market Cross House, Cooper Street,
Chichester, West Sussex, PO19 1EB, England

A division of
Simon & Schuster International Group
A Paramount Communications Company

© Ellis Horwood Limited, 1990

All rights reserved. No part of this publication may be
reproduced, stored in a retrieval system, or transmitted,
in any form, or by any means, electronic, mechanical,
photocopying, recording or otherwise, without the prior
permission, in writing, of the publisher

Printed and bound in Great Britain
by Bookcraft (Bath) Limited, Midsomer Norton, Avon

British Library Cataloguing in Publication Data

Farrashkhalvat, M.
Tensor methods for engineers.
1. Tensor analysis
I. Title II. Miles, J. P. III. Series
515.63
ISBN 0–13–904079–X

Library of Congress Cataloging-in-Publication Data

Farrashkhalvat, M.
Tensor methods for engineers / M. Farrashkhalvat,
J. P. Miles.
  p. cm. — (Ellis Horwood series in mathematics and its
applications)
ISBN 0–13–904079–X
1. Engineering mathematics. 2. Calculus of tensors.
I. Miles, J. P. II. Title. III. Series.
TA347.T4F37 1990
620′.001′51563–dc20   90–35578
                                                      CIP

Table of Contents

Preface vii

Chapter 1   Vectors   **1**
1.1   Basic concepts 1
1.2   Operations on vectors 2
1.3   Rectangular cartesian co-ordinate systems 6
1.4   Transformations between rectangular cartesian systems. . . . 19
1.5   General cartesian co-ordinate systems. 22
1.6   Vector fields. 27
1.7   Curvilinear co-ordinate systems 34

Chapter 2   Matrix algebra and determinants   **41**
2.1   Matrices 41
2.2   Determinants 49
2.3   Matrix inversion and diagonalization 53

Chapter 3   Tensor analysis   **63**
3.1   Introduction. 63
3.2   Cartesian tensors 64
3.3   Generalized tensors 86

Chapter 4   Covariant differentiation   **111**
4.1   Introduction. 111
4.2   Differentiation of base vectors and Christoffel symbols . . 112
4.3   Examples of Christoffel symbols. 114
4.4   Other properties of Christoffel symbols 116
4.5   Covariant derivatives of vector fields 118

4.6	Div and curl.	120
4.7	Covariant derivatives of higher-order tensors	122
4.8	Special cases.	124
4.9	Second derivatives.	125

Chapter 5 Orthogonal curvilinear co-ordinates and physical components   **127**
5.1	General theory.	127
5.2	Examples of orthogonal curvilinear co-ordinates	131
5.3	Application to particle kinematics	145

Chapter 6 Applications in continuum mechanics   **147**
6.1	Introduction.	147
6.2	Applications.	147
6.3	Transport equations	169

Chapter 7 Turbulence equations   **180**
7.1	Introduction.	180
7.2	Examples.	181

Appendix: Chain rules.   **204**

Answers to problems   **207**

References   **210**

Preface

It is our experience that engineers needing to solve standard mathematical equations of physics and engineering frequently find it difficult, when dealing with a particular boundary-value problem which naturally suggests a certain curvilinear co-ordinate system, to derive the appropriate form of the equations. This is particularly the case when the equations have the complexity of, for example, the Navier–Stokes equations for viscous fluid flow. One consequence of these difficulties is an overdependence on equations quoted in textbooks and research papers. Misprints (and also mistakes) inevitably occur from time to time and may be reproduced and perpetuated if engineers are unable to check their equations.

One standard approach to the derivation of the equations appropriate to a given curvilinear co-ordinate system is to apply physical principles (such as conservation of mass, momentum and energy in continuum mechanics) to an infinitesimal curvilinear element. We have no objection to such methods in principle. Indeed, the approach followed in this book in the examples given in Chapters 6 and 7 takes the form of the equations with respect to a set of rectangular cartesian co-ordinates as given, and this form may well have been established by considering an infinitesimal rectangular element (with sides parallel to the cartesian axes). However, once the rectangular cartesian form of the equations has been obtained, we believe that generalized tensor methods offer the most *systematic* way of deriving the form of the equations with respect to a given curvilinear co-ordinate system. This belief has motivated us to write this book.

To apply generalized tensor methods with confidence, the concept of a tensor must be thoroughly understood, and Chapter 3 contains the central theory. We feel that it is highly desirable to acquire a good grasp of cartesian tensors before embarking on the study of generalized tensors, and Chapter 3 has been structured accordingly. Our attention throughout is restricted to tensors in euclidean spaces, which we believe are of most interest to engineers. Unfortunately, this rules out the two-dimensional non-euclidean (curved) spaces which arise, for example, in the theory of shells in solid mechanics.

The first two chapters contain introductory material on vectors and matrices, Chapter 1 starting from a basic level but soon moving on to a discussion of vector fields, curvilinear co-ordinates, and covariant and contravariant components. The integral theorems of vector field theory have been excluded but may be found in standard references (Aris 1962, Spiegel 1959). The treatment of matrix theory in Chapter 2 is restricted to those aspects which are relevant to the remainder of the book.

Chapter 4 concerns the differentiation of vector and tensor fields and avoids any discussion of the question of parallelism since non-euclidean geometries have been excluded from consideration. In Chapter 5 we discuss orthogonal curvilinear co-ordinate systems and the concept of physical components of vectors and tensors, before giving a number of examples of such systems. We have not attempted to present any discussion of physical components in non-orthogonal co-ordinate systems.

Chapters 6 and 7 contain a considerable number of examples of equations from continuum mechanics which are transformed from a rectangular cartesian co-ordinate system to a curvilinear one. The transformations are carried out for the most part in much more elaborate detail than would normally be found in textbooks. The connections between different examples are often indicated, but our aim was not to write a book on continuum mechanics (see, for example, Hunter 1983). Neither was the book intended as a reference for continuum mechanics equations with respect to various co-ordinate systems, although it may in practice have a limited role as such. Rather, the objective is to make clear the method of tensor transformation, so that the reader will be able to apply it as required in a completely systematic way. We have considered only transformations from one spatial co-ordinate system to another. Time-dependent co-ordinate transformations have not been discussed. Thus we give no examples involving moving boundaries (such as fluid flow in the combustion chamber of an engine), in which the solution domain varies with time and time-dependent transformations would be appropriate. The final chapter contains applications of tensor transformations in the theory of turbulence.

The choice of notation in tensor theory is fraught with difficulties. Here we have used y_i or (y_1, y_2, y_3) consistently to refer to a set of rectangular cartesian co-ordinates (except for some of the examples in which it seemed clearer to use (x, y, z)), while x^k or $\{x^1, x^2, x^3\}$, with the superscript, is used for generalized co-ordinates. Cylindrical polar co-ordinates are denoted by $\{\rho, \phi, z\}$ to distinguish them from spherical polar co-ordinates $\{r, \theta, \phi\}$, except for applications involving the scalar density function, which is denoted by ρ. We have avoided the complexities of invariant notation for different types of tensor products, which did not seem essential here. The abbreviations LHS and RHS are used for left-hand side and right-hand side, respectively.

We have not thought it necessary to discuss the implications of the left- and right-handedness of general co-ordinate systems on the definitions of generalized tensors. Thus the concepts of *polar* and *axial* vectors, pseudoscalars, and also of *relative* tensors, are not mentioned here.

In conclusion, we would like to record our thanks to our colleague Dr. A. Kaye for a number of valuable suggestions and to Mrs. Sandra Kershaw for her considerable patience and excellent typing.

1

Vectors

1.1 BASIC CONCEPTS

Physical quantities which are completely specified in terms of a single number are called **scalar quantities**. The number will depend on a scale of measurement and may be positive, negative or zero but must not depend on any choice of spatial co-ordinate system. Examples of scalar quantities are the mass of a physical object, the pressure, density and temperature of a gas, and the work done by a force. Direct comparison of such quantities is possible if they have the same physical dimension. We can say that two scalars are equal if they have the same physical dimensions and if measurement of the quantities in the same system of units gives the same magnitude (absolute value) and sign.

Vector quantities, on the other hand, are physical quantities which require both a magnitude and a direction for their complete specification. It is convenient to represent a vector symbolically by an arrow (a 'directed line segment') whose length and direction are the same, respectively, as the magnitude (in a chosen system of units) and direction of the vector quantity. Bold-face letters **a**, **b**, etc., are typically used as symbolic representations for vectors, and the magnitude of **a** is conventionally denoted by $|\mathbf{a}|$. Examples of vector quantities are the displacement, velocity, acceleration and momentum of a particle. Such quantities are clearly dependent for their specification on the position and motion of the spatial frame of reference. Other vector quantities, such as the gravitational force acting on a particle, may be regarded as existing independently of any frame of reference. Two vectors **a** and **b** with the same physical dimensions and with magnitude measured in the same system of units are said to be equal ($\mathbf{a} = \mathbf{b}$) if they have the same magnitude and direction.

Other types of physical quantities besides scalars and vectors arise in physics and engineering. For example, theoretical analysis of the stresses acting on a material element in the interior of an elastic body loaded by surface forces led to the concept of the **stress tensor**, a physical quantity which has nine components (usually reducing to six independent components). In the theory of turbulence the time averaging of the Navier–Stokes equation gives rise to a double velocity correlation known as the **Reynolds stress**. These quantities are examples of **tensors**, and the theory of tensors and its applications will be discussed from Chapter 3 onwards. First, however, we present the basic mathematics of vector analysis. We place some emphasis on the two alternative ways of representing (and manipulating) vectors—firstly in terms of arrows (directed line segments), as mentioned above, and secondly in terms of components with respect to a given co-ordinate system. Our primary concern is to explain the techniques involved; mathematical rigour is a secondary matter.

1.2 OPERATIONS ON VECTORS

1.2.1 Addition and subtraction

If two forces $\mathbf{F}_1$ and $\mathbf{F}_2$ act simultaneously on a particle, their effect is equivalent to that of a single force, called the **resultant** of $\mathbf{F}_1$ and $\mathbf{F}_2$, acting on the particle. It is well known that the resultant may be obtained by using the so-called **parallelogram law** of the composition of forces. The mathematical sum $\mathbf{a} + \mathbf{b}$ of two vectors $\mathbf{a}$ and $\mathbf{b}$ is defined in a way which is consistent with the parallelogram law.

Let the vectors be represented by arrows each having a 'terminal point' (at the head of the arrow) and an 'initial point'. We assume that for present purposes we can move arrows parallel to themselves (thus preserving their magnitude and direction) with complete freedom in space if we so wish. (Thus either the terminal point or the initial point, but not both, can be chosen arbitrarily.) We place the arrows $\mathbf{a}$ and $\mathbf{b}$ so that the initial point of $\mathbf{b}$ coincides with the terminal point of $\mathbf{a}$. Then the sum $\mathbf{a} + \mathbf{b}$ is a vector represented by the arrow extending from the initial point of $\mathbf{a}$ to the terminal point of $\mathbf{b}$ (Fig. 1.1).

This definition of addition is evidently consistent with the definition of equality, i.e. if $\mathbf{a} = \mathbf{a}'$ and $\mathbf{b} = \mathbf{b}'$ then $\mathbf{a} + \mathbf{b} = \mathbf{a}' + \mathbf{b}'$. Since opposite sides of the parallelogram in Fig. 1.1 are equal in magnitude and have the same direction, vector addition must be **commutative**, i.e. $\mathbf{a} + \mathbf{b} = \mathbf{b} + \mathbf{a}$. Fig. 1.2 shows that the vector addition is also **associative**, i.e. $(\mathbf{a} + \mathbf{b}) + \mathbf{c} = \mathbf{a} + (\mathbf{b} + \mathbf{c})$. Thus there is no ambiguity in writing $\mathbf{a} + \mathbf{b} + \mathbf{c}$ without parentheses for the sum of three vectors.

If we have three vectors $\mathbf{a}, \mathbf{b}, \mathbf{c}$ with $\mathbf{a} + \mathbf{b} = \mathbf{c}$, then we can consistently write $\mathbf{a} = \mathbf{c} - \mathbf{b} = \mathbf{c} + (-\mathbf{b})$ provided that we define the vector $(-\mathbf{b})$ to be the vector with

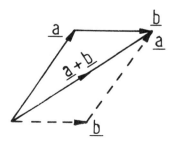

Fig. 1.1.

Sec. 1.2]	**Operations on vectors**	3

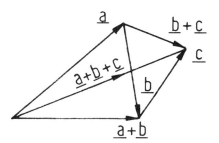

Fig. 1.2.

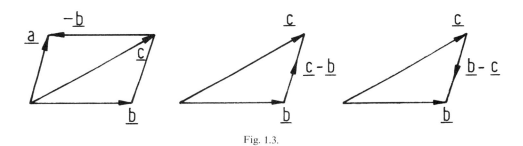

Fig. 1.3.

the same magnitude as **b** but with the opposite direction (Fig. 1.3). Note that we have defined the difference of two vectors $\mathbf{c} - \mathbf{b} = \mathbf{c} + (-\mathbf{b})$, and that $\mathbf{c} - \mathbf{b} = -(\mathbf{b} - \mathbf{c})$.

The magnitude $|\mathbf{a}|$ of vector **a** is a scalar quantity. From the definition of $-\mathbf{b}$, we have $|-\mathbf{b}| = |\mathbf{b}|$. We may also deduce from Fig. 1.1 the triangle inequality for vector addition

$$|\mathbf{a} + \mathbf{b}| \leqslant |\mathbf{a}| + |\mathbf{b}|,$$

based on the observation that the length of one side of a triangle cannot be greater than the sum of the lengths of the other two sides.

It is mathematically convenient to define a **zero vector 0** with zero magnitude and direction unspecified, satisfying $\mathbf{a} - \mathbf{a} = \mathbf{0}$, $\mathbf{a} + \mathbf{0} = \mathbf{a}$, etc., for any vector **a**.

Note that the diagrams and theory as presented so far are 'co-ordinate free', in that no mention has yet been made of co-ordinate systems of reference.

1.2.2 Multiplication of a vector by a scalar

A scalar here is taken to mean a real number. Multiplying a vector **a** by a scalar λ produces a vector $\lambda\mathbf{a}$ with magnitude equal to the product of the magnitude of **a** and the absolute value of λ; it has the same direction as **a** if $\lambda > 0$ but the opposite direction if $\lambda < 0$. It follows that

$$-\mathbf{a} = (-1)\mathbf{a},$$

with $-\mathbf{a}$ as defined above. More generally, we have

$$|\lambda \mathbf{a}| = |\lambda| |\mathbf{a}|$$

4 Vectors [Ch. 1

and, if λ and μ are any scalars, and **a** and **b** any vectors,

$$\lambda(\mu\mathbf{a}) = (\lambda\mu)\mathbf{a},$$

$$\lambda(\mathbf{a} + \mathbf{b}) = \lambda\mathbf{a} + \lambda\mathbf{b}$$

and

$$(\lambda + \mu)\mathbf{a} = \lambda\mathbf{a} + \mu\mathbf{a}.$$

These results can be made convincing by drawing appropriate simple diagrams.

A **unit vector** is one whose magnitude is unity (in some system of units).

Two non-zero vectors **a** and **b** are **parallel** (i.e. they have the same direction, though not necessarily the same **sense**) if

$$\mathbf{a} = \lambda\mathbf{b}$$

for some scalar λ.

1.2.3 The scalar product

One form of product of two vectors **a** and **b** may be generated by first moving one of the vectors parallel to itself so that the initial points of the arrows coincide. If θ is then the angle between the vectors, the **scalar product** $\mathbf{a} \cdot \mathbf{b}$ is defined to be the real number

$$\mathbf{a} \cdot \mathbf{b} = |\mathbf{a}||\mathbf{b}|\cos\theta.$$

This is equal to $|\mathbf{a}|(|\mathbf{b}|\cos\theta)$, i.e. the product of the magnitude of **a** and the 'projection' ($|\mathbf{b}|\cos\theta$) of **b** onto the direction of **a**. It is similarly equal to the product of the magnitude of **b** and the projection of **a** onto the direction of **b** (Fig. 1.4).

The scalar product is clearly **commutative**, i.e. $\mathbf{a} \cdot \mathbf{b} = \mathbf{b} \cdot \mathbf{a}$. Moreover, the operations of scalar product and vector addition satisfy the **distributive** rule; for any vectors **a, b, c**,

$$\mathbf{a} \cdot (\mathbf{b} + \mathbf{c}) = \mathbf{a} \cdot \mathbf{b} + \mathbf{a} \cdot \mathbf{c}.$$

This may be seen from Fig. 1.5 by considering the projections of the vectors **b**, **c** and **b** + **c** onto the direction of **a**.

If the scalar product of two non-zero vectors **a** and **b** is zero, the angle θ between their directions must be $\pi/2$, i.e. the vectors are perpendicular to each other. We may call them **orthogonal** vectors.

For a direct physical application, the work done by a force is most conveniently

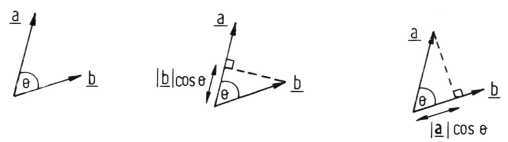

Fig. 1.4.

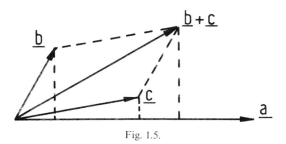

Fig. 1.5.

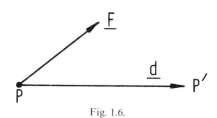

Fig. 1.6.

defined in terms of a scalar product. If a constant force **F** acts on a particle which is displaced in a straight line from a point P to a point P', the displacement may be considered as a vector **d** (an arrow with initial point P and terminal point P'), and the work done by **F** is the scalar product **F**·**d** (Fig. 1.6).

Note that, since the angle between a vector **a** and itself is zero, we have

$$\mathbf{a}\cdot\mathbf{a} = |\mathbf{a}|^2.$$

1.2.4 The vector product

Another way of combining two vectors **a** and **b** is the **vector product a × b**. This may formally be defined as follows. We can again move one of the vectors parallel to itself so that the initial points coincide. Suppose the angle between the vectors is θ (taking $0 \leqslant \theta \leqslant \pi$). Then

(1) **a** × **b** is a vector with magnitude given by

$$|\mathbf{a} \times \mathbf{b}| = |\mathbf{a}||\mathbf{b}|\sin\theta.$$

(This magnitude is equal to the area of the parallelogram 'spanned' by **a** and **b**.)
(2) The direction of **a** × **b** is perpendicular to both **a** and **b**, i.e. to the plane formed by **a** and **b** when their initial points coincide.
(3) To specify the sense of **a** × **b**, a 'right-handed' convention is used. The vector **a** × **b** points in the direction from which a rotation from the direction of **a** to the direction of **b** (through the angle θ) would appear anti-clockwise. Thus the vector product points in the direction of advance of a right-handed screw turned from the direction of **a** towards the direction of **b**, through the (smaller) angle $\theta \leqslant \pi$ (Fig. 1.7). (When $\theta = \pi$ the direction of rotation is immaterial since $\sin \pi = 0$ and hence the magnitude of **a** × **b** will be zero, i.e. **a** × **b** = **0** in this case.)

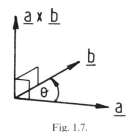

Fig. 1.7.

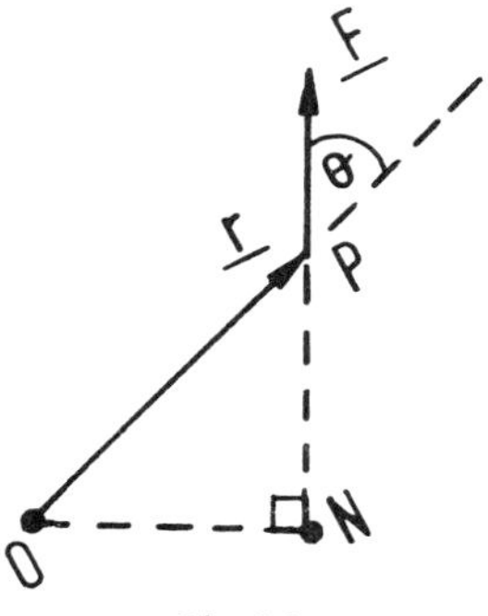

Fig. 1.8.

It follows that

$$\mathbf{b} \times \mathbf{a} = -\mathbf{a} \times \mathbf{b},$$

i.e. the vector product is not commutative. However, the distributive law

$$\mathbf{a} \times (\mathbf{b} + \mathbf{c}) = \mathbf{a} \times \mathbf{b} + \mathbf{a} \times \mathbf{c}$$

is satisfied for any vectors **a**, **b**, **c**. To demonstrate this result in co-ordinate-free terms is (perhaps surprisingly) a little complicated, and so we defer proof. Two non-zero vectors **a** and **b** are parallel ($\theta = 0$ or π) if $\mathbf{a} \times \mathbf{b} = \mathbf{0}$.

In mechanics the moment of a force **F** acting at a point P is most conveniently expressed as a vector product. The moment of **F** about a point O is equal to $\mathbf{r} \times \mathbf{F}$ where **r** is the **position vector** of P with respect to O, i.e. an arrow with initial point O and terminal point P. The magnitude of the moment vector $\mathbf{r} \times \mathbf{F}$ is equal to $|\mathbf{F}| \times \text{OP} \times \sin\theta = |\mathbf{F}| \times \text{ON}$, where ON is the shortest distance between O and the line of action of **F** (Fig. 1.8).

1.3 RECTANGULAR CARTESIAN CO-ORDINATE SYSTEMS

1.3.1 Base vectors and components of vectors

It is important to retain the co-ordinate-free picture of vector quantities (and also higher-order tensor quantities) but at the same time to acknowledge the indispensable service that co-ordinate systems can provide.

Sec. 1.3]   Rectangular cartesian co-ordinate systems   7

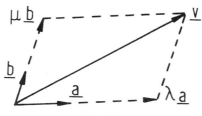

Fig. 1.9.

Let us first consider vectors in a two-dimensional space (a plane). Given two non-zero non-parallel vectors **a** and **b**, any vector **v** can be expressed as

$$\mathbf{v} = \lambda\mathbf{a} + \mu\mathbf{b},$$

where λ and μ are appropriate scalars; i.e. if all vectors have the same initial point, **v** can be represented as the diagonal of a parallelogram whose sides are parallel to **a** and **b** (Fig. 1.9).

Extending this result to three dimensions, suppose we are given three non-zero non-parallel vectors **a**, **b** and **c**, which have the property that, after having been moved parallel to themselves so that they have the same initial point, they do not lie in the same plane. Then any vector **v** can be represented as the long diagonal of a parallelepiped whose edges are parallel to **a**, **b** and **c**. We have

$$\mathbf{v} = \lambda\mathbf{a} + \mu\mathbf{b} + \nu\mathbf{c},$$

for some appropriate scalars λ, μ, ν, where $\lambda\mathbf{a}$, a scalar multiple λ of **a**, is a vector along one edge of the parallelepiped, etc. (Fig. 1.10). Thus, given the 'base vectors' **a**, **b** and **c**, an arbitrary vector **v** is completely (and uniquely) specified by a set of three real numbers $\{\lambda, \mu, \nu\}$. The set of vectors $\{\mathbf{a}, \mathbf{b}, \mathbf{c}\}$ is said to constitute a **basis** for the three-dimensional space.

A particularly important set of base vectors follows when a rectangular cartesian co-ordinate system has been established. In three dimensions this choice of co-ordinate system involves a choice of origin, i.e. a point O in space, and set of three mutually perpendicular axes Oy_1, Oy_2 and Oy_3 along which the three co-ordinates y_1, y_2 and y_3 of a point in space can be measured. Our choice of the symbols $\{y_1, y_2, y_3\}$ as co-ordinates, rather than $\{x, y, z\}$, conforms more naturally with the theoretical

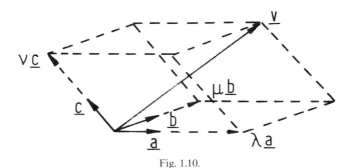

Fig. 1.10.

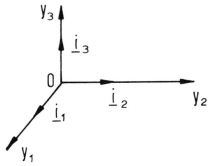

Fig. 1.11.

developments in this book. The corresponding base vectors are unit vectors denoted by $\mathbf{i}_1, \mathbf{i}_2, \mathbf{i}_3$ (rather than the commonly used $\mathbf{i}, \mathbf{j}, \mathbf{k}$) along the positive directions of the axes (Fig. 1.11). We usually assume, moreover, that the co-ordinate system is a 'right-handed' one, i.e. a right-handed screw rotated from the positive y_1-axis towards the positive y_2 axis through the angle $90°$ would advance in the positive y_3 direction. The following identities are then satisfied:

$$\left.\begin{array}{l}\mathbf{i}_1 \cdot \mathbf{i}_1 = \mathbf{i}_2 \cdot \mathbf{i}_2 = \mathbf{i}_3 \cdot \mathbf{i}_3 = 1, \\ \mathbf{i}_1 \cdot \mathbf{i}_2 = \mathbf{i}_2 \cdot \mathbf{i}_3 = \mathbf{i}_3 \cdot \mathbf{i}_1 = 0,\end{array}\right\} \quad (1.1)$$

$$\mathbf{i}_1 \times \mathbf{i}_2 = \mathbf{i}_3, \quad \mathbf{i}_2 \times \mathbf{i}_3 = \mathbf{i}_1, \quad \mathbf{i}_3 \times \mathbf{i}_1 = \mathbf{i}_2. \quad (1.2)$$

A basis satisfying these equations is called **orthonormal**.

An arbitrary vector **a** can be moved parallel to itself so that its initial point coincides with O, its terminal point then coinciding with the point P, say. The line OP then occupies the position of the long diagonal of a rectangular box with edges parallel to the co-ordinate axes (Fig. 1.12), and the vector **a** is the sum of three vectors in the directions of $\mathbf{i}_1, \mathbf{i}_2, \mathbf{i}_3$. If the point P has cartesian co-ordinates (a_1, a_2, a_3), these vectors are $a_1\mathbf{i}_1$, $a_2\mathbf{i}_2$ and $a_3\mathbf{i}_3$, and

$$\mathbf{a} = a_1\mathbf{i}_1 + a_2\mathbf{i}_2 + a_3\mathbf{i}_3. \quad (1.3)$$

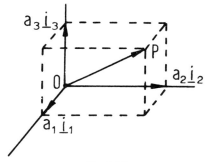

Fig. 1.12.

Sec. 1.3] Rectangular cartesian co-ordinate systems

The real numbers (a_1, a_2, a_3) are called the **components** of the vector **a** with respect to the given co-ordinate system. Use of the usual distance formula, based on Pythagoras's theorem, shows that

$$|\mathbf{a}| = \sqrt{a_1^2 + a_2^2 + a_3^2}.$$

Vectors are commonly expressed, using the component representation, as rows or columns, i.e.

$$\mathbf{a} = [a_1 \quad a_2 \quad a_3] \quad \text{or} \quad \mathbf{a} = \begin{bmatrix} a_1 \\ a_2 \\ a_3 \end{bmatrix}.$$

Vector additions can be carried out in this form, e.g. if $\mathbf{a} = a_1 \mathbf{i}_1 + a_2 \mathbf{i}_2 + a_3 \mathbf{i}_3$ and $\mathbf{b} = b_1 \mathbf{i}_1 + b_2 \mathbf{i}_2 + b_3 \mathbf{i}_3$ then

(i) $\mathbf{a} + \mathbf{b} = (a_1 + b_1)\mathbf{i}_1 + (a_2 + b_2)\mathbf{i}_2 + (a_3 + b_3)\mathbf{i}_3$ or
(ii) $[a_1 \quad a_2 \quad a_3] + [b_1 \quad b_2 \quad b_3] = [a_1 + b_1 \quad a_2 + b_2 \quad a_3 + b_3]$ or
(iii) $\begin{bmatrix} a_1 \\ a_2 \\ a_3 \end{bmatrix} + \begin{bmatrix} b_1 \\ b_2 \\ b_3 \end{bmatrix} = \begin{bmatrix} a_1 + b_1 \\ a_2 + b_2 \\ a_3 + b_3 \end{bmatrix}.$

Multiplication of vectors by scalars can also be carried out easily in component form:

$$\lambda \mathbf{a} = \lambda a_1 \mathbf{i}_1 + \lambda a_2 \mathbf{i}_2 + \lambda a_3 \mathbf{i}_3, \quad \text{etc.}$$

The components of **a** may be 'generically' denoted by the symbol a_i, where the index i, here appearing in the lower position as a suffix (rather than a superscript, as in a^i), may take any of the values 1, 2 or 3. However, this symbol is often used to represent the vector itself; thus we may speak of 'the vector a_i'.

Points P in space with co-ordinates (y_1, y_2, y_3) may also be specified by the **position vector**, an arrow with initial point O and terminal point P. We have already referred to this vector as **r**, and

$$\mathbf{r} = y_1 \mathbf{i}_1 + y_2 \mathbf{i}_2 + y_3 \mathbf{i}_3 = \sum_{j=1}^{3} y_j \mathbf{i}_j, \tag{1.4}$$

with the usual notation for summation.

Example

If $\mathbf{a} = \mathbf{i}_1 + 2\mathbf{i}_2 + 3\mathbf{i}_3$ and $\mathbf{b} = -\mathbf{i}_2 + 4\mathbf{i}_3$, calculate the magnitude of $2\mathbf{a} - \mathbf{b}$.

Solution

We have
$$2\mathbf{a} = 2\mathbf{i}_1 + 4\mathbf{i}_2 + 6\mathbf{i}_3$$
and
$$2\mathbf{a} - \mathbf{b} = 2\mathbf{i}_1 + 5\mathbf{i}_2 + 2\mathbf{i}_3.$$
Hence
$$|2\mathbf{a} - \mathbf{b}| = \sqrt{2^2 + 5^2 + 2^2} = \sqrt{33}.$$

1.3.2 Scalar products

We can make use of the distributive rule $\mathbf{a} \cdot (\mathbf{b} + \mathbf{c}) = \mathbf{a} \cdot \mathbf{b} + \mathbf{a} \cdot \mathbf{c}$, with vectors expressed in component form, to obtain another formula for the scalar product. For vectors $\mathbf{a} = a_1\mathbf{i}_1 + a_2\mathbf{i}_2 + a_3\mathbf{i}_3$ and $\mathbf{b} = b_1\mathbf{i}_1 + b_2\mathbf{i}_2 + b_3\mathbf{i}_3$, we have

$$\begin{aligned}
\mathbf{a} \cdot \mathbf{b} &= (a_1\mathbf{i}_1 + a_2\mathbf{i}_2 + a_3\mathbf{i}_3) \cdot (b_1\mathbf{i}_1 + b_2\mathbf{i}_2 + b_3\mathbf{i}_3) \\
&= a_1 b_1 \mathbf{i}_1 \cdot \mathbf{i}_1 + a_1 b_2 \mathbf{i}_1 \cdot \mathbf{i}_2 + a_1 b_3 \mathbf{i}_1 \cdot \mathbf{i}_3 \\
&\quad + a_2 b_1 \mathbf{i}_2 \cdot \mathbf{i}_1 + a_2 b_2 \mathbf{i}_2 \cdot \mathbf{i}_2 + a_2 b_3 \mathbf{i}_2 \cdot \mathbf{i}_3 \\
&\quad + a_3 b_1 \mathbf{i}_3 \cdot \mathbf{i}_1 + a_3 b_2 \mathbf{i}_3 \cdot \mathbf{i}_2 + a_3 b_3 \mathbf{i}_3 \cdot \mathbf{i}_3 \\
&= a_1 b_1 + a_2 b_2 + a_3 b_3 \qquad \text{in view of (1.1)} \\
&= \sum_{j=1}^{3} a_j b_j,
\end{aligned} \qquad (1.5)$$

again using the standard notation for summation.

For the case $\mathbf{a} = \mathbf{b}$ we obtain

$$\mathbf{a} \cdot \mathbf{a} = a_1^2 + a_2^2 + a_3^2 = \sum_{j=1}^{3} a_j a_j. \qquad (1.6)$$

Since $\mathbf{a} \cdot \mathbf{a} = |\mathbf{a}|^2$, we recover the formula

$$|\mathbf{a}| = \sqrt{a_1^2 + a_2^2 + a_3^2}.$$

Example

Calculate the cosine of the angle between the vectors $\mathbf{a}$ and $\mathbf{b}$ if $\mathbf{a} = \mathbf{i}_1 + \mathbf{i}_2 + 2\mathbf{i}_3$ and $\mathbf{b} = -\mathbf{i}_1 + \lambda\mathbf{i}_3$, and find the value of λ if $\mathbf{a}$ is perpendicular to $\mathbf{b}$.

Solution

$$\mathbf{a} \cdot \mathbf{b} = (1)(-1) + (1)(0) + (2)(\lambda) = 2\lambda - 1.$$

Also, $\mathbf{a} \cdot \mathbf{b} = |\mathbf{a}||\mathbf{b}|\cos\theta$, where θ is the angle between the vectors. Now

$$|\mathbf{a}| = \sqrt{1^2 + 1^2 + 2^2} = \sqrt{6}$$

and

$$|\mathbf{b}| = \sqrt{(-1)^2 + 0^2 + \lambda^2} = \sqrt{1 + \lambda^2}.$$

Hence

$$\cos\theta = \frac{\mathbf{a} \cdot \mathbf{b}}{|\mathbf{a}||\mathbf{b}|} = \frac{2\lambda - 1}{\sqrt{6}\sqrt{1 + \lambda^2}}.$$

If the vectors are perpendicular, then $\cos\theta = 0$, and it follows that $\lambda = \tfrac{1}{2}$.

If $\mathbf{a} = a_1\mathbf{i}_1 + a_2\mathbf{i}_2 + a_3\mathbf{i}_3$, we obtain

$$\left.\begin{aligned}
\mathbf{a} \cdot \mathbf{i}_1 &= a_1\mathbf{i}_1 \cdot \mathbf{i}_1 + a_2\mathbf{i}_2 \cdot \mathbf{i}_1 + a_3\mathbf{i}_3 \cdot \mathbf{i}_1 = a_1. \\
\text{Similarly,} \quad \mathbf{a} \cdot \mathbf{i}_2 &= a_2 \quad \text{and} \quad \mathbf{a} \cdot \mathbf{i}_3 = a_3.
\end{aligned}\right\} \qquad (1.7)$$

Thus the components of $\mathbf{a}$ are the projections of the vector $\mathbf{a}$ in the directions of the

Sec. 1.3]   **Rectangular cartesian co-ordinate systems**   11

co-ordinate axes, as is evident from Fig. 1.12. Equations (1.7) may be succinctly written as

$$\mathbf{a} \cdot \mathbf{i}_j = a_j, \tag{1.8}$$

where the suffix j can take any value from 1 to 3. The vector **a** can be written as

$$\mathbf{a} = (\mathbf{a} \cdot \mathbf{i}_1)\mathbf{i}_1 + (\mathbf{a} \cdot \mathbf{i}_2)\mathbf{i}_2 + (\mathbf{a} \cdot \mathbf{i}_3)\mathbf{i}_3 = \sum_{j=1}^{3} (\mathbf{a} \cdot \mathbf{i}_j)\mathbf{i}_j. \tag{1.9}$$

We can also write

$$\mathbf{a} = |\mathbf{a}|\hat{\mathbf{a}},$$

where $\hat{\mathbf{a}}$ is a unit vector in the direction **a**, i.e.

$$\hat{\mathbf{a}} = \frac{a_1}{|\mathbf{a}|}\mathbf{i}_1 + \frac{a_2}{|\mathbf{a}|}\mathbf{i}_2 + \frac{a_3}{|\mathbf{a}|}\mathbf{i}_3$$
$$= n_1\mathbf{i}_1 + n_2\mathbf{i}_2 + n_3\mathbf{i}_3, \tag{1.10}$$

where n_1, n_2, n_3 are the **direction cosines** of **a**; they are the cosines of the angles that the direction of **a** makes with each of the co-ordinate axes. Since $\hat{\mathbf{a}}$ is a unit vector,

$$n_1^2 + n_2^2 + n_3^2 = 1.$$

It is convenient here to introduce a set of nine quantities labelled δ_{ij}, where the suffixes i and j can take any value from 1 to 3, defined as follows:

$$\delta_{ij} = \begin{cases} 1, & \text{if } i \text{ and } j \text{ are the same,} \\ 0, & \text{otherwise,} \end{cases} \tag{1.11}$$

i.e. $\delta_{11} = \delta_{22} = \delta_{33} = 1$, $\delta_{12} = \delta_{21} = \delta_{23} = \delta_{32} = \delta_{31} = \delta_{13} = 0$. We can immediately economize on the six equations (1.1) required to write out the mutual scalar products of the base vectors by writing a single equation

$$\mathbf{i}_j \cdot \mathbf{i}_k = \delta_{jk}. \tag{1.12}$$

We call δ_{ij} the '**Kronecker delta**'. Any letters whatsoever may be used for the suffixes; δ_{pq}, for example, stands for the same set of nine quantities, and equation (1.12) may equally well be written as

$$\mathbf{i}_p \cdot \mathbf{i}_q = \delta_{pq}.$$

Suffixes appearing once on each side of an equation can take any value from 1 to 3, and are called **free suffixes**; further details follow in section 1.3.4.

1.3.3 Vector products and the alternating symbol

If $\mathbf{a} = a_1\mathbf{i}_1 + a_2\mathbf{i}_2 + a_3\mathbf{i}_3$, $\mathbf{b} = b_1\mathbf{i}_1 + b_2\mathbf{i}_2 + b_3\mathbf{i}_3$, and $\mathbf{a} \times \mathbf{b} = \mathbf{c}$, where $\mathbf{c} = c_1\mathbf{i}_1 + c_2\mathbf{i}_2 + c_3\mathbf{i}_3$, we know that **c** is perpendicular to both **a** and **b**, and hence

$$\mathbf{c} \cdot \mathbf{a} = c_1 a_1 + c_2 a_2 + c_3 a_3 = 0$$

and

$$\mathbf{c} \cdot \mathbf{b} = c_1 b_1 + c_2 b_2 + c_3 b_3 = 0.$$

Eliminating c_3 from these two equations gives

$$c_1(a_3b_1 - a_1b_3) = c_2(a_2b_3 - a_3b_2).$$

We may deduce that the ratios of the components of **c** satisfy

$$c_1:c_2:c_3 = (a_2b_3 - a_3b_2):(a_3b_1 - a_1b_3):(a_1b_2 - a_2b_1).$$

Thus

$$\left.\begin{aligned} c_1 &= \lambda(a_2b_3 - a_3b_2), \\ c_2 &= \lambda(a_3b_1 - a_1b_3), \\ c_3 &= \lambda(a_1b_2 - a_2b_1), \end{aligned}\right\} \tag{1.13}$$

for some scalar λ.

Moreover,

$$|\mathbf{c}|^2 = |\mathbf{a}|^2|\mathbf{b}|^2 \sin^2 \theta,$$

where θ is the angle between **a** and **b**; therefore

$$\begin{aligned} |\mathbf{c}|^2 &= |\mathbf{a}|^2|\mathbf{b}|^2(1 - \cos^2 \theta) = |\mathbf{a}|^2|\mathbf{b}|^2 - |\mathbf{a}||\mathbf{b}|^2 \cos^2 \theta \\ &= |\mathbf{a}|^2|\mathbf{b}|^2 - (\mathbf{a} \cdot \mathbf{b})^2 \\ &= (a_1^2 + a_2^2 + a_3^2)(b_1^2 + b_2^2 + b_3^2) - (a_1b_1 + a_2b_2 + a_3b_3)^2 \\ &= (a_2b_3 - a_3b_2)^2 + (a_3b_1 - a_1b_3)^2 + (a_1b_2 - a_2b_1)^2 \end{aligned}$$

after a little algebra. But $|c|^2 = c_1^2 + c_2^2 + c_3^2$, by (1.6). After substituting from (1.13), we deduce that $\lambda = \pm 1$. By considering particular cases, e.g. $\mathbf{a} = \mathbf{i}_1$ and $\mathbf{b} = \mathbf{i}_2$, it is clear that the solution $\lambda = +1$ corresponds to the right-handed convention for the vector product, whereas $\lambda = -1$ would correspond to an opposite (left-handed) convention.

Thus we have shown that

$$\mathbf{a} \times \mathbf{b} = (a_2b_3 - a_3b_2)\mathbf{i}_1 + (a_3b_1 - a_1b_3)\mathbf{i}_2 + (a_1b_2 - a_2b_1)\mathbf{i}_3 \tag{1.14}$$

in component form.

For arbitrary vectors **u**, **v**, **w** we can now deduce the distributive law

$$\begin{aligned} \mathbf{u} \times (\mathbf{v} + \mathbf{w}) &= [u_2(v_3 + w_3) - u_3(v_2 + w_2)]\mathbf{i}_1 + [u_3(v_1 + w_1) - u_1(v_3 + w_3)]\mathbf{i}_2 \\ &\quad + [u_1(v_2 + w_2) - u_2(v_1 + w_1)]\mathbf{i}_3 \\ &= [(u_2v_3 - u_3v_2)\mathbf{i}_1 + (u_3v_1 - u_1v_3)\mathbf{i}_2 + (u_1v_2 - u_2v_1)\mathbf{i}_3] \\ &\quad + [(u_2w_3 - u_3w_2)\mathbf{i}_1 + (u_3w_1 - u_1w_3)\mathbf{i}_2 + (u_1w_2 - u_2w_1)\mathbf{i}_3] \\ &= \mathbf{u} \times \mathbf{v} + \mathbf{u} \times \mathbf{w} \end{aligned}$$

in co-ordinate-free form.

Of course, if we had already known the distributive law, we could have deduced (1.14) from

$$\begin{aligned} \mathbf{a} \times \mathbf{b} &= (a_1\mathbf{i}_1 + a_2\mathbf{i}_2 + a_3\mathbf{i}_3) \times (b_1\mathbf{i}_1 + b_2\mathbf{i}_2 + b_3\mathbf{i}_3) \\ &= a_1b_1\mathbf{i}_1 \times \mathbf{i}_1 + a_1b_2\mathbf{i}_1 \times \mathbf{i}_2 + a_1b_3\mathbf{i}_1 \times \mathbf{i}_3 \\ &\quad + a_2b_1\mathbf{i}_2 \times \mathbf{i}_1 + a_2b_2\mathbf{i}_2 \times \mathbf{i}_2 + a_2b_3\mathbf{i}_2 \times \mathbf{i}_3 \end{aligned}$$

$$+ a_3b_1\mathbf{i}_3 \times \mathbf{i}_1 + a_3b_2\mathbf{i}_3 \times \mathbf{i}_2 + a_3b_3\mathbf{i}_3 \times \mathbf{i}_3$$
$$= (a_2b_3 - a_3b_2)\mathbf{i}_1 + (a_3b_1 - a_1b_3)\mathbf{i}_2 + (a_1b_2 - a_2b_1)\mathbf{i}_3$$

because of (1.2).

If $\mathbf{a} \times \mathbf{b} = \mathbf{c}$, we could write

$$c_i = a_j b_k - a_k b_j,$$

where the suffixes i, j, k must be a cyclic permutation of the numbers $1, 2, 3$. (The possible permutations are $\{1, 2, 3\}$, $\{2, 3, 1\}$ and $\{3, 1, 2\}$.) However, it turns out to be more satisfactory to introduce a set of 27 quantities labelled e_{ijk}, where the suffixes i, j, k can take any value from 1 to 3, defined as follows:

$$e_{ijk} = \begin{cases} 0, & \text{if any of } i, j, k \text{ are the same.} \\ +1, & \text{if } \{i, j, k\} \text{ is a cyclic permutation of } \{1, 2, 3\}, \\ -1, & \text{otherwise (i.e. if } \{i, j, k\} \text{ is an 'anti-cyclic' permutation of } \{1, 2, 3\}). \end{cases}$$
(1.15)

We then have

$$c_i = \sum_{j=1}^{3} \sum_{k=1}^{3} e_{ijk} a_j b_k. \qquad (1.16)$$

Note that only six of the 27 possible e_{ijk} values are non-zero. These are $e_{123} = e_{231} = e_{312} = 1$, $e_{132} = e_{321} = e_{213} = -1$. It follows that only two of the nine terms in the summation of (1.16) fail to vanish, e.g.

$$c_1 = \sum_{j=1}^{3} \sum_{k=1}^{3} e_{1jk} a_j b_k$$
$$= e_{123} a_2 b_3 + e_{132} a_3 b_2$$
$$= (+1)a_2 b_2 + (-1)a_3 b_2$$
$$= a_2 b_3 - a_3 b_2.$$

For the moment we call e_{ijk} the **'alternating symbol'**, although later we show that with respect to rectangular cartesian systems it has tensor properties and merits the name of **alternating tensor**. Note that

$$e_{jik} = -e_{ijk}, \quad \text{etc.,}$$

i.e. e_{ijk} is **skew symmetric** (or anti-symmetric) in its suffixes.

1.3.4 The summation convention and the permutation identity

We consider first the set of quantities

$$A_{jklm} = \sum_{i=1}^{3} e_{ijk} e_{ilm},$$

which arise in the next section. Here the suffixes j, k, l, m can take any value from 1 to 3, whereas summation takes place over the suffix i. Thus the quantities A_{jklm} are, explicitly,

$$e_{1jk} e_{1lm} + e_{2jk} e_{2lm} + e_{3jk} e_{3lm},$$

with $3^4 = 81$ possible choices of j, k, l and m. Of these choices, those in which j and k are identical or l and m are identical will give a zero contribution in every term. The result can only be non-zero if both

(i) j and k take different values and
(ii) l and m take different values.

Moreover, if j and k are different, then a non-zero result can follow only if *either* $l = j$ and $m = k$, *or* if $l = k$ and $m = j$. For example, if $j = 2$ and $k = 3$, the first term $e_{1jk}e_{1lm}$ is the only one of the three which can be non-zero (since $e_{223} = e_{323} = 0$), and this will be so only if $l = 2$ and $m = 3$, giving $e_{123}e_{123} = (+1)^2 = 1$, or if $l = 3$ and $m = 2$, giving $e_{123}e_{132} = (+1)(-1) = -1$. Thus we are led to the following result:

$$A_{jklm} = \begin{cases} +1, & \text{if } j \neq k, l = j \text{ and } m = k, \\ -1, & \text{if } j \neq k, l = k \text{ and } m = j, \\ 0, & \text{otherwise.} \end{cases}$$

Another way of writing these 81 possible quantities is

$$A_{jklm} = \delta_{jl}\delta_{km} - \delta_{jm}\delta_{kl}.$$

To see this, we can check the various possibilities as follows.

(a) If $j \neq k$, $l = j$ and $m = k$, then the RHS $= 1 - 0 = 1$.
(b) If $j \neq k$, $l = k$ and $m = j$, the RHS $= 0 - 1 = -1$.
(c) If $j \neq k$, any other choice of l and m apart from those in (a) and (b) will make the RHS $= 0 - 0 = 0$.
(d) If $j = k$, the only choice of l and m which does not give $0 - 0$ on the RHS is $l = m = j = k$, and this gives $1 - 1 = 0$ for the RHS.

Thus we have established the identity (for all possible choices of j, k, l, m)

$$\sum_{i=1}^{3} e_{ijk}e_{ilm} = \delta_{jl}\delta_{km} - \delta_{jm}\delta_{kl}.$$

This is called the **permutation identity**. It is not too difficult to memorize and proves useful in establishing standard identities.

This is a convenient point at which to introduce the **summation convention** (due to Einstein). According to this the summation sign in the above identity may safely be omitted and summation over the *repeated* suffix i automatically assumed. Thus we write the permutation identity simply as

$$e_{ijk}e_{ilm} = \delta_{jl}\delta_{km} - \delta_{jm}\delta_{kl}. \tag{1.17}$$

The following points are intended to serve as a guide to the summation convention.

(i) Each expression $e_{ijk}e_{ilm}, \delta_{jl}\delta_{km}$ and $\delta_{jm}\delta_{kl}$ featuring in our identity (equation) contains certain indices i, j, k, l, m, and one appears twice (here i).
(ii) Summation from 1 to 3 over suffixes appearing twice is automatically assumed; such suffixes may be called **repeated** (or **dummy**) **suffixes** and can be replaced by any other letter not used elsewhere in the expression, e.g.

$$e_{ijk}e_{ilm} = e_{pjk}e_{plm}.$$

(iii) Suffixes which appear once only in each expression are called **free suffixes**, and can take any value from 1 to 3; for consistency, the same free suffixes must appear once in each expression in the given equation. Terms separated by plus or minus signs, such as $\delta_{jl}\delta_{km}$ and $\delta_{jm}\delta_{kl}$ on the RHS of (1.17), are to be regarded as distinct expressions for present purposes, and the free suffixes j,k,l,m can legitimately appear once in each. In an equation such as (1.17) involving a number of expressions, the same free suffixes must be used consistently in each. Equations such as

$$a_i + b_j = 0$$

or

$$a_i = \delta_{ij} + b_k$$

are clearly not meaningful in the present context.

(iv) A dummy suffix must not be used more than twice in the same expression (e.g. e_{iii} is not admissible), but more than one dummy suffix may be used. For example, the expression $\delta_{ii}\delta_{jj}$ stands for

$$\sum_{i=1}^{3} \sum_{j=1}^{3} \delta_{ii}\delta_{jj} = \delta_{11}\delta_{11} + \delta_{22}\delta_{11} + \delta_{33}\delta_{11} + \delta_{11}\delta_{22} + \delta_{22}\delta_{22}$$
$$+ \delta_{33}\delta_{22} + \delta_{11}\delta_{33} + \delta_{22}\delta_{33} + \delta_{33}\delta_{33}.$$

The summation convention offers considerable economy to the sometimes rather complicated vector and tensor expressions which appear in this book and will henceforth be taken for granted (with reminders from time to time). If we want to use a suffix twice *without* summation, we shall state explicitly that the convention is not being observed. The symbol (n.s.) standing for 'no summation' will be used. When we come to consider more general co-ordinate systems the summation convention will need modification, as will be explained later, but for the time being we write the scalar product $\mathbf{a}\cdot\mathbf{b}$ compactly in component form, from (1.5), as

$$\mathbf{a}\cdot\mathbf{b} = a_i b_i, \tag{1.18}$$

and the vector product $\mathbf{a} \times \mathbf{b}$, from (1.16), as

$$(\mathbf{a} \times \mathbf{b})_i = e_{ijk} a_j b_k, \quad \text{or} \quad \mathbf{a} \times \mathbf{b} = e_{pqr} a_q b_r \mathbf{i}_p. \tag{1.19}$$

Note that in the last expression the indices p,q,r are all repeated, so that we have omitted three summation signs; without the summation convention, we have

$$\mathbf{a} \times \mathbf{b} = \sum_{p=1}^{3} \sum_{q=1}^{3} \sum_{r=1}^{3} e_{pqr} a_q b_r \mathbf{i}_p.$$

Further illustrations

(1) Consider the expression $a_k b_k c_j$. Since j is here a free suffix, the expression has three possible values depending on which number from 1 to 3 is assigned to j. On the other hand, k is a dummy suffix and is automatically summed from 1 to 3. For $j=1$, the expression is thus

$$a_1 b_1 c_1 + a_2 b_2 c_1 + a_3 b_3 c_1 = (a_1 b_1 + a_2 b_2 + a_3 b_3)c_1 = (\mathbf{a}\cdot\mathbf{b})c_1.$$

Similarly, for $j = 2$, it becomes $(\mathbf{a} \cdot \mathbf{b})c_2$ and, for $j = 3$, it becomes $(\mathbf{a} \cdot \mathbf{b})c_3$. Since $\mathbf{a} \cdot \mathbf{b}$ is a scalar, the expression is just the j component of the vector $(\mathbf{a} \cdot \mathbf{b})\mathbf{c}$.

(2) Consider the expression $a_i b_i c_i$. According to the summation convention this is unacceptable since the repeated suffix i occurs three times. If such an expression arises in the course of mathematical calculations, the rules of the summation convention must have been violated at some stage.

(3) The order of terms in an expression can be varied, e.g.

$$a_i T_{ij} b_j = b_j T_{ij} a_i,$$

since both expressions are equal to

$$a_1 T_{11} b_1 + a_1 T_{12} b_2 + a_1 T_{13} b_3 + a_2 T_{21} b_1 + a_2 T_{22} b_2 + a_2 T_{23} b_3$$
$$+ a_3 T_{31} b_1 + a_3 T_{32} b_2 + a_3 T_{33} b_3.$$

The resulting expression is independent of the order in which the summations over i and j are carried out.

(4) Substitution of one expression into another may require a change of symbol for the suffixes employed in order to avoid a suffix appearing more than twice. For example, if we have

$$u_i = A_{ij} v_j \quad \text{and} \quad v_i = B_{ij} w_j,$$

then to express u_i in terms of w_i we need to substitute for v_j and, from the second equation,

$$v_j = B_{jk} w_k,$$

writing j in place of i and k in place of j. Then we have

$$u_i = A_{ij} B_{jk} w_k = C_{ik} w_k, \quad \text{say},$$

where $C_{ik} = A_{ij} B_{jk}$. We could also write, with the same consistency,

$$u_i = C_{ij} w_j \quad \text{with} \quad C_{ij} = A_{ik} B_{kj}.$$

(5) Use of the summation convention leads to a convenient way of regarding the action of the Kronecker delta on components of vectors. If we consider the component a_j of a vector $\mathbf{a}$ and then the expression $\delta_{ij} a_j$, with automatic summation over j, the result is

$$\sum_{j=1}^{3} \delta_{ij} a_j = \delta_{i1} a_1 + \delta_{i2} a_2 + \delta_{i3} a_3.$$

If $i = 1$, only the first term is non-zero; if $i = 2$, only the second term; etc. Thus we deduce that the RHS is precisely equal to a_i for any choice of i. Hence

$$\delta_{ij} a_j = a_i,$$

and the action of δ_{ij} has been effectively to change the j of a_j into an i. The Kronecker delta is sometimes called the **substitution operator** for this reason. We also have results such as

$$\delta_{ij} \delta_{jk} = \delta_{ik},$$
$$\delta_{ij} \delta_{kl} a_j b_l = a_i b_k, \quad \text{etc.}$$

(6) Suppose that $S_{ij} = S_{ji}$, i.e. S_{ij} represents a set of nine quantities which is **symmetric** in its suffixes, while $T_{ij} = -T_{ji}$, i.e. T_{ij} is a set of nine quantities **skew symmetric** in its suffixes. Then

$$S_{ij}T_{ij} = 0,$$

where summation over both i and j is implied. This result may be obtained by first considering the identity

$$S_{ij}T_{ij} = S_{ji}T_{ji}$$

in which, by the properties of dummy suffixes, we may write j in place of i and vice versa. Hence

$$S_{ij}T_{ij} = S_{ji}T_{ji} = -S_{ij}T_{ij},$$

using the symmetry properties. Thus $2S_{ij}T_{ij} = 0$, and the result follows.

1.3.5 Triple products

Given three vectors $\mathbf{a}, \mathbf{b}$ and $\mathbf{c}$, the **scalar triple product** $(\mathbf{a} \times \mathbf{b})\cdot\mathbf{c}$ can be formed. If the vectors are moved parallel to themselves so that they have the same initial point (Fig. 1.13), the scalar triple product is simply equal to the volume V of the parallelepiped spanned by the three vectors with a positive or negative sign depending on whether the angle between $\mathbf{a} \times \mathbf{b}$ and $\mathbf{c}$ is acute or obtuse. To see this, note that

$$(\mathbf{a} \times \mathbf{b})\cdot\mathbf{c} = |\mathbf{a} \times \mathbf{b}|h,$$

where h is the projection of $\mathbf{c}$ in the direction of $\mathbf{a} \times \mathbf{b}$, i.e. the 'height' of the parallelepiped with the parallelogram spanned by $\mathbf{a}$ and $\mathbf{b}$ as base, while $|\mathbf{a} \times \mathbf{b}|$ is equal to the area of the parallelogram.

It follows that

$$(\mathbf{a} \times \mathbf{b})\cdot\mathbf{c} = (\mathbf{b} \times \mathbf{c})\cdot\mathbf{a} = (\mathbf{c} \times \mathbf{a})\cdot\mathbf{b}$$
$$= \mathbf{a}\cdot(\mathbf{b} \times \mathbf{c}) = \mathbf{b}\cdot(\mathbf{c} \times \mathbf{a}) = \mathbf{c}\cdot(\mathbf{a} \times \mathbf{b}).$$

Under anti-cyclic permutations, however, a negative sign is introduced, i.e.

$$(\mathbf{b} \times \mathbf{a})\cdot\mathbf{c} = -(\mathbf{a} \times \mathbf{b})\cdot\mathbf{c}, \quad \text{etc.}$$

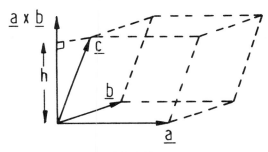

Fig. 1.13.

In the component form

$$(\mathbf{a} \times \mathbf{b})_i = e_{ijk} a_j b_k,$$

and so

$$\begin{aligned}
(\mathbf{a} \times \mathbf{b}) \cdot \mathbf{c} &= (\mathbf{a} \times \mathbf{b})_i c_i \\
&= e_{ijk} a_j b_k c_i \\
&= e_{jki} a_j b_k c_i
\end{aligned}$$

(since $e_{ijk} = e_{jki}$, with cyclic permutation of i, j, k)

$$= e_{ijk} a_i b_j c_k, \tag{1.20}$$

substituting i in the place of j, j for k, and k for i, as is permissible for dummy suffixes, and avoiding the appearance of any dummy suffix in more than two places. Evaluating the summation explicitly gives, since only six of the 27 components of e_{ijk} are non-zero,

$$(\mathbf{a} \times \mathbf{b}) \cdot \mathbf{c} = a_1 b_2 c_3 + a_2 b_3 c_1 + a_3 b_1 c_2 - a_1 b_3 c_2 - a_3 b_2 c_1 - a_2 b_1 c_3$$

in component form.

The scalar triple product vanishes when the vectors $\mathbf{a}, \mathbf{b}$ and $\mathbf{c}$ are coplanar (after moving them, if necessary, to have the same initial point), since the volume of the corresponding parallelepiped shrinks to zero.

The **vector triple product** $\mathbf{a} \times (\mathbf{b} \times \mathbf{c})$ can be formed. If $\mathbf{a} \times (\mathbf{b} \times \mathbf{c}) = \mathbf{d}$, we have, from (1.14),

$$\begin{aligned}
d_1 &= a_2 (\mathbf{b} \times \mathbf{c})_3 - a_3 (\mathbf{b} \times \mathbf{c})_2 \\
&= a_2 (b_1 c_2 - b_2 c_1) - a_3 (b_3 c_1 - b_1 c_3) \\
&= (a_2 c_2 b_1 + a_3 c_3 b_1) - (a_2 b_2 c_1 + a_3 b_3 c_1) \\
&= (a_2 c_2 + a_3 c_3) b_1 - (a_2 b_2 + a_3 b_3) c_1 \\
&= (a_1 c_1 + a_2 c_2 + a_3 c_3) b_1 - (a_1 b_1 + a_2 b_2 + a_3 b_3) c_1,
\end{aligned}$$

adding in the zero contribution $(a_1 c_1 b_1 - a_1 b_1 c_1)$, i.e.

$$d_1 = (\mathbf{a} \cdot \mathbf{c}) b_1 - (\mathbf{a} \cdot \mathbf{b}) c_1.$$

Similarly, we can show that

$$d_2 = (\mathbf{a} \cdot \mathbf{c}) b_2 - (\mathbf{a} \cdot \mathbf{b}) c_2$$

and

$$d_3 = (\mathbf{a} \cdot \mathbf{c}) b_3 - (\mathbf{a} \cdot \mathbf{b}) c_3.$$

These equations can be summarized in a single vector equation

$$\mathbf{d} = (\mathbf{a} \cdot \mathbf{c}) \mathbf{b} - (\mathbf{a} \cdot \mathbf{b}) \mathbf{c},$$

giving the identity

$$\mathbf{a} \times (\mathbf{b} \times \mathbf{c}) = (\mathbf{a} \cdot \mathbf{c}) \mathbf{b} - (\mathbf{a} \cdot \mathbf{b}) \mathbf{c}. \tag{1.21}$$

This proof may be contrasted with one which makes use of the suffix method together with the permutation identity. We know that $\mathbf{a} \times (\mathbf{b} \times \mathbf{c})$ is a vector with the i component given by

$$\begin{aligned}
[\mathbf{a} \times (\mathbf{b} \times \mathbf{c})]_i &= e_{ijk} a_j (\mathbf{b} \times \mathbf{c})_k \\
&= e_{ijk} a_j (e_{klm} b_l c_m).
\end{aligned}$$

Note that we must introduce new dummy suffixes as required, to avoid the same repeated suffix appearing more than twice. The resulting expression involves four dummy suffixes (with four corresponding summations understood) and the free suffix i. We can rearrange the order of terms to give

$$e_{ijk}e_{klm}a_jb_lc_m$$
$$= e_{kij}e_{klm}a_jb_lc_m$$

permuting the suffixes cyclically in the first alternating symbol

$$= (\delta_{il}\delta_{jm} - \delta_{im}\delta_{jl})a_jb_lc_m$$

by the permutation identity (1.17)

$$= \delta_{il}\delta_{jm}a_jb_lc_m - \delta_{im}\delta_{jl}a_jb_lc_m$$
$$= a_mb_ic_m - a_jb_jc_i$$

by the substitution property of the Kronecker deltas

$$= (a_mc_m)b_i - (a_jb_j)c_i$$
$$= (\mathbf{a}\cdot\mathbf{c})b_i - (\mathbf{a}\cdot\mathbf{b})c_i$$
$$= [(\mathbf{a}\cdot\mathbf{c})\mathbf{b} - (\mathbf{a}\cdot\mathbf{b})\mathbf{c}]_i.$$

The identity of each i component implies the identity of the corresponding vectors

$$\mathbf{a} \times (\mathbf{b} \times \mathbf{c}) = (\mathbf{a}\cdot\mathbf{c})\mathbf{b} - (\mathbf{a}\cdot\mathbf{b})\mathbf{c}.$$

Similarly, we can show that

$$(\mathbf{a} \times \mathbf{b}) \times \mathbf{c} = (\mathbf{a}\cdot\mathbf{c})\mathbf{b} - (\mathbf{b}\cdot\mathbf{c})\mathbf{a}.$$

Note that the implicit introduction of a co-ordinate system, with corresponding components for the vectors $\mathbf{a}, \mathbf{b}$ and $\mathbf{c}$, proves to be a useful device for establishing the co-ordinate-free identity (1.21).

Problems

1.3.1 Prove the following identities:

(i) $e_{ijk}e_{ljk} = 2\delta_{il}$;
(ii) $e_{ijk}e_{ijk} = 6$.

1.3.2 Use the permutation identity to prove the following identities:

(i) $(\mathbf{a} \times \mathbf{b})\cdot(\mathbf{c} \times \mathbf{d}) = (\mathbf{a}\cdot\mathbf{c})(\mathbf{b}\cdot\mathbf{d}) - (\mathbf{a}\cdot\mathbf{d})(\mathbf{b}\cdot\mathbf{c})$;
(ii) $(\mathbf{a} \times \mathbf{b}) \times (\mathbf{c} \times \mathbf{d}) = [\mathbf{a}\cdot(\mathbf{c} \times \mathbf{d})]\mathbf{b} - [\mathbf{b}\cdot(\mathbf{c} \times \mathbf{d})]\mathbf{a} = [(\mathbf{a} \times \mathbf{b})\cdot\mathbf{d}]\mathbf{c} - [(\mathbf{a} \times \mathbf{b})\cdot\mathbf{d}]\mathbf{c}$.

1.4 TRANSFORMATIONS BETWEEN RECTANGULAR CARTESIAN SYSTEMS

1.4.1 Change of basis

A vector $\mathbf{u}$ has components u_i relative to a rectangular cartesian system $\mathbf{i}_1, \mathbf{i}_2, \mathbf{i}_3$, but an entirely different set of components, which we label u_i', with respect to a differently orientated rectangular system given by unit vectors $\mathbf{i}_1', \mathbf{i}_2', \mathbf{i}_3'$. The relation between

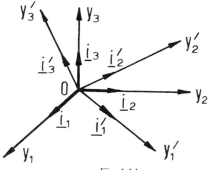

Fig. 1.14.

the two sets of components of **u** can be derived once we have established a way of specifying the orientation of the 'new' base vectors $\{\mathbf{i}_1', \mathbf{i}_2', \mathbf{i}_3'\}$ with respect to the 'old' ones $\{\mathbf{i}_1, \mathbf{i}_2, \mathbf{i}_3\}$. We can suppose that both systems have the same origin O, with axes $Oy_1 y_2 y_3$ and $Oy_1' y_2' y_3'$ (Fig. 1.14). A convenient way of fixing the position of the new axes in terms of the old ones is to specify the direction cosines of the new unit vectors with respect to the axes $Oy_1 y_2 y_3$.

We let l_{jp} denote the cosine of the angle between the jth axis of the new system and the pth axis of the old system. Then the direction cosines of the vector $\mathbf{i}_j'$ with respect to the old system are l_{j1}, l_{j2} and l_{j3}; it follows that

$$\mathbf{i}_j' = l_{j1}\mathbf{i}_1 + l_{j2}\mathbf{i}_2 + l_{j3}\mathbf{i}_3 \qquad (j=1,2,3)$$

as in (1.10).

Using the summation convention (with summation over p),

$$\mathbf{i}_j' = l_{jp}\mathbf{i}_p. \tag{1.22}$$

Clearly

$$l_{jp} = \mathbf{i}_j' \cdot \mathbf{i}_p.$$

(Note that in this equation both j and p are free suffixes.)

Similar expressions exist for the vectors $\mathbf{i}_1, \mathbf{i}_2, \mathbf{i}_3$ of the old system in terms of the new system. We have

$$\mathbf{i}_p = l_{1p}\mathbf{i}_1' + l_{2p}\mathbf{i}_2' + l_{3p}\mathbf{i}_3',$$

since the coefficients l_{jp} still represent the cosines of the angle between $\mathbf{i}_p$ and $\mathbf{i}_j'$. Thus

$$\mathbf{i}_p = l_{jp}\mathbf{i}_j', \tag{1.23}$$

where j here is a dummy suffix and p a free suffix.

Combining the two equations (1.22) and (1.23) gives

$$\mathbf{i}_j' = l_{jp}\mathbf{i}_p = l_{jp}(l_{kp}\mathbf{i}_k'). \tag{1.24}$$

Here we have to introduce a new suffix k for the dummy suffix in (1.23), since j is already in use as a free suffix.

We can write (1.24) as

$$l_{jp}l_{kp}\mathbf{i}_k' = \mathbf{i}_j' = \delta_{jk}\mathbf{i}_k',$$

and identifying coefficients of $\mathbf{i}_k'$ on both sides gives

$$l_{jp}l_{kp} = \delta_{jk}, \tag{1.25}$$

where j and k are now free suffixes and p a dummy. This gives nine equations, of which only the following six are essentially different:

$$\left. \begin{array}{l} l_{1p}l_{1p} = l_{11}{}^2 + l_{12}{}^2 + l_{13}{}^2 = \delta_{11} = 1, \\ l_{2p}l_{2p} = l_{21}{}^2 + l_{22}{}^2 + l_{23}{}^2 = \delta_{22} = 1, \\ l_{3p}l_{3p} = l_{31}{}^2 + l_{32}{}^2 + l_{33}{}^2 = \delta_{33} = 1, \\ l_{1p}l_{2p} = l_{11}l_{21} + l_{12}l_{22} + l_{13}l_{23} = \delta_{12} = 0, \\ l_{2p}l_{3p} = l_{21}l_{31} + l_{22}l_{32} + l_{23}l_{33} = \delta_{23} = 0, \\ l_{3p}l_{1p} = l_{31}l_{11} + l_{32}l_{12} + l_{33}l_{13} = \delta_{31} = 0. \end{array} \right\} \tag{1.26}$$

An alternative combination of (1.22) and (1.23) gives

$$\mathbf{i}_p = l_{jp}\mathbf{i}_j' = l_{jp}l_{jq}\mathbf{i}_q = \delta_{pq}\mathbf{i}_q,$$

and thus

$$l_{jp}l_{jq} = \delta_{pq}, \tag{1.27}$$

which gives the six equations

$$\left. \begin{array}{l} l_{11}{}^2 + l_{21}{}^2 + l_{31}{}^2 = 1, \\ l_{12}{}^2 + l_{22}{}^2 + l_{32}{}^2 = 1, \\ l_{13}{}^2 + l_{23}{}^2 + l_{33}{}^2 = 1, \\ l_{11}l_{12} + l_{21}l_{22} + l_{31}l_{32} = 0, \\ l_{12}l_{13} + l_{22}l_{23} + l_{32}l_{33} = 0, \\ l_{13}l_{11} + l_{23}l_{21} + l_{33}l_{31} = 0. \end{array} \right\} \tag{1.28}$$

Equations (1.28) are not independent of (1.26) and can in principle be derived from them, as will be shown later. The nine quantities l_{jp} are thus essentially related by the six equations (1.26).

As an example we consider a transformation in which the unit vectors $\mathbf{i}_1, \mathbf{i}_2$ are rotated through an angle θ about the direction $\mathbf{i}_3$ to reach the new directions $\mathbf{i}_1'$ and $\mathbf{i}_2'$, $\mathbf{i}_3'$ being unchanged in direction (Fig. 1.15). Then equations (1.22) become

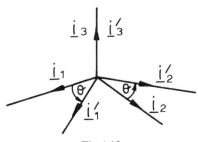

Fig. 1.15.

$$\left.\begin{array}{l}\mathbf{i}_1' = \cos\theta\,\mathbf{i}_1 + \sin\theta\,\mathbf{i}_2 + 0\mathbf{i}_3, \\ \mathbf{i}_2' = -\sin\theta\,\mathbf{i}_1 + \cos\theta\,\mathbf{i}_2 + 0\mathbf{i}_3, \\ \mathbf{i}_3' = 0\mathbf{i}_1 + 0\mathbf{i}_2 + 1\mathbf{i}_3,\end{array}\right\} \quad (1.29)$$

the coefficients satisfying (1.26) and (1.28).

1.4.2 Transformation of vector components

A vector **u** has the decomposition

$$\mathbf{u} = u_1\mathbf{i}_1 + u_2\mathbf{i}_2 + u_3\mathbf{i}_3 = u_p\mathbf{i}_p \quad (1.30)$$

into components with respect to the base vectors $\mathbf{i}_1, \mathbf{i}_2, \mathbf{i}_3$, and also the decomposition

$$\mathbf{u} = u_1'\mathbf{i}_1' + u_2'\mathbf{i}_2' + u_3'\mathbf{i}_3' = u_j'\mathbf{i}_j'$$

with respect to the 'new' system. Since the systems are related by (1.22), we have

$$\mathbf{u} = u_j'\mathbf{i}_j' = u_j'l_{jp}\mathbf{i}_p.$$

Identifying components here with those in (1.30), we have

$$u_p = u_j'l_{jp}, \quad (1.31)$$

which expresses the 'old' components in terms of the 'new' ones.

Similarly, using (1.23),

$$\mathbf{u} = u_p\mathbf{i}_p = u_p(l_{jp}\mathbf{i}_j') = u_j'\mathbf{i}_j',$$

and hence

$$u_j' = u_p l_{jp} = l_{jp} u_p. \quad (1.32)$$

These transformation equations are easily remembered and are fundamental to the definitions of tensors we give in Chapter 3.

1.5 GENERAL CARTESIAN CO-ORDINATE SYSTEMS

1.5.1 Non-rectangular systems and reciprocal bases

Components of vectors are commonly given, if required, with respect to some chosen rectangular cartesian system of axes. There are, however, certain problems in which the use of non-rectangular systems is convenient; these arise, for example, in the theory of crystal lattices. We have already seen in section 1.3.1 that an arbitrary vector in three dimensions may be expressed in terms of three given non-coplanar vectors. Such 'base' vectors do not have to be unit vectors, even if they are mutually orthogonal.

Suppose that in three-dimensional space we have a set of base vectors, which we write as $\mathbf{e}_1, \mathbf{e}_2, \mathbf{e}_3$, in general neither unit vectors nor mutually orthogonal. According to section 1.3.5, the condition for them to be non-coplanar (when moved parallel to themselves, if necessary, to have the same initial point) is that their scalar triple product $(\mathbf{e}_1 \times \mathbf{e}_2)\cdot\mathbf{e}_3$ should be non-zero. The set of vectors is then said to constitute a **basis** for the three-dimensional 'space' of vectors, and an arbitrary vector **v** can be expressed uniquely in the form

$$\mathbf{v} = v^1 \mathbf{e}_1 + v^2 \mathbf{e}_2 + v^3 \mathbf{e}_3 = v^i \mathbf{e}_i, \qquad (1.33)$$

where v^1, v^2 and v^3 can be regarded as the components of $\mathbf{v}$ on the given basis $\{\mathbf{e}_1, \mathbf{e}_2, \mathbf{e}_3\}$, and $v^1 \mathbf{e}_1, v^2 \mathbf{e}_2, v^3 \mathbf{e}_3$ are the vector components of $\mathbf{v}$.

The use of superscripts here rather than suffixes in representing the components of $\mathbf{v}$ is an important feature of the analysis of vectors and tensors with respect to general bases and co-ordinate systems. It is important, of course, not to confuse v^2 and v^3 with powers of v^1. (The square and cube of v^1 would always be written here as $(v^1)^2$ and $(v^1)^3$, respectively.)

Associated with the given basis is another important basis. This is a set of three vectors, written as $\mathbf{e}^1, \mathbf{e}^2, \mathbf{e}^3$, which satisfy the conditions

$$\mathbf{e}^i \cdot \mathbf{e}_j = \begin{cases} 0, & \text{if } i \neq j, \\ 1, & \text{if } i = j, \end{cases} \qquad (1.34)$$

where i and j take any values from 1 to 3. Thus the vector $\mathbf{e}^1$ is orthogonal to both $\mathbf{e}_2$ and $\mathbf{e}_3$, and its magnitude is such as to make $\mathbf{e}^1 \cdot \mathbf{e}_1 = 1$, i.e.

$$|\mathbf{e}^1| = \frac{1}{|\mathbf{e}_1| \cos(\mathbf{e}^1, \mathbf{e}_1)},$$

where $\cos(\mathbf{e}^1, \mathbf{e}_1)$ denotes the cosine of the angle between $\mathbf{e}^1$ and $\mathbf{e}_1$. Since $|\mathbf{e}^1|$ and $|\mathbf{e}_1|$ are both positive, $\cos(\mathbf{e}^1, \mathbf{e}_1)$ must be positive, which means that the angle between $\mathbf{e}^1$ and $\mathbf{e}_1$ must be acute. Similar considerations apply to the pairs $\mathbf{e}^2$ and $\mathbf{e}_2$, and $\mathbf{e}^3$ and $\mathbf{e}_3$.

Typical configurations are illustrated in Fig. 1.16.

The basis $\{\mathbf{e}^1, \mathbf{e}^2, \mathbf{e}^3\}$ is called a **reciprocal basis**, and the relationship between the bases is naturally symmetric, according to (1.34).

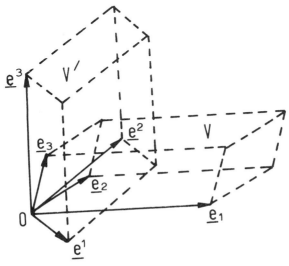

Fig. 1.16.

Now the vector product $\mathbf{e}_2 \times \mathbf{e}_3$ is perpendicular to both $\mathbf{e}_2$ and $\mathbf{e}_3$ and so must be parallel to $\mathbf{e}^1$. Hence $\mathbf{e}^1 = \lambda(\mathbf{e}_2 \times \mathbf{e}_3)$ for some scalar λ. By taking scalar products of both sides with $\mathbf{e}_1$,

$$\mathbf{e}^1 \cdot \mathbf{e}_1 = 1 = \lambda \mathbf{e}_1 \cdot (\mathbf{e}_2 \times \mathbf{e}_3).$$

We may recall that the scalar triple product $\mathbf{e}_1 \cdot (\mathbf{e}_2 \times \mathbf{e}_3)$ is equal to the volume V of the parallelepiped spanned by the base vectors $\mathbf{e}_1, \mathbf{e}_2$ and $\mathbf{e}_3$. (If the scalar triple product turns out to be negative, we can reverse the sign by interchanging the suffixes on $\mathbf{e}_2$ and $\mathbf{e}_3$, i.e. relabelling $\mathbf{e}_2$ as $\mathbf{e}_3$ and vice versa.) Hence $\lambda = 1/V$, and it follows that

$$\mathbf{e}^1 = \frac{\mathbf{e}_2 \times \mathbf{e}_3}{V}.$$

Similarly,

$$\mathbf{e}^2 = \frac{\mathbf{e}_3 \times \mathbf{e}_1}{V} \quad \text{and} \quad \mathbf{e}^3 = \frac{\mathbf{e}_1 \times \mathbf{e}_2}{V}.$$

(1.35)

We also have reciprocal relations

$$\mathbf{e}_1 = \frac{\mathbf{e}^2 \times \mathbf{e}^3}{V'}, \quad \mathbf{e}_2 = \frac{\mathbf{e}^3 \times \mathbf{e}^1}{V'} \quad \text{and} \quad \mathbf{e}_3 = \frac{\mathbf{e}^1 \times \mathbf{e}^2}{V'},$$

where V' is the volume of the parallelepiped spanned by $\mathbf{e}^1, \mathbf{e}^2$ and $\mathbf{e}^3$.

In the particular case of a right-handed rectangular cartesian set of unit vectors $\{\mathbf{i}_1, \mathbf{i}_2, \mathbf{i}_3\}$, the reciprocal basis consists of exactly the same vectors, in view of (1.12). For a mutually orthogonal set of non-unit base vectors $\{\mathbf{e}_1, \mathbf{e}_2, \mathbf{e}_3\}$, we have $\mathbf{e}^1 = \mathbf{e}_1/|\mathbf{e}_1|^2$, $\mathbf{e}^2 = \mathbf{e}_2/|\mathbf{e}_2|^2$, since, for example, we then have

$$\mathbf{e}^1 \cdot \mathbf{e}_1 = \frac{\mathbf{e}_1 \cdot \mathbf{e}_1}{|\mathbf{e}_1|^2} = \frac{|\mathbf{e}_1|^2}{|\mathbf{e}_1|^2} = 1.$$

Thus $|\mathbf{e}^1| = 1/|\mathbf{e}_1|$, etc. This situation arises when considering orthogonal curvilinear co-ordinate systems.

In general, it follows from the results

$$\mathbf{e}_1 = \frac{\mathbf{e}^2 \times \mathbf{e}^3}{V'} \quad \text{and} \quad \mathbf{e}^3 = \frac{\mathbf{e}_1 \times \mathbf{e}_2}{V}$$

that

$$\mathbf{e}_1 = \frac{\mathbf{e}^2 \times (\mathbf{e}_1 \times \mathbf{e}_2)}{V'V} = \frac{(\mathbf{e}^2 \cdot \mathbf{e}_2)\mathbf{e}_1 - (\mathbf{e}^2 \cdot \mathbf{e}_1)\mathbf{e}_2}{V'V} \quad \text{by (1.21)}$$

$$= \frac{1\mathbf{e}_1 - 0\mathbf{e}_2}{V'V} \quad \text{by (1.34)}$$

$$= \frac{\mathbf{e}_1}{V'V}.$$

Hence $V'V = 1$, i.e. $V' = 1/V$. (1.36)

It turns out here to be convenient to write (1.34) as

$$\mathbf{e}^i \cdot \mathbf{e}_j = \delta^i_j,$$

1.5.2 Covariant and contravariant vector components

Taking $\{\mathbf{e}_1, \mathbf{e}_2, \mathbf{e}_3\}$ as basis, and $\{\mathbf{e}^1, \mathbf{e}^2, \mathbf{e}^3\}$ as the reciprocal basis, we can express an arbitrary vector $\mathbf{v}$ in terms of component vectors parallel to $\mathbf{e}_1, \mathbf{e}_2, \mathbf{e}_3$, as in (1.33). But $\mathbf{v}$ can also be expressed in terms of component vectors parallel to $\mathbf{e}^1, \mathbf{e}^2$ and $\mathbf{e}^3$ as

$$\mathbf{v} = v_1 \mathbf{e}^1 + v_2 \mathbf{e}^2 + v_3 \mathbf{e}^3 = v_i \mathbf{e}^i. \tag{1.37}$$

We call the components $\{v_1, v_2, v_3\}$ the **covariant** components of $\mathbf{v}$, i.e. the components associated with the reciprocal basis, whereas the components $\{v^1, v^2, v^3\}$ are called the **contravariant** components of $\mathbf{v}$. The distinction between covariant and contravariant components is a rather arbitrary one here, inasmuch as the reciprocal relation between the bases is symmetric. In fact, the distinction appears to lie merely in whether the attached index is a subscript or a superscript. When we come to considering curvilinear co-ordinate systems, however, we shall find that the distinction occurs quite naturally.

By taking scalar products of both sides of (1.33) with one of the reciprocal base vectors $\mathbf{e}^j$, we obtain

$$\mathbf{v} \cdot \mathbf{e}^j = v^i \mathbf{e}_i \cdot \mathbf{e}^j = v^i \delta_i^j = v^j,$$

since δ_i^j still has the substitution operator properties.

Similarly,

$$\mathbf{v} \cdot \mathbf{e}_j = v_i \mathbf{e}^i \cdot \mathbf{e}_j = v_i \delta_j^i = v_j.$$

Thus the covariant and contravariant components of $\mathbf{v}$ are given by the scalar products of $\mathbf{v}$ with the corresponding base vectors:

$$v_i = \mathbf{v} \cdot \mathbf{e}_i \quad \text{and} \quad v^i = \mathbf{v} \cdot \mathbf{e}^i. \tag{1.38}$$

We can conveniently illustrate the components of $\mathbf{v}$ in a two-dimensional space, with base vectors $\{\mathbf{e}_1, \mathbf{e}_2\}$ and reciprocal vectors $\{\mathbf{e}^1, \mathbf{e}^2\}$ (Fig. 1.17).

An orthonormal basis is self-reciprocal, and so there was no need to distinguish between covariant and contravariant components in the previous sections. Moreover, with respect to such a basis the components of a vector have the same physical

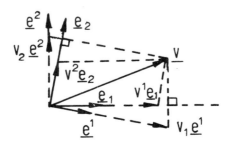

Fig. 1.17.

dimensions as the vector itself. This is not necessarily true of the covariant and contravariant components of a vector with respect to a general basis, as we shall discuss later. In section 1.4, suffixes alone are required; in this section both subscripts and superscripts are required, often in the same expression.

Important remark

It is important to note that, *when we are dealing with general co-ordinate systems,* the use of superscripts leads to a further constraint on the use of the summation convention. When repeated indices occur in an expression, as, for example, i in (1.33) and (1.37), it is vital to ensure that one is a subscript and the other a superscript, i.e. they must occur at different levels. It is also necessary to ensure that any free index in an equation must occur in each expression at the same level, as either a subscript or a superscript.

Thus, expressions such as $a^i b^i$ or $u_j v_j$ will not (or should not) arise in general systems, but $a^i b_i$ or $a_i b^i$ will. Equations such as

$$a^i = b_i u^k v_k$$

are invalid because, although the k occurs correctly, the i indices appear on different levels.

Previously stated rules about letters not occurring more than twice still apply.

1.5.3 Relations between general cartesian and rectangular cartesian systems

It may be convenient to express the base vectors $\mathbf{e}_i$ of a general cartesian system in terms of base vectors $\mathbf{i}_j$ of a given rectangular cartesian system. Suppose that

$$\mathbf{e}_1 = L_1^{\cdot 1}\mathbf{i}_1 + L_1^{\cdot 2}\mathbf{i}_2 + L_1^{\cdot 3}\mathbf{i}_3,$$
$$\mathbf{e}_2 = L_2^{\cdot 1}\mathbf{i}_1 + L_2^{\cdot 2}\mathbf{i}_2 + L_2^{\cdot 3}\mathbf{i}_3,$$

and
$$\mathbf{e}_3 = L_3^{\cdot 1}\mathbf{i}_1 + L_3^{\cdot 2}\mathbf{i}_2 + L_3^{\cdot 3}\mathbf{i}_3,$$

for scalars $L_1^{\cdot 1}, \ldots, L_3^{\cdot 3}$. We have written the indices of the coefficients $L_i^{\cdot j}$ in this way (with one subscript and one superscript) so that the index form of these equations

$$\mathbf{e}_i = L_i^{\cdot j}\mathbf{i}_j \qquad (1.39)$$

is consistent with the use of the summation convention for general systems as explained above, i.e. the repeated index j occurs once on the top and once on the bottom. The dot above the index i implies that i is to be regarded as the first index (and j the second). The significance of this should become apparent in section 2.1.3.

Suppose also that the reciprocal base vectors are given by

$$\mathbf{e}^i = M_{\cdot j}^{i}\mathbf{i}_j, \qquad (1.40)$$

for some coefficients $M_{\cdot j}^{i}$. Here we must admit to a lack of consistency in the use of the repeated suffix j, which occurs twice in the lower position in this equation. In fact, it is sometimes difficult to be completely consistent when dealing with a mixture of rectangular cartesian and general systems. We could restore consistency by raising the final j to a superscript, in the knowledge that an orthonormal basis is self-reciprocal, and thus $\mathbf{i}^j = \mathbf{i}_j$. It is not so convenient to raise the first j, as we shall see. Again, the dot below the index i serves to place the two indices in order (i being first and j second).

If we write

$$\mathbf{e}^i = M^i_{\cdot j}\mathbf{i}^j,$$

we find

$$\mathbf{e}^i \cdot \mathbf{e}_j = \delta^i_j = (M^i_{\cdot k}\mathbf{i}^k) \cdot (L_j^{\cdot l}\mathbf{i}_l)$$
$$= M^i_{\cdot k}L_j^{\cdot l}(\mathbf{i}^k \cdot \mathbf{i}_l) = M^i_{\cdot k}L_j^{\cdot l}\delta^k_l,$$

i.e.

$$M^i_{\cdot k}L_j^{\cdot k} = \delta^i_j, \tag{1.41}$$

which gives a set of nine equations relating the coefficients $M^i_{\cdot j}$ and $L_i^{\cdot j}$.

Suppose the rectangular system and the general cartesian system have the same origin. Then the position vector $\mathbf{r}$ of a point in space may be expressed as

$$\mathbf{r} = x^i\mathbf{e}_i = y_j\mathbf{i}_j,$$

where (y_1, y_2, y_3) are rectangular cartesian co-ordinates and (x^1, x^2, x^3) are general *cartesian* co-ordinates. The relationship between the co-ordinates may be obtained as follows:

$$x^i\mathbf{e}_i = x^iL_i^{\cdot j}\mathbf{i}_j = y_j\mathbf{i}_j.$$

Hence, equating coefficients, $y_j = L_i^{\cdot j}x^i$, where, to restore consistency, it would be necessary to raise the suffix of y_j to a superscript. Let us do this. Then we have

$$y^j = L_i^{\cdot j}x^i. \tag{1.42}$$

It is now possible to obtain an expression for x^i in terms of y^j by multiplying both sides by $M^k_{\cdot j}$, implying summation over j, and using (1.41). We get

$$M^k_{\cdot j}y^j = M^k_{\cdot j}L_i^{\cdot j}x^i = \delta^k_i x^i = x^k,$$

i.e.

$$x^i = M^i_{\cdot j}y^j. \tag{1.43}$$

1.6 VECTOR FIELDS

1.6.1 Scalar fields

Examples of physical scalar quantities given above were pressure, density and temperature in a gas. For many purposes it is adequate to regard a given volume of gas as homogeneous, with pressure, density and temperature taking certain averaged values. In some situations it is important to take account of inhomogeneities and the variation in these quantities throughout the gas. Then the quantities must be regarded as functions of position, associated either with a point of space (the 'Eulerian' description) or with a particle of the gas (the 'Lagrangian' description). In either case the quantities are called scalar **fields**.

Suppose Φ represents a scalar quantity associated with position in space, and suppose that we set up a rectangular cartesian set of axes $Oy_1y_2y_3$ in space. Then Φ is a function of the co-ordinates (y_1, y_2, y_3). We may wish to consider the variation in Φ throughout space. A good deal of information is provided by the partial derivatives of Φ with respect to the co-ordinates, which we can assemble into a row vector. This is the **gradient vector**, written grad Φ, i.e.

$$\text{grad } \Phi = \left[\frac{\partial \Phi}{\partial y_1} \quad \frac{\partial \Phi}{\partial y_2} \quad \frac{\partial \Phi}{\partial y_3} \right]. \tag{1.44}$$

In suffix notation, the i component of grad Φ is $(\text{grad } \Phi)_i = \partial \Phi / \partial y_i$.

It is not immediately apparent that it is meaningful to regard grad Φ as a physical vector, with a corresponding arrow having magnitude and direction. However, if we consider the usual formula for the first-order increment $\delta \Phi$ in Φ when y_1, y_2 and y_3 are increased by the small quantities $\delta y_1, \delta y_2$ and δy_3, i.e.

$$\delta \Phi = \frac{\partial \Phi}{\partial y_1} \delta y_1 + \frac{\partial \Phi}{\partial y_2} \delta y_2 + \frac{\partial \Phi}{\partial y_3} \delta y_3, \tag{1.45}$$

which according to the summation convention can be expressed as

$$\delta \Phi = \frac{\partial \Phi}{\partial y_i} \delta y_i,$$

then, by (1.18), we can regard the RHS as the scalar product of grad Φ and the increment vector $\delta \mathbf{r}$ of the position vector. The resulting scalar product is the scalar quantity $\delta \Phi$, and this gives a little more credibility to the vector status of grad Φ. (A more rigorous demonstration will be given in section 3.2.1.)

Suppose that the increment in the position vector corresponds to a short step of magnitude δs from a point P to a neighbouring point P' (Fig. 1.18). Then

$$\delta \mathbf{r} = \mathbf{n} \, \delta s,$$

where $\mathbf{n}$ is a unit vector in the direction PP', and the components (n_1, n_2, n_3) of $\mathbf{n}$ are direction cosines which specify the direction of the step from P to P'. The increment

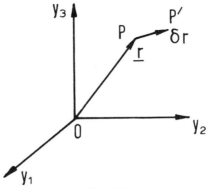

Fig. 1.18.

formula becomes

$$\partial \Phi = \frac{\partial \Phi}{\partial y_i} n_i \, \delta s,$$

and in the limit as $\delta s \to 0$ we obtain the exact formula

$$\frac{d\Phi}{ds} = \frac{\partial \Phi}{\partial y_i} n_i$$

for the rate of change in Φ with respect to distance in the direction PP′, the rate of change being evaluated at P. This is called the **directional derivative** of Φ at P, and we have

$$\frac{d\Phi}{ds} = \text{grad } \Phi \cdot \mathbf{n}. \tag{1.46}$$

For varying $\mathbf{n}$, the RHS of (1.46) has its greatest value when $\mathbf{n}$ is parallel to the direction of grad Φ, so that the angle between $\mathbf{n}$ and grad Φ is zero. Thus at the given point P the direction of grad Φ gives the direction in which Φ increases most rapidly in space.

We can visualize our three-dimensional space filled with contour surfaces, on each of which Φ takes a constant value. If the step $\delta \mathbf{r}$ at P is tangential to the contour surface passing through P, then the first-order increment in Φ will be zero, and so will be the directional derivative. This will happen only if the vector grad Φ is orthogonal to each direction $\mathbf{n}$ tangential to the contour surface. The implication is that the direction of grad Φ at P must be the same as the direction of the normal to the contour surface passing through P. The magnitude of grad Φ is equal to the greatest rate of increase in Φ in space at P.

We see that grad Φ is a vector associated with each point P of space; such a vector function is called a **vector field**.

Example 1

Suppose

$$\Phi = y_1^2 + y_2^2 + y_3^2 = |\mathbf{r}|^2.$$

then

$$\frac{\partial \Phi}{\partial y_1} = 2y_1, \qquad \frac{\partial \Phi}{\partial y_2} = 2y_2, \qquad \frac{\partial \Phi}{\partial y_3} = 2y_3,$$

and

$$\text{grad } \Phi = [2y_1 \quad 2y_2 \quad 2y_3]$$
$$= 2[y_1 \quad y_2 \quad y_3] = 2\mathbf{r}.$$

Example 2

If $\Phi = \frac{1}{2} a_{ij} y_i y_j$, which according to the summation convention is equivalent to

$$\Phi = \tfrac{1}{2} a_{11} y_1^2 + \tfrac{1}{2} a_{22} y_2^2 + \tfrac{1}{2} a_{33} y_3^2 + \tfrac{1}{2}(a_{12} + a_{21}) y_1 y_2$$
$$+ \tfrac{1}{2}(a_{23} + a_{32}) y_2 y_3 + \tfrac{1}{2}(a_{31} + a_{13}) y_1 y_3,$$

where the a_{ij} are a set of nine constants a_{11}, a_{12}, etc., then

$$\frac{\partial \Phi}{\partial y_k} = \tfrac{1}{2} a_{ij} \frac{\partial y_i}{\partial y_k} y_j + \tfrac{1}{2} a_{ij} y_i \frac{\partial y_j}{\partial y_k}$$

using the product rule for differentiation; hence

$$\frac{\partial \Phi}{\partial y_k} = \tfrac{1}{2} a_{ij} \delta_{ik} y_j + \tfrac{1}{2} a_{ij} y_i \delta_{jk},$$

since the y_i are an independent set of co-ordinates. Thus

$$\frac{\partial \Phi}{\partial y_k} = \tfrac{1}{2} a_{kj} y_j + \tfrac{1}{2} a_{ik} y_i$$
$$= \tfrac{1}{2} a_{kj} y_j + \tfrac{1}{2} a_{jk} y_j.$$

This may be written

$$\frac{\partial \Phi}{\partial y_i} = \tfrac{1}{2} a_{ij} y_j + \tfrac{1}{2} a_{ji} y_j = \tfrac{1}{2}(a_{ij} + a_{ji}) y_j.$$

If the quantities a_{ij} are symmetric, with $a_{ij} = a_{ji}$, then

$$\frac{\partial \Phi}{\partial y_i} = a_{ij} y_j.$$

Note that this example is a generalization of Example 1, in which we may put $a_{ij} = 2\delta_{ij}$, giving

$$\Phi = \delta_{ij} y_i y_j = y_i y_i = y_1^2 + y_2^2 + y_3^2,$$

$$(\text{grad }\Phi)_i = \frac{\partial \Phi}{\partial y_i} = 2\delta_{ij} y_j = 2y_i,$$

and hence

$$\text{grad } \Phi = 2\mathbf{r}.$$

For a further example, suppose that $a_{11} = 3$, $a_{22} = 2$, $a_{33} = 1$, $a_{12} = a_{21} = 1$, $a_{23} = a_{32} = a_{31} = a_{13} = 0$. Then the equation $\Phi = 1$ represents the ellipsoidal surface

$$\tfrac{1}{2}(3y_1^2 + 2y_1 y_2 + 2y_2^2 + y_3^2) = 1,$$

and we know that the normal to the surface at a point (y_1, y_2, y_3) is the direction of grad Φ, which is parallel to the vector with components $a_{ij} y_j$, i.e. $(3y_1 + y_2, y_1 + 2y_2, y_3)$.

Example 3

Suppose the temperature T in a solid at the point with cartesian co-ordinates (y_1, y_2, y_3) is given by the expression

$$T = y_1 - y_2^2 + y_1 y_2 y_3 + 273.$$

At the point with co-ordinates $(1, -2, 3)$, the value of T is 264. We can calculate the direction in which T increases (with respect to distance) most rapidly at P. This

Sec. 1.6]   **Vector fields**   31

direction is parallel to grad T, where

$$\text{grad } T = (1 + y_2 y_3, -2y_2 + y_1 y_3, y_1 y_2),$$

and this is equal to $(-5, 7, 3)$ at P. The corresponding rate of increase (the directional derivative in this direction) is equal to the magnitude of grad T at P, which is $\sqrt{5^2 + 7^2 + 3^2} = \sqrt{83}$.

1.6.2 div and curl

A vector field associates a vector, say **u**, with each point $P(y_1, y_2, y_3)$ in some region of three-dimensional space. Each component of the vector is thus a function of the co-ordinates (y_1, y_2, y_3) with respect to some reference system,

$$u_i = u_i(y_1, y_2, y_3).$$

The **divergence** of the vector field is a scalar quantity, denoted by div **u**, associated with each point (and thereby constituting a scalar field). It is given by

$$\text{div } \mathbf{u} = \frac{\partial u_1}{\partial y_1} + \frac{\partial u_2}{\partial y_2} + \frac{\partial u_3}{\partial y_3} = \frac{\partial u_k}{\partial y_k}. \tag{1.47}$$

This definition is heavily dependent on the co-ordinate system, and it is certainly not obvious that div **u** is a true scalar. To show this involves demonstrating that equation (1.47) would yield the same result if a different reference system (with a different origin, orientation of axes, and hence different components of **u**) were adopted. We defer a proof that div **u** is a scalar until Chapter 3 but, as an example, we may note here the fundamental concept of the velocity field in fluid mechanics, whereby a velocity vector **v** of a fluid particle may be associated with a point in space occupied by the fluid at some given instant of time. It can be shown that div **v** then represents the rate at which the volume of a material element is locally expanding (per unit reference volume) and is consequently independent of the choice of reference system.

It is also possible to derive a vector field curl **u** from a vector field **u** by the formula

$$(\text{curl } \mathbf{u})_i = e_{ijk} \frac{\partial u_k}{\partial y_j}, \tag{1.48}$$

i.e.

$$\text{curl } \mathbf{u} = \left[\frac{\partial u_3}{\partial y_2} - \frac{\partial u_2}{\partial y_3}, \frac{\partial u_1}{\partial y_3} - \frac{\partial u_3}{\partial y_1}, \frac{\partial u_2}{\partial y_1} - \frac{\partial u_1}{\partial y_2} \right].$$

Again the vector nature of curl **u** is not obvious, although it is well-known in fluid mechanics that, when **u** is the velocity field **v**, curl **v** gives a vector measure of the local rate of rotation of a material element.

We should perhaps emphasize that the definitions (1.44), (1.47) and (1.48) are given specifically in terms of rectangular cartesian co-ordinate systems.

Example 1

Given the following vector fields, we can calculate the divergence and curl.

(a) $\mathbf{u} = y_1 \mathbf{i}_1 + y_2 \mathbf{i}_2 + y_3 \mathbf{i}_3 = \mathbf{r}$, the position vector.

$$\operatorname{div} \mathbf{u} = \frac{\partial y_1}{\partial y_1} + \frac{\partial y_2}{\partial y_2} + \frac{\partial y_3}{\partial y_3} = 1 + 1 + 1 = 3.$$

$$\operatorname{curl} \mathbf{u} = \left[\frac{\partial y_3}{\partial y_2} - \frac{\partial y_2}{\partial y_3} \quad \frac{\partial y_1}{\partial y_3} - \frac{\partial y_3}{\partial y_1} \quad \frac{\partial y_2}{\partial y_1} - \frac{\partial y_1}{\partial y_2} \right] = \mathbf{0}.$$

(b)
$$\mathbf{u} = \frac{1}{r}\mathbf{r},$$

where $r = |\mathbf{r}|$. Thus

$$\mathbf{u} = \frac{y_1}{r}\mathbf{i}_1 + \frac{y_2}{r}\mathbf{i}_2 + \frac{y_3}{r}\mathbf{i}_3.$$

Note that $r^2 = y_1^2 + y_2^2 + y_3^2$. Hence we obtain

$$2r \frac{\partial r}{\partial y_1} = 2y_1 \quad \text{and} \quad \frac{\partial r}{\partial y_1} = \frac{y_1}{r}.$$

Therefore

$$\operatorname{div} \mathbf{u} = \frac{\partial}{\partial y_1}\left(\frac{y_1}{r}\right) + \frac{\partial}{\partial y_2}\left(\frac{y_2}{r}\right) + \frac{\partial}{\partial y_3}\left(\frac{y_3}{r}\right)$$

$$= \left(\frac{1}{r} - \frac{y_1}{r^2}\frac{\partial r}{\partial y_1}\right) + \left(\frac{1}{r} - \frac{y_2}{r^2}\frac{\partial r}{\partial y_2}\right) + \left(\frac{1}{r} - \frac{y_3}{r^2}\frac{\partial r}{\partial y_3}\right)$$

$$= \left(\frac{1}{r} - \frac{y_1^2}{r^3}\right) + \left(\frac{1}{r} - \frac{y_2^2}{r^3}\right) + \left(\frac{1}{r} - \frac{y_3^2}{r^3}\right)$$

$$= \frac{3}{r} - \frac{y_1^2 + y_2^2 + y_3^2}{r^3} = \frac{3}{r} - \frac{r^2}{r^3} = \frac{2}{r}$$

and
$$(\operatorname{curl} \mathbf{u})_1 = \frac{\partial}{\partial y_2}\left(\frac{y_3}{r}\right) - \frac{\partial}{\partial y_3}\left(\frac{y_2}{r}\right)$$

$$= -\frac{y_3 y_2}{r^3} + \frac{y_2 y_3}{r^3} = 0.$$

Similarly for the other components of curl **u**. Hence

$$\operatorname{curl} \mathbf{u} = \mathbf{0}.$$

(c) $\mathbf{u} = y_1^2 \mathbf{i}_1 + y_1 y_2 \mathbf{i}_2 + y_1 y_3 \mathbf{i}_3$.
We obtain

$$\operatorname{div} \mathbf{u} = 4y_1 \quad \text{and} \quad \operatorname{curl} \mathbf{u} = (0 \quad -y_3 \quad y_2).$$

Example 2

We can compare the above calculations in Example 1, (a) and (b), with those using the suffix notation.
In (a) we have $u_i = y_i$. Hence

$$\operatorname{div} \mathbf{u} = \frac{\partial u_i}{\partial y_i} = \frac{\partial y_i}{\partial y_i} = \delta_{ii} = 1 + 1 + 1 = 3,$$

and
$$(\operatorname{curl} \mathbf{u})_i = e_{ijk}\frac{\partial u_k}{\partial y_j} = e_{ijk}\frac{\partial y_k}{\partial y_j} = e_{ijk}\delta_{kj} = e_{ijj} = 0.$$

Moreover in (b) we also have a convenient suffix representation of $\mathbf{u}$, with $u_i = y_i/r$. Hence

$$\operatorname{div} \mathbf{u} = \frac{\partial u_i}{\partial y_i} = \frac{\partial}{\partial y_i}\left(\frac{y_i}{r}\right) = \frac{\delta_{ii}}{r} - \frac{y_i}{r^2}\frac{\partial r}{\partial y_i}$$

$$= \frac{3}{r} - \frac{y_i}{r^2}\frac{y_i}{r} = \frac{3}{r} - \frac{y_i y_i}{r^3} = \frac{3}{r} - \frac{r^2}{r^3} = \frac{2}{r},$$

and
$$(\operatorname{curl} \mathbf{u})_i = e_{ijk}\frac{\partial}{\partial y_j}\left(\frac{y_k}{r}\right) = e_{ijk}\left(\frac{\delta_{jk}}{r} - \frac{y_k y_j}{r^3}\right)$$

$$= \frac{e_{ijj}}{r} - \frac{e_{ijk} y_k y_j}{r^3} = 0,$$

since e_{ijk} is skew symmetric and $y_k y_j$ symmetric in k and j, drawing a similar conclusion to that in note (6) on p. 17.

1.6.3 Some standard identities

Suffix notation can be used to establish the following identities involving arbitrary scalar fields Φ, vector fields $\mathbf{u}$ and $\mathbf{v}$, and the operators grad, div and curl.

(a) $\operatorname{curl}(\operatorname{grad} \Phi) = \mathbf{0}$,
(b) $\operatorname{div}(\operatorname{curl} \mathbf{u}) = 0$,
(c) $\operatorname{div}(\Phi \mathbf{u}) = \Phi \operatorname{div} \mathbf{u} + \mathbf{u} \cdot \operatorname{grad} \Phi$, (1.49)
(d) $\operatorname{curl}(\Phi \mathbf{u}) = \Phi \operatorname{curl} \mathbf{u} + \operatorname{grad} \Phi \times \mathbf{u}$,
(e) $\operatorname{div}(\mathbf{u} \times \mathbf{v}) = \operatorname{curl} \mathbf{u} \cdot \mathbf{v} - \mathbf{u} \cdot \operatorname{curl} \mathbf{v}.$

As an example, we demonstrate (d). We have

$$(\operatorname{curl}(\Phi \mathbf{u}))_i = e_{ijk}\frac{\partial}{\partial y_j}(\Phi u_k)$$

$$= e_{ijk}\left(\Phi\frac{\partial u_k}{\partial y_j} + \frac{\partial \Phi}{\partial y_j}u_k\right)$$

$$= \Phi e_{ijk}\frac{\partial u_k}{\partial y_j} + e_{ijk}\frac{\partial \Phi}{\partial y_j}u_k$$

$$= \Phi(\operatorname{curl} \mathbf{u})_i + e_{ijk}(\operatorname{grad} \Phi)_j u_k$$

$$= (\Phi \operatorname{curl} \mathbf{u})_i + (\operatorname{grad} \Phi \times \mathbf{u})_i$$

$$= (\Phi \operatorname{curl} \mathbf{u} + \operatorname{grad} \Phi \times \mathbf{u})_i.$$

Equality of each i component implies equality of the vectors on each side, yielding (d).
The reader should also verify the other identities above.

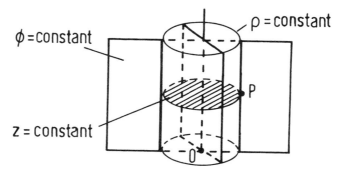

Fig. 1.19.

1.7 CURVILINEAR CO-ORDINATE SYSTEMS

1.7.1 (Covariant) base vectors

A great range of co-ordinate systems is available in three-dimensional space to specify the position of a point. In particular problems under investigation the geometry of the situation will tend to indicate what systems might be most helpful; circular cylinders suggest cylindrical polar co-ordinates, spheres spherical polar co-ordinates, etc. Suppose we designate the co-ordinates by $\{x^1, x^2, x^3\}$, standing possibly for cylindrical polars $\{\rho, \phi, z\}$, or spherical polars $\{r, \theta, \phi\}$, or some other set of co-ordinates. The reason for using superscripts rather than subscripts will become clear later. Space will be filled by 'co-ordinate surfaces' $x^1 = $ constant, $x^2 = $ constant, $x^3 = $ constant, and through any point P in space (except for certain 'singular' points) there will pass three curves, on each of which one co-ordinate alone will vary.

For cylindrical polars the surfaces are

(i) cylinders $\rho = $ constant,
(ii) planes containing the z axis with $\phi = $ constant and
(iii) planes orthogonal to the z-axis with $z = $ constant.

Through any point P (not on the z-axis) there pass three curves, two of which are straight lines with z and ρ varying, respectively, and the third the circle $\rho = z = $ constant (Fig. 1.19).

For spherical polars the surfaces are

(i) spheres $r = $ constant,
(ii) right circular cones with vertex at the origin O and semi-angle θ, defined by $\theta = $ constant
(iii) planes $\phi = $ constant.

Through any point P there pass two curves on the surface $r = $ constant, which we may think of as a circle of latitude, on which ϕ alone varies, and a circle of longitude, on which θ varies; r alone varies along the straight line OP (Fig. 1.20). In these examples the co-ordinate surfaces passing through P and the associated curves intersect at right angles, and in fact cylindrical and spherical polars are standard examples of **orthogonal** curvilinear co-ordinates.

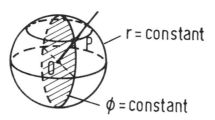

Fig. 1.20.

The three co-ordinate curves passing through P define three directions at P, tangential to the curves, and it is convenient to establish a set of base vectors at P with these directions. A natural set of base vectors can be obtained by taking the partial derivatives of the position vector **r** with respect to each co-ordinate in turn. Considering $\partial \mathbf{r}/\partial x^1$ first, this is the rate of change at our point P in **r** with respect to x^1, holding x^2 and x^3 fixed, i.e. it is the rate of change in **r** with respect to x^1 as a point moves along the x^1-co-ordinate curve at P. The limit of $\delta \mathbf{r}/\delta x^1$ along this curve as $\delta x^1 \to 0$ is a vector tangent to the curve at P. We shall write this vector as $\mathbf{g}_1$, so that, defining $\mathbf{g}_2$ and $\mathbf{g}_3$ similarly, we have a set of base vectors

$$\mathbf{g}_i = \frac{\partial \mathbf{r}}{\partial x^i}, \qquad i = 1, 2, 3. \tag{1.50}$$

In general these vectors will not be unit vectors. In fact they will be unit vectors only if the corresponding x^i measures distance along the co-ordinate curve. We sometimes refer to $\mathbf{g}_1, \mathbf{g}_2, \mathbf{g}_3$ as **covariant base vectors**.

The components of $\mathbf{g}_1, \mathbf{g}_2$ and $\mathbf{g}_3$ with respect to some fixed (background) set of rectangular cartesian axes $Oy_1 y_2 y_3$ can be evaluated. Since (y_1, y_2, y_3) are the cartesian components of **r**, the cartesian components of $\mathbf{g}_i$ can be written

$$(\mathbf{g}_i)_j = \frac{\partial y_j}{\partial x^i}, \qquad j = 1, 2, 3. \tag{1.51}$$

For cylindrical polars, with the usual choice of axes, Oy_3 coinciding with the z-axis, we have

$$\left. \begin{array}{l} y_1 = \rho \cos \phi, \\ y_2 = \rho \sin \phi, \\ y_3 = z. \end{array} \right\} \tag{1.52}$$

Then

$$\begin{aligned} \mathbf{g}_1 &= \left[\frac{\partial y_1}{\partial r} \quad \frac{\partial y_2}{\partial r} \quad \frac{\partial y_3}{\partial r} \right] = [\cos \phi \quad \sin \phi \quad 0] \\ &= \cos \phi \, \mathbf{i}_1 + \sin \phi \, \mathbf{i}_2, \\ \mathbf{g}_2 &= \left[\frac{\partial y_1}{\partial \phi} \quad \frac{\partial y_2}{\partial \phi} \quad \frac{\partial y_3}{\partial \phi} \right] = [-\rho \sin \phi \quad \rho \cos \phi \quad 0] \end{aligned} \right\}$$

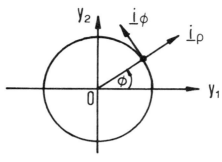

Fig. 1.21.

$$\begin{aligned}
&= \rho(-\sin\phi\, \mathbf{i}_1 + \cos\phi\, \mathbf{i}_2), \\
\mathbf{g}_3 &= \left[\frac{\partial y_1}{\delta z}\ \frac{\partial y_2}{\partial z}\ \frac{\partial y_3}{\partial z}\right] = [\ 0\ \ 0\ \ 1] \\
&= \mathbf{i}_3.
\end{aligned} \quad (1.53)$$

Thus $\mathbf{g}_1$ and $\mathbf{g}_3$ are unit vectors, but $|\mathbf{g}_2| = \rho$. In two dimensions we can define unit vectors $\mathbf{i}_\rho$ and $\mathbf{i}_\phi$ in the radial and circumferential directions with respect to the polar co-ordinates ρ and ϕ (Fig. 1.21). Then we have $\mathbf{g}_1 = \mathbf{i}_\rho$, $\mathbf{g}_2 = \rho\mathbf{i}_\phi$. Note that the directions of $\mathbf{g}_1$ and $\mathbf{g}_2$ vary with the position of P in space, as does the magnitude of $\mathbf{g}_2$.

The natural base vectors for a spherical polar co-ordinate system can be found in a similar manner. With the usual choice of rectangular cartesian axes, Oy_1y_2,y_3, with Oy_1y_2 forming an equatorial plane and Oy_1y_3 a meridional plane, we have the relations

$$\left.\begin{aligned}
y_1 &= r\sin\theta\cos\phi, \\
y_2 &= r\sin\theta\sin\phi, \\
y_3 &= r\cos\theta
\end{aligned}\right\} \quad (1.54)$$

between the spherical polar and cartesian co-ordinates of a point. Then the partial derivatives give cartesian components

$$\begin{aligned}
\mathbf{g}_1 &= \left[\frac{\partial y_1}{\partial r}\ \frac{\partial y_2}{\partial r}\ \frac{\partial y_3}{\partial r}\right] = [\sin\theta\cos\phi\ \ \sin\theta\sin\phi\ \ \cos\theta], \\
\mathbf{g}_2 &= \left[\frac{\partial y_1}{\partial \theta}\ \frac{\partial y_2}{\partial \theta}\ \frac{\partial y_3}{\partial \theta}\right] = [r\cos\theta\cos\phi\ \ r\cos\theta\sin\phi\ \ -r\sin\theta], \\
\mathbf{g}_3 &= \left[\frac{\partial y_1}{\partial \phi}\ \frac{\partial y_2}{\partial \phi}\ \frac{\partial y_3}{\partial \phi}\right] = [-r\sin\theta\sin\phi\ \ r\sin\theta\cos\phi\ \ 0].
\end{aligned} \quad (1.55)$$

As expected, these vectors are mutually orthogonal, with

$$\mathbf{g}_1\cdot\mathbf{g}_2 = \mathbf{g}_2\cdot\mathbf{g}_3 = \mathbf{g}_3\cdot\mathbf{g}_1 = 0.$$

Only $\mathbf{g}_1$ is a unit vector. We have

$$|\mathbf{g}_1| = 1,\quad |\mathbf{g}_2| = r,\quad |\mathbf{g}_3| = r\sin\theta.$$

Of course, when we are considering the axis of points for which the spherical polar co-ordinate θ is zero, then value of the co-ordinate ϕ for such points is indeterminate. Moreover there will also be no uniquely determined set of base vectors $\{\mathbf{g}_1, \mathbf{g}_2, \mathbf{g}_3\}$ at such points. Curvilinear co-ordinate systems generally have such singularities, and care may sometimes be required in dealing with them.

For a spherical polar co-ordinate system, if we make the obvious correspondence between a co-ordinate surface $r = $ constant and the surface of the earth, with the point at $\theta = 0$ representing the North Pole, then the base vectors at a point P on the spherical surface will point in directions vertically upwards (along a line joining the centre of the sphere to P), southwards and eastwards, respectively. If we represent unit vectors in these directions as $\mathbf{i}_r$, $\mathbf{i}_\theta$ and $\mathbf{i}_\phi$, then we have

$$\mathbf{g}_1 = \mathbf{i}_r, \quad \mathbf{g}_2 = r\mathbf{i}_\theta, \quad \mathbf{g}_3 = r\sin\theta\, \mathbf{i}_\phi. \tag{1.56}$$

In general, the difference $d\mathbf{r}$ in the position vectors of two points infinitesimally close together is given by the first-order formula

$$d\mathbf{r} = \frac{\partial \mathbf{r}}{\partial x^1} dx^1 + \frac{\partial \mathbf{r}}{\partial x^2} dx^2 + \frac{\partial \mathbf{r}}{\partial x^3} dx^3,$$

where dx^1, dx^2, dx^3 are the differences in the corresponding co-ordinates of the two points. Because of the way the base vectors have been defined, this expression can be written as

$$d\mathbf{r} = \mathbf{g}_1\, dx^1 + \mathbf{g}_2\, dx^2 + \mathbf{g}_3\, dx^3 = \mathbf{g}_i\, dx^i. \tag{1.57}$$

If $\mathbf{u}(x^1, x^2, x^3)$ is a vector field, associating the vector $\mathbf{u}$ with any point in space with curvilinear co-ordinates (x^1, x^2, x^3), we can express $\mathbf{u}$ in terms of the base vectors at that point, i.e.

$$\mathbf{u} = u^1 \mathbf{g}_1 + u^2 \mathbf{g}_2 + u^3 \mathbf{g}_3 = u^i \mathbf{g}_i. \tag{1.58}$$

The components $\{u^1, u^2, u^3\}$ are the **contravariant components** of $\mathbf{u}$ with respect to the given co-ordinate system.

1.7.2 Reciprocal (contravariant) base vectors

For a given set of curvilinear co-ordinates $\{x^1, x^2, x^3\}$ we now have at any point in space (except for certain singular points) a set of natural base vectors $\{\mathbf{g}_1, \mathbf{g}_2, \mathbf{g}_3\}$ with $\mathbf{g}_i = \partial \mathbf{r}/\partial x^i$. It is again convenient to define a set of **reciprocal base vectors** $\{\mathbf{g}^1, \mathbf{g}^2, \mathbf{g}^3\}$ at each point in space, satisfying the conditions (1.34), i.e.

$$\mathbf{g}^i \cdot \mathbf{g}_j = \delta^i_j. \tag{1.59}$$

Moreover, we have

$$\mathbf{g}^1 = \frac{1}{V} \mathbf{g}_2 \times \mathbf{g}_3, \quad \mathbf{g}^2 = \frac{1}{V} \mathbf{g}_3 \times \mathbf{g}_1, \quad \mathbf{g}^3 = \frac{1}{V} \mathbf{g}_1 \times \mathbf{g}_2,$$

where $V = \mathbf{g}_1 \cdot (\mathbf{g}_2 \times \mathbf{g}_3)$, as in (1.35).

A vector $\mathbf{u}$ at a point in space then has the representation

$$\mathbf{u} = u_1 \mathbf{g}^1 + u_2 \mathbf{g}^2 + u_3 \mathbf{g}^3 = u_i \mathbf{g}^i, \tag{1.60}$$

where $\{u_1, u_2, u_3\}$ are the **covariant components** of $\mathbf{u}$.

We have also the equations

$$u_i = \mathbf{u} \cdot \mathbf{g}_i, \qquad u^i = \mathbf{u} \cdot \mathbf{g}^i. \tag{1.61}$$

The vectors $\mathbf{g}^1, \mathbf{g}^2, \mathbf{g}^3$ will sometimes be called **contravariant base vectors**.

Since $\mathbf{g}^1$ is orthogonal to $\mathbf{g}_2$ and $\mathbf{g}_3$, which are in the directions of the co-ordinate curves on which x^2 and x^3, respectively, vary, it must be orthogonal to the plane containing the tangents to these curves. This plane is just the tangent plane to the surface $x^1 = $ constant at the point in question. Hence $\mathbf{g}^1$ is in the direction of the **normal** to the surface $x^1 = $ constant. This means that the unit normal to this surface is the vector $\mathbf{g}^1/|\mathbf{g}^1|$, and therefore that the covariant components of the unit normal are $(1/|\mathbf{g}^1|, 0, 0)$.

If the curvilinear co-ordinate system is orthogonal, as are cylindrical and spherical polars, pairs of base vectors $\mathbf{g}_1$ and $\mathbf{g}^1$, etc., will be parallel, but in general not equal in magnitude.

For cylindrical polars, with $x^1 = \rho$, $x^2 = \phi$, $x^3 = z$, we have

$$\mathbf{g}^1 = \mathbf{g}_1, \qquad \mathbf{g}^2 = \frac{1}{\rho^2}\mathbf{g}_2, \qquad \mathbf{g}^3 = \mathbf{g}_3, \tag{1.62}$$

since, for example,

$$\mathbf{g}^2 \cdot \mathbf{g}_2 = \frac{1}{\rho^2}\mathbf{g}_2 \cdot \mathbf{g}_2 = \frac{1}{\rho^2}(\rho \mathbf{i}_\phi) \cdot (\rho \mathbf{i}_\phi) = \mathbf{i}_\phi \cdot \mathbf{i}_\phi = 1,$$

as required. (Note here the use of the superscript in ρ^2 as an *exponent*, without ambiguity, since indices are not relevant for the quantity ρ.)

For spherical polars, with $x^1 = r$, $x^2 = \theta$, $x^3 = \phi$, we get

$$\mathbf{g}^1 = \mathbf{g}_1, \qquad \mathbf{g}^2 = \frac{1}{r^2}\mathbf{g}_2, \qquad \mathbf{g}^3 = \frac{1}{r^2 \sin^2 \theta}\mathbf{g}_3. \tag{1.63}$$

Certain formulas established in rectangular cartesian co-ordinate systems need to be modified when extended to curvilinear co-ordinate systems. For example, consider the formula for the scalar product of two vectors $\mathbf{u}$ and $\mathbf{v}$. In a general co-ordinate system, this becomes

$$\mathbf{u} \cdot \mathbf{v} = (u^i \mathbf{g}_i) \cdot (v_j \mathbf{g}^j) = u^i v_j \mathbf{g}_i \cdot \mathbf{g}^j = u^i v_j \delta_i^j = u^i v_i,$$

i.e. it is the sum of product of corresponding *contravariant* components of $\mathbf{u}$ and *covariant* components of $\mathbf{v}$. Since the scalar product is symmetric, we have

$$\mathbf{u} \cdot \mathbf{v} = u^i v_i = u_i v^i. \tag{1.64}$$

1.7.3 Relation between covariant and contravariant components

For an arbitrary curvilinear co-ordinate system a vector $\mathbf{u}$ has covariant components u_i and contravariant components u^i. Since $\mathbf{u} = u_k \mathbf{g}^k$, we have, by (1.61),

$$\begin{aligned} u^i &= \mathbf{u} \cdot \mathbf{g}^i = (u_k \mathbf{g}^k) \cdot \mathbf{g}^i \\ &= u_k (\mathbf{g}^k \cdot \mathbf{g}^i), \end{aligned}$$

and, similarly,

$$u_i = \mathbf{u} \cdot \mathbf{g}_i = (u^k \mathbf{g}_k) \cdot \mathbf{g}_i = u^k(\mathbf{g}_k \cdot \mathbf{g}_i).$$

It is convenient to introduce the symbols g^{ik} and g_{ik} for the mutual scalar products of base vectors:

and
$$\left.\begin{array}{l} g_{ik} = \mathbf{g}_i \cdot \mathbf{g}_k, \\ g^{ik} = \mathbf{g}^i \cdot \mathbf{g}^k. \end{array}\right\} \quad (1.65)$$

Note that $g_{ik} = g_{ki}$ and $g^{ik} = g^{ki}$ because of the symmetry of the scalar product. Then we have the following convenient relations between the covariant and contravariant components of **u**:

$$u^i = g^{ik} u_k, \quad (1.66)$$

$$u_i = g_{ik} u^k. \quad (1.67)$$

It is natural to speak of the actions of g^{ik} and g_{ik} on the RHS of (1.66) and (1.67) as **raising** the index of **u** and **lowering** the index respectively.

The quantities g_{ij} are related to the distance between two points infinitesimally close together. From (1.57) we have $d\mathbf{r} = \mathbf{g}_i\, dx^i$ for the infinitesimal line segment joining the two points. The distance between the points ds is equal to the magnitude of $d\mathbf{r}$. Hence

$$ds^2 = d\mathbf{r} \cdot d\mathbf{r} = (\mathbf{g}_i\, dx^i) \cdot (\mathbf{g}_j\, dx^j)$$

(ensuring that no dummy suffix appears more than twice)

$$= \mathbf{g}_i \cdot \mathbf{g}_j\, dx^i\, dx^j$$
$$= g_{ij}\, dx^i\, dx^j. \quad (1.68)$$

Explicitly,

$$ds^2 = g_{11}(dx^1)^2 + g_{22}(dx^2)^2 + g_{33}(dx^3)^2 + 2g_{12}\, dx^1\, dx^2 + 2g_{23}\, dx^2\, dx^3 + 2g_{31}\, dx^3\, dx^1,$$

since $g_{12} = g_{21}$, etc. Thus g_{ij} plays an important part in the measurement of distance with respect to the given curvilinear co-ordinates.

In cylindrical polars, we have

$$g_{11} = 1, \quad g_{22} = \rho^2, \quad g_{33} = 1, \quad g_{12} = g_{23} = g_{31} = 0,$$

with

$$ds^2 = (d\rho)^2 + \rho^2(d\phi)^2 + (dz)^2, \quad (1.69)$$

and, in spherical polars,

$$g_{11} = 1, \quad g_{22} = r^2, \quad g_{33} = r^2 \sin\theta, \quad g_{12} = g_{23} = g_{31} = 0,$$

with

$$ds^2 = (dr)^2 + r^2(d\theta)^2 + r^2 \sin^2\theta\, (d\phi)^2. \quad (1.70)$$

It is sometimes convenient to relate the components of various quantities with respect to a curvilinear system to those with respect to a (background) rectangular cartesian system of co-ordinates. For the base vectors $\mathbf{g}_i$, we have cartesian components

$\partial y_j/\partial x^i$, by (1.51). It follows that

$$g_{ij} = \mathbf{g}_i \cdot \mathbf{g}_j = (\mathbf{g}_i)_k (\mathbf{g}_j)_k \qquad \text{summed over } k$$

$$= \frac{\partial y_k}{\partial x^i} \frac{\partial y_k}{\partial x^j}. \qquad (1.71)$$

We have already seen that $\mathbf{g}^i$ is orthogonal to the surface $x^i = $ constant, but the normal to a surface $\Phi = $ constant is in the direction of the gradient vector grad Φ, which according to section 1.6.1 has cartesian components $\partial \Phi/\partial y_j$. Hence the cartesian components of $\mathbf{g}^i$ must be proportional to $\partial x^i/\partial y_j$. Thus

$$(\mathbf{g}^i)_j = \lambda \frac{\partial x^i}{\partial y_j},$$

for some λ. Now we have

$$\mathbf{g}^i \cdot \mathbf{g}_j = (\mathbf{g}^i)_k (\mathbf{g}_j)_k \qquad \text{summing over cartesian components}$$

$$= \lambda \frac{\partial x^i}{\partial y_k} \frac{\partial y_k}{\partial x^j} = \lambda \frac{\partial x^i}{\partial x^j} = \lambda \delta^i_j, \qquad \text{by the chain rule.}$$

Hence $\lambda = 1$, by (1.59), and we have cartesian components

$$(\mathbf{g}^i)_j = \frac{\partial x^i}{\partial y_j}. \qquad (1.72)$$

We may deduce the following expression for g^{ij}:

$$g^{ij} = \frac{\partial x^i}{\partial y_k} \frac{\partial x^j}{\partial y_k}. \qquad (1.73)$$

It follows that

$$g_{ij} g^{kj} = \frac{\partial y_l}{\partial x^i} \frac{\partial y_l}{\partial x^j} \frac{\partial x^k}{\partial y_m} \frac{\partial x^j}{\partial y_m}$$

$$= \frac{\partial y_l}{\partial x^i} \frac{\partial x^k}{\partial y_m} \left(\frac{\partial y_l}{\partial x^j} \frac{\partial x^j}{\partial y_m} \right) = \frac{\partial y_l}{\partial x^i} \frac{\partial x^k}{\partial y_m} \delta_{lm}$$

$$= \frac{\partial y_l}{\partial x^i} \frac{\partial x^k}{\partial y_l} = \delta^k_i \qquad (1.74)$$

with two applications of the chain rule.

Note also that from (1.65) we may deduce the relations

and
$$\left. \begin{array}{l} \mathbf{g}_i = g_{ij} \mathbf{g}^j \\ \mathbf{g}^i = g^{ij} \mathbf{g}_j. \end{array} \right\} \qquad (1.75)$$

2

Matrix algebra and determinants

2.1 MATRICES

2.1.1 Definitions

In this chapter we give a brief account of basic matrix algebra, concentrating on the particular aspects which are of relevance to the tensor analysis in later chapters.

A **matrix** is a rectangular array with m rows and n columns (we call this an $m \times n$ matrix) whose basic **elements**, or **entries**, are numbers. For the purposes of this book we concern ourselves mainly with real numbers, although occasionally the possibility of complex numbers may arise. An example of a 4×3 matrix is

$$\begin{bmatrix} 3 & 8 & -2 \\ 17 & 6 & 5 \\ 8 & -9 & 1 \\ 10 & 1 & 0 \end{bmatrix},$$

where the array of elements is enclosed in square brackets. A general 4×3 matrix could be written as

$$\begin{bmatrix} a_{11} & a_{12} & a_{13} \\ a_{21} & a_{22} & a_{23} \\ a_{31} & a_{32} & a_{33} \\ a_{41} & a_{42} & a_{43} \end{bmatrix},$$

where the general element is a_{ij}. Note that the two subscripts (i and j) of the general element signify the position of the element in the array, the first subscript referring to the number of the row and the second to the column in which the element appears. It is usual to represent the complete matrix by a capital letter, such as A.

If the number of rows is equal to the number of columns, the matrix is called **square**.

A matrix with only one row, such as the 1×3 matrix

$$[1 \quad 5 \quad -7],$$

is called a **row vector**. Similarly, a matrix with one column, such as the 2×1 matrix

$$\begin{bmatrix} 4 \\ 3 \end{bmatrix}$$

is called a **column vector**. Row vectors and column vectors will generally be distinguished from other matrices by the use of bold-face symbols such as **u** (as used for vectors in Chapter 1). In fact, when we arrange the three components (u_1, u_2, u_3) of a vector **u** as a matrix, we shall normally choose the 3×1 array of a column vector.

A square matrix $(n \times n)$ is **symmetric** if the element a_{ij} satisfies $a_{ij} = a_{ji}$ for all relevant i and j. For a 3×3 matrix, this implies $a_{12} = a_{21}$, $a_{23} = a_{32}$ and $a_{31} = a_{13}$. Examples of symmetric matrices are

$$\begin{bmatrix} 5 & 6 \\ 6 & 5 \end{bmatrix} \quad \text{and} \quad \begin{bmatrix} 1 & 3 & -2 \\ 3 & 0 & 2 \\ -2 & 2 & 5 \end{bmatrix}.$$

The mth row of such a matrix is identical with the mth column; so interchange of rows and columns leaves the matrix unchanged.

In general this operation of interchanging rows and columns on an $m \times n$ matrix A produces an $n \times m$ matrix called the **transpose** of A and denoted by A^T. Thus, if

$$A = \begin{bmatrix} a_{11} & a_{12} & \cdots & a_{1n} \\ a_{21} & \cdot & \cdots & \cdot \\ \vdots & & & \vdots \\ a_{m1} & \cdot & \cdots & a_{mn} \end{bmatrix},$$

we obtain

$$A^T = \begin{bmatrix} a_{11} & a_{21} & \cdots & a_{m1} \\ a_{12} & \cdot & \cdots & \cdot \\ \vdots & & & \vdots \\ a_{1n} & \cdot & \cdots & a_{mn} \end{bmatrix}. \tag{2.1}$$

We may say that, if the i–j entry of A is a_{ij} (paying due regard to the position of the subscripts), the i–j entry of A^T is a_{ji}. As an example, if

$$A = \begin{bmatrix} 5 & 6 & -1 \\ 9 & 10 & 2 \end{bmatrix},$$

then

$$A^T = \begin{bmatrix} 5 & 9 \\ 6 & 10 \\ -1 & 2 \end{bmatrix}.$$

Naturally, $(A^T)^T = A$, for any matrix A.

A square matrix is called **skew symmetric** (or **anti-symmetric**) if the transposition procedure produces a matrix whose elements are unchanged except for a change of sign, i.e. A is skew symmetric if $a_{ji} = -a_{ij}$. It follows that the diagonal elements a_{11},

Sec. 2.1] **Matrices** 43

a_{22}, etc., must all be zero, since they are not affected by transposition, e.g. $a_{11} = -a_{11} \rightarrow a_{11} = 0$. An example of a skew-symmetric matrix is

$$\begin{bmatrix} 0 & 1 & -3 \\ -1 & 0 & 4 \\ 3 & -4 & 0 \end{bmatrix}.$$

2.1.2 Basic operations

2.1.2.1 Addition and scalar multiplication

The operations of addition and scalar multiplication of matrices are a rather obvious extension of the similar operations on vectors discussed in Chapter 1. The definition of matrix multiplication, on the other hand, is not quite so straightforward, but one could perhaps compare the level of complexity with that involved in defining scalar and vector products of vectors, which may seem a little obscure on first acquaintance. All these operations play a part in tensor analysis.

We must begin with the notion of **equality**. Two matrices A and B are said to be equal if and only if each matrix has the same number of rows, say m, and the same number of columns, say n, and if all corresponding elements are equal. Thus they must both be $m \times n$, and, if their elements are denoted by a_{ij} and b_{ij}, then $a_{ij} = b_{ij}$ for all relevant i $(i = 1, 2, \ldots, m)$ and j $(j = 1, 2, \ldots, n)$.

Suppose now we have two matrices A and B, not necessarily equal. We can define their sum $A + B$ by adding corresponding elements, but this operation makes sense only if A and B are the same size, i.e. only if they both have m rows and n columns, for some m and n. If their elements are a_{ij} and b_{ij}, respectively, the sum is another $m \times n$ matrix C with elements c_{ij} given by

$$c_{ij} = a_{ij} + b_{ij}.$$

For example, we have

$$\begin{bmatrix} 5 & 6 & -1 \\ 9 & 10 & 2 \end{bmatrix} + \begin{bmatrix} 3 & -2 & 1 \\ -2 & 4 & 6 \end{bmatrix} = \begin{bmatrix} 8 & 4 & 0 \\ 7 & 14 & 8 \end{bmatrix}.$$

Clearly the **commutative** and **associative** rules,

$$A + B = B + A$$

and

$$(A + B) + C = A + (B + C),$$

respectively, are satisfied for matrix addition, where A, B, C are arbitrary matrices, provided that they are all $m \times n$ for some m and some n.

For any $m \times n$ matrix A and scalar λ we can define the **scalar multiple** λA to be another $m \times n$ matrix obtained from A by multiplying all the elements of A by λ. If the elements of A are a_{ij}, the i–j entry of λA is just λa_{ij}.

We can now define $-A$ as $(-1)A$. This is the matrix with entries

$$\begin{bmatrix} -a_{11} & -a_{12} & \cdots & -a_{1n} \\ -a_{21} & \cdot & \cdots & \cdot \\ \vdots & & & \vdots \\ -a_{m1} & \cdot & \cdots & -a_{mn} \end{bmatrix}.$$

We can then regard subtraction of matrices $A - B$ as equivalent to the sum $A + (-B)$.

2.1.2.2 Matrix multiplications

The **product** AB of two matrices A and B is defined if and only if the number of columns of A is equal to the number of rows of B. Thus, if A is $m \times n$, for some m and n, then B must be $n \times p$, for some p. The matrices are then said to be **conformable**. The resulting product AB is an $m \times p$ matrix C, say, whose i–j entry c_{ij} is obtained by multiplying each element of the ith row of A by its corresponding element in the jth column of B and adding the results together. We may write, formally,

$$c_{ij} = \sum_{k=1}^{n} a_{ik} b_{kj}. \tag{2.2}$$

As an example, suppose A is the 2×3 matrix

$$\begin{bmatrix} 1 & 4 & 3 \\ 2 & 5 & 1 \end{bmatrix}$$

and B is the 3×2 matrix

$$\begin{bmatrix} 6 & 9 \\ 8 & 0 \\ 7 & 1 \end{bmatrix}.$$

Then AB is the 2×2 matrix

$$\begin{bmatrix} 1 \times 6 + 4 \times 8 + 3 \times 7 & 1 \times 9 + 4 \times 0 + 3 \times 1 \\ 2 \times 6 + 5 \times 8 + 1 \times 7 & 2 \times 9 + 5 \times 0 + 1 \times 1 \end{bmatrix} = \begin{bmatrix} 59 & 12 \\ 49 & 19 \end{bmatrix}.$$

In this case we can also form the product AB, since the number of columns of B is equal to the number of rows of A. We obtain

$$BA = \begin{bmatrix} 6 \times 1 + 9 \times 2 & 6 \times 4 + 9 \times 5 & 6 \times 3 + 9 \times 1 \\ 8 \times 1 + 0 \times 2 & 8 \times 4 + 0 \times 5 & 8 \times 3 + 0 \times 1 \\ 7 \times 1 + 1 \times 2 & 7 \times 4 + 1 \times 5 & 7 \times 3 + 1 \times 1 \end{bmatrix} = \begin{bmatrix} 24 & 69 & 27 \\ 8 & 32 & 24 \\ 9 & 33 & 22 \end{bmatrix}.$$

Note that $BA \neq AB$. In fact, BA and AB are not even matrices of the same size. Matrix multiplication is thus clearly not commutative in general. The $n \times n$ matrix whose diagonal elements are all equal to unity and whose off-diagonal elements are all zero plays an important part in matrix multiplication. We shall denote this matrix as I, irrespective of size, although in the applications in this book it will normally be 3×3. Thus

$$I = \begin{bmatrix} 1 & 0 & 0 & \cdots & 0 \\ 0 & 1 & \cdot & \cdots & 0 \\ 0 & \cdot & \cdot & & \cdot \\ \vdots & & & \ddots & \vdots \\ 0 & \cdot & \cdot & \cdots & 1 \end{bmatrix}, \tag{2.3}$$

and it is not difficult to see that, if A is $m \times n$ and B is $n \times m$, we have

$$AI = A \quad \text{and} \quad IB = B. \tag{2.4}$$

So I behaves like unity for the purposes of matrix multiplication, and we call it the **unit** (or **identity**) $n \times n$ matrix. In the 3×3 case the elements of I are equal to the components δ_{ij} of the Kronecker delta (1.11). By (2.2), equations (2.4) are then equivalent to the equations

$$\sum_{k=1}^{3} a_{ik}\delta_{kj} = a_{ij} \quad \text{and} \quad \sum_{k=1}^{3} \delta_{ik}b_{kj} = b_{ij},$$

which may remind the reader of the substitution operator properties of δ_{ij}.

A fundamental property of matrix multiplication is that it is **associative**, i.e. if A, B and C are arbitrary $m \times n$, $n \times p$ and $p \times q$ matrices, respectively, then we can form the matrix products $(AB)C$ and $A(BC)$, and we always have

$$(AB)C = A(BC). \tag{2.5}$$

The proof of (2.5) is technically straightforward but requires some facility in the manipulation of terms involving suffixes. Thus the i–j entry of $(AB)C$ is, using (2.2),

$$\sum_{k=1}^{p} (AB)_{ik} c_{kj}$$

since AB is $m \times p$ and C is $p \times q$. This may be written, using (2.2) again,

$$\sum_{k=1}^{p} \left(\sum_{l=1}^{n} a_{il} b_{lk} \right) c_{kj}$$

$$= \sum_{l=1}^{n} \left(\sum_{k=1}^{p} a_{il} b_{lk} c_{kj} \right) \quad \text{(changing the order in which the terms are summed)}$$

$$= \sum_{l=1}^{n} a_{il} \left(\sum_{k=1}^{p} b_{lk} c_{kj} \right)$$

$$= \sum_{l=1}^{n} a_{il} (BC)_{lj} = (A(BC))_{ij}.$$

Equality of each i–j entry ensures that

$$(AB)C = A(BC).$$

Hence we can write the triple product simply as ABC without any ambiguity.

If any step in this proof seems obscure to the reader, it would be sensible to check each step by considering a particular example, choosing specific values (low for convenience) of m, n, p and q, and evaluating the matrix products in the proof explicitly. (See problem 2.1.1)

The summation convention introduced in Chapter 1 may be safely used in matrix multiplication, summation signs thereby being omitted, provided that the repeated subscripts are summed over the appropriate range of values. Thus (2.2) may be written $c_{ij} = a_{ik}b_{kj}$ with k summed from 1 to n, and the triple product ABC above has the i–j entry $a_{il}b_{lk}c_{kj}$ where l is summed from 1 to n and k from 1 to p. Note that it is essential when using the summation convention for each repeated index in an expression representing a matrix product to appear in adjacent positions, as do the l and the k in $a_{il}b_{lk}c_{kj}$, which represents the i–j entry of the product ABC. Having established this expression, however, we are free to vary the order of terms, e.g.

$$a_{il}b_{lk}c_{kj} = b_{lk}a_{il}c_{kj} = c_{kj}b_{lk}a_{il}, \quad \text{etc.,}$$

for the obvious reason that the product of several numbers does not depend on the order in which they are multiplied.

In this book we are principally concerned with three-dimensional problems for which summation is carried out from 1 to 3 (or 1 to 2 for plane problems). The important matrix products here are products of square 3×3 matrices and products $A\mathbf{u}$ of 3×3 matrices A and 3×1 column vectors $\mathbf{u}$. The matrix equation

$$A\mathbf{u} = \mathbf{v},$$

where $\mathbf{v}$ is another 3×1 column vector, may be written as

$$a_{ij}u_j = v_i,$$

using the summation convention.

The matrix product $\mathbf{u}^T A \mathbf{u}$ represents a 1×1 matrix, which, by discarding the matrix brackets, we can regard as equivalent to a single number. In terms of the entries u_i and a_{ij} of $\mathbf{u}$ and A, respectively, this number is

$$\begin{aligned} u_i a_{ij} u_j &= a_{ij} u_i u_j \\ &= a_{11}u_1^2 + a_{12}u_1u_2 + a_{13}u_1u_3 + a_{21}u_2u_1 \\ &\quad + a_{22}u_2^2 + a_{23}u_2u_3 + a_{31}u_3u_1 + a_{32}u_3u_2 + a_{33}u_3^2 \\ &= a_{11}u_1^2 + a_{22}u_2^2 + a_{33}u_3^2 + (a_{12}+a_{21})u_1u_2 \\ &\quad + (a_{23}+a_{32})u_2u_3 + (a_{31}+a_{13})u_3u_1. \end{aligned}$$

This expression may also be obtained by direct matrix multiplication

$$\begin{bmatrix} u_1 & u_2 & u_3 \end{bmatrix} \begin{bmatrix} a_{11} & a_{12} & a_{13} \\ a_{21} & a_{22} & a_{23} \\ a_{31} & a_{32} & a_{33} \end{bmatrix} \begin{bmatrix} u_1 \\ u_2 \\ u_3 \end{bmatrix}.$$

The operations of matrix addition and multiplication are connected through the **distributive law**, which states that, if A is any $m \times n$ matrix and B and C are any $n \times p$ matrices, we must have

$$A(B+C) = AB + AC. \tag{2.6}$$

This follows directly from consideration of the identity

$$\sum_{k=1}^n a_{ik}(b_{kj} + c_{kj}) = \sum_{k=1}^n a_{ik}b_{kj} + \sum_{k=1}^n a_{ik}c_{kj},$$

or, equivalently, using the summation convention,

$$a_{ik}(b_{kj} + c_{kj}) = a_{ik}b_{kj} + a_{ik}c_{kj}.$$

2.1.2.3 Matrix multiplication and transposition

If the matrices A and B are $m \times n$ and $n \times p$, respectively, the product AB is $m \times p$. The transpose of the product $(AB)^T$ is therefore $p \times m$. Now B^T is $p \times n$ and A^T is

$n \times m$; thus B^T and A^T are conformable and their product $B^T A^T$ is also $p \times m$. In fact we can prove that $(AB)^T$ and $B^T A^T$ are identical for any conformable A and B; thus

$$(AB)^T = B^T A^T. \tag{2.7}$$

To show this, consider the i–j entry of $(AB)^T$, which equals the j–i entry of AB. This is $\sum_{k=1}^{n} a_{jk} b_{ki}$, by (2.2), which may be written as $\sum_{k=1}^{n} b_{ki} a_{jk}$, changing the order in which multiplication is carried out in each term in the summation. Now we note that b_{ki} is the i–k entry of B^T and a_{jk} is the k–j entry of A^T. It follows that $\sum_{k=1}^{n} b_{ki} a_{jk}$ is the i–j entry of $B^T A^T$. Identity of i–j entries establishes the result.

Example 1

Transposing both sides of the matrix equation

$$A\mathbf{u} = \mathbf{v},$$

where A is 3×3 and $\mathbf{u}$ and $\mathbf{v}$ are 3×1 column vectors, yields

$$\mathbf{u}^T A^T = \mathbf{v}^T,$$

where $\mathbf{u}^T$ and $\mathbf{v}^T$ are 1×3 row vectors.

Example 2

The matrix product $\mathbf{u}^T A \mathbf{u}$, where A is 3×3 and $\mathbf{u}$ is 3×1, is 1×1, and transposition leaves it unchanged. Transposing a matrix product is equivalent to writing down the matrices in reverse order and transposing each one (see problem 2.1.2). In this case we obtain

$$\mathbf{u}^T A \mathbf{u} = (\mathbf{u}^T A \mathbf{u})^T = \mathbf{u}^T A^T (\mathbf{u}^T)^T = \mathbf{u}^T A^T \mathbf{u}.$$

Note that in the special case where A is a *skew-symmetric matrix*, satisfying $A^T = -A$, we have

$$\mathbf{u}^T A \mathbf{u} = \mathbf{u}^T (-A) \mathbf{u} = -\mathbf{u}^T A \mathbf{u}.$$

Hence $2\mathbf{u}^T A \mathbf{u} = 0$, and we conclude that $\mathbf{u}^T A \mathbf{u} = 0$.

Problems

2.1.1. If A, B and C are arbitrary 2×2 matrices, of the form

$$\begin{bmatrix} a_{11} & a_{12} \\ a_{21} & a_{22} \end{bmatrix}, \begin{bmatrix} b_{11} & b_{12} \\ b_{21} & b_{22} \end{bmatrix}, \begin{bmatrix} c_{11} & c_{12} \\ c_{21} & c_{22} \end{bmatrix},$$

verify by explicit calculation that

$$(AB)C = A(BC).$$

2.1.2. Making use of the identity (2.7), establish the result

$$(ABC)^T = C^T B^T A^T$$

for arbitrary A, B, C, conformable for multiplication.

2.1.3. Show that for any $m \times n$ matrix $A, (AA^T)^T = AA^T$. Deduce that AA^T must be a symmetric $m \times m$ matrix, and similarly that $A^T A$ must be a symmetric $n \times n$ matrix. Evaluate these matrices, and thus verify the result, when

$$A = \begin{bmatrix} 5 & 6 & -1 \\ 9 & 10 & 2 \end{bmatrix}.$$

2.1.4. Express the 'quadratic form'

$$u_1^2 + 3u_2^2 - u_3^2 + 4u_1 u_2 - 8u_2 u_3$$

in the form $\mathbf{u}^T A \mathbf{u}$, where A is a symmetric 3×3 matrix and

$$\mathbf{u} = [u_1 \quad u_2 \quad u_3]^T.$$

2.1.3 Applications

Here we indicate how some of the results of Chapter 1 may be expressed in matrix form.

First, the component form of the scalar product (1.5) may be written as $\mathbf{a}^T \mathbf{b}$, remembering that $\mathbf{a}$ now stands for the 3×1 column vector of rectangular cartesian components of the 'absolute' vector $\mathbf{a}$, and hence that $\mathbf{a}^T$ is the 1×3 row vector of components. Multiplying the 1×3 row vector $\mathbf{a}^T$ into the 3×1 column vector $\mathbf{b}$ produces the 1×1 array whose single entry is the scalar product.

Next, it is convenient to represent the 3×3 array of direction cosines l_{jp} defined in section 1.4.1 by the matrix R. Then equation (1.25) becomes in matrix terms

$$RR^T = I, \tag{2.8}$$

while (1.27) is equivalent to

$$R^T R = I. \tag{2.9}$$

Note that the requirement in matrix multiplication that the repeated suffixes appear in adjacent positions leads to the replacement of R by R^T in the precise positions shown in (2.8) and (2.9). Moreover, taking the transpose of both sides in (2.8) does not lead to (2.9) but merely leaves both sides unchanged. We shall see in section 2.3.2 that (2.9) *can* be deduced from (2.8), but that the new operation of matrix inversion is required.

The transformation equations (1.31) and (1.32) may be expressed as

$$\mathbf{u} = R^T \mathbf{u}' \tag{2.10}$$

and

$$\mathbf{u}' = R\mathbf{u}. \tag{2.11}$$

Some care is needed in handling (1.31), since, although the repeated suffix j occurs in adjacent positions, the matrix equation $\mathbf{u} = \mathbf{u}' R$ would not be correct; in fact the 3×1 column vector of transformed components $\mathbf{u}'$ and the 3×3 matrix R are not even conformable for multiplication as they stand. On the other hand, the equation $\mathbf{u}^T = (\mathbf{u}')^T R$ is correct and may be obtained by transposing both sides of (2.10).

Finally, if we represent the arrays of coefficients $L_i{}^j$ and $M^i{}_j$ in Section 1.5.3 by the matrices L and M, i referring to the row and j to the column in each case, equations

(1.41), (1.42) and (1.43) may be written

and
$$\left.\begin{array}{r}ML^{\mathrm{T}}=I,\\ \mathbf{y}=L^{\mathrm{T}}\mathbf{x}^{\mathrm{c}},\\ \mathbf{x}^{\mathrm{c}}=M\mathbf{y},\end{array}\right\} \quad (2.12)$$

where $\mathbf{x}^{\mathrm{c}}$ and $\mathbf{y}$ are the column vectors of the general and rectangular cartesian components of the position vector, respectively.

2.2 DETERMINANTS

2.2.1 Definitions

The **determinant** of a 2×2 matrix A, where

$$A = \begin{bmatrix} a_{11} & a_{12} \\ a_{21} & a_{22} \end{bmatrix},$$

is simply the scalar quantity $a_{11}a_{22} - a_{12}a_{21}$ and is denoted here by det A. Alternatively it can be represented by the array

$$\begin{vmatrix} a_{11} & a_{12} \\ a_{21} & a_{22} \end{vmatrix},$$

which may be distinguished from the matrix A itself by the use of the enclosing vertical lines rather than square brackets. For example,

$$\begin{vmatrix} 7 & 4 \\ 2 & 3 \end{vmatrix} = 7 \times 3 - 4 \times 2 = 13.$$

Note that an interchange of row or of column changes the sign of the determinant. Thus

$$\begin{vmatrix} 2 & 3 \\ 7 & 4 \end{vmatrix} = \begin{vmatrix} 4 & 7 \\ 3 & 2 \end{vmatrix} = 8 - 21 = -13.$$

We now define the determinant det A of a 3×3 matrix

$$A = \begin{bmatrix} a_{11} & a_{12} & a_{13} \\ a_{21} & a_{22} & a_{23} \\ a_{31} & a_{32} & a_{33} \end{bmatrix}.$$

First we define the **minors** M_{ij} of the matrix. The minor M_{ij} is the determinant of the 2×2 matrix obtained by deleting the ith row and jth column of A. Thus

$$M_{11} = \begin{vmatrix} a_{22} & a_{23} \\ a_{32} & a_{33} \end{vmatrix}, \quad M_{12} = \begin{vmatrix} a_{21} & a_{23} \\ a_{31} & a_{33} \end{vmatrix}, \quad \text{etc.}$$

Then

$$\det A = \begin{vmatrix} a_{11} & a_{12} & a_{13} \\ a_{21} & a_{22} & a_{23} \\ a_{31} & a_{32} & a_{33} \end{vmatrix} = a_{11}M_{11} - a_{12}M_{12} + a_{13}M_{13}. \quad (2.13)$$

(The determinant of a 3×3 matrix is called a **third-order determinant**.) The expression

(2.13) for det A (expanding by the first row of A) is equivalent to that obtained by expanding by the first column,

$$\det A = a_{11}M_{11} - a_{21}M_{21} + a_{31}M_{31}, \tag{2.14}$$

as may be easily checked. Written out in full, these expressions become

$$\det A = a_{11}(a_{22}a_{33} - a_{23}a_{32}) - a_{12}(a_{21}a_{33} - a_{23}a_{31}) + a_{13}(a_{21}a_{32} - a_{22}a_{31})$$
$$= a_{11}a_{22}a_{33} + a_{12}a_{23}a_{31} + a_{13}a_{21}a_{32} - a_{11}a_{23}a_{32} - a_{12}a_{21}a_{33} - a_{13}a_{22}a_{31}. \tag{2.15}$$

For example,

$$\begin{vmatrix} 1 & 2 & 3 \\ 4 & 5 & 6 \\ 7 & 8 & 9 \end{vmatrix} = 1(45-48) - 2(36-42) + 3(32-35) = -3 + 12 - 9 = 0.$$

The expressions (2.13) and (2.14) may be rewritten so as to eliminate minus signs by introducing the **cofactors** A_{ij} of the matrix. These are defined in terms of the minors by

$$A_{ij} = (-1)^{i+j} M_{ij},$$

so that $A_{ij} = M_{ij}$ if the sum $i+j$ is even and $A_{ij} = -M_{ij}$ if this sum is odd. We then have

$$\det A = a_{11}A_{11} + a_{12}A_{12} + a_{13}A_{13} = a_{11}A_{11} + a_{21}A_{21} + a_{31}A_{31}. \tag{2.16}$$

The expression (2.15) may be written in terms of the alternating symbol e_{ijk}, defined by (1.15), as

$$\det A = \sum_{p=1}^{3} \sum_{q=1}^{3} \sum_{r=1}^{3} e_{pqr} a_{1p} a_{2q} a_{3r}$$
$$= e_{pqr} a_{1p} a_{2q} a_{3r} \qquad \text{by the summation convention.} \tag{2.17}$$

It follows that

$$\det A = e_{pqr} a_{2q} a_{1p} a_{3r}$$
$$= e_{qpr} a_{2p} a_{1q} a_{3r} \qquad \text{writing } p \text{ for } q \text{ and } q \text{ for } p \text{ (dummy suffixes)}$$
$$= -e_{pqr} a_{2p} a_{1q} a_{3r} \qquad \text{using the anti-symmetric properties of } e_{ijk}$$
$$= -\begin{vmatrix} a_{21} & a_{22} & a_{23} \\ a_{11} & a_{12} & a_{13} \\ a_{31} & a_{32} & a_{33} \end{vmatrix}.$$

We have thus illustrated one of the basic properties of determinants, that interchange of rows (here the first and the second) changes the *sign* of the determinant. A useful consequence of this result is that, if a matrix has two identical rows, its determinant must be zero.

Rewriting (2.15) as

$$\det A = a_{11}a_{22}a_{33} + a_{21}a_{32}a_{13} + a_{31}a_{12}a_{23} - a_{11}a_{32}a_{23} - a_{31}a_{22}a_{13} - a_{21}a_{12}a_{33},$$

we observe that we have the further identity

$$\det A = \sum_{p=1}^{3}\sum_{q=1}^{3}\sum_{r=1}^{3} e_{pqr} a_{p1} a_{q2} a_{r3} = e_{pqr} a_{p1} a_{q2} a_{r3}. \tag{2.18}$$

It follows immediately that the above observations on interchange of *rows* apply equally well to interchange of *columns*.

We should emphasize that the expressions (2.13)–(2.18) are valid only for third-order determinants. It is possible to extend these results to the general case of a determinant of an $n \times n$ matrix. The properties of determinants to be discussed next, in particular (2.21) and (2.22), remain true for higher-order determinants.

Note that the determinant of the unit 3×3 matrix I is equal to 1. This remains true for the $n \times n$ unit matrix.

A final observation arises from comparing (2.17) and (2.18) with the expression (1.20) for the scalar triple product of three vectors. It follows from (2.17) that, if we regard each *row* of a 3×3 matrix as comprising the three components of a vector in three-dimensional space with respect to a rectangular cartesian reference system, then the determinant is equal to the scalar triple product of the three vectors defined by the three rows. Similarly, from (2.18), the *columns* may be regarded as comprising the components of vectors, and their scalar triple product has the same value, the determinant of the matrix.

2.2.2 Properties of third-order determinants

We have seen that the interchange of rows or columns changes the sign of a determinant. It also follows from (2.17) and (2.18), since $e_{pqr} = e_{qrp} = e_{rpq}$, that the value of a determinant is unchanged if the position of rows or columns is varied in a cyclic manner, e.g.

$$\begin{vmatrix} 1 & 9 & 8 \\ 4 & 5 & 2 \\ 7 & 3 & 6 \end{vmatrix} = \begin{vmatrix} 4 & 5 & 2 \\ 7 & 3 & 6 \\ 1 & 9 & 8 \end{vmatrix} = \begin{vmatrix} 7 & 3 & 6 \\ 1 & 9 & 8 \\ 4 & 5 & 2 \end{vmatrix} = -250.$$

Hence, by considering the two determinants

$$\begin{vmatrix} a_{i1} & a_{i2} & a_{i3} \\ a_{j1} & a_{j2} & a_{j3} \\ a_{k1} & a_{k2} & a_{k3} \end{vmatrix} \quad \text{and} \quad \begin{vmatrix} a_{1i} & a_{1j} & a_{1k} \\ a_{2i} & a_{2j} & a_{2k} \\ a_{3i} & a_{3j} & a_{3k} \end{vmatrix}.$$

we may deduce the interesting identities

$$e_{pqr} a_{ip} a_{jq} a_{kr} = (\det A) e_{ijk} \tag{2.19}$$

and
$$e_{pqr} a_{pi} a_{qj} a_{rk} = (\det A) e_{ijk}, \tag{2.20}$$

since the RHSs are equal to

(i) $\det A$ if i,j,k are in cyclic order $1,2,3$, corresponding to a cyclic interchange of rows or columns,
(ii) $-\det A$ if i,j,k are in the anti-cyclic order $3,2,1$ (which would be produced by interchanging two rows or columns) and
(iii) 0 if any of i,j,k are the same (corresponding to two identical rows or columns).

Now consider the matrix product AB, where A and B are both 3×3 matrices with elements a_{ij} and b_{ij}. Then AB has the i–p element $a_{ij}b_{jp}$ (using the summation convention). Hence

$$\det(AB) = e_{pqr}(a_{1j}b_{jp})(a_{2k}b_{kq})(a_{3l}b_{lr}) \qquad \text{(by (2.17))},$$

where we have needed to introduce new dummy suffixes k, l. Rearranging the terms,

$$\det(AB) = (e_{pqr}b_{jp}b_{kq}b_{lr})(a_{1j}a_{2k}a_{3l})$$
$$= (\det B)e_{jkl}(a_{1j}a_{2k}a_{3l}) \qquad \text{(by (2.19))}$$
$$= (\det B)(e_{jkl}a_{1j}a_{2k}a_{3l}) = (\det B)(\det A) \qquad \text{(by (2.17))}.$$

We have thus established the fundamental result

$$\det(AB) = \det A \det B \qquad (2.21)$$

for 3×3 matrices. This result may also be verified for 2×2 matrices, and can be extended to the general $n \times n$ case, once a definition of nth-order determinants has been supplied, although we shall not require this general result here.

For 3×3 matrices A, B and C, it follows immediately from (2.21) that

$$\det(ABC) = \det A \det(BC) = \det A \det B \det C. \qquad (2.22)$$

Another basic property of determinants is that, for any A,

$$\det(A^{\mathrm{T}}) = \det A, \qquad (2.23)$$

i.e. the value of a determinant is unchanged when rows and columns are interchanged. This result may be seen by noting that, from (2.17),

$$\det(A^{\mathrm{T}}) = e_{pqr}a_{p1}a_{q2}a_{r3},$$

which, by (2.18), is the same as $\det A$.

Other properties of determinants, which frequently simplify their evaluation, will not be needed in this book. These include the following:

(a) adding a scalar multiple of any row (column) to another row (column) does not affect the value of the determinant;
(b) a row (column) in which each element contains a common factor may be factorized and the common factor taken outside the determinant, e.g.

$$\begin{vmatrix} 6 & 12 & 18 \\ 5 & 1 & 2 \\ -1 & 0 & 4 \end{vmatrix} = 6 \begin{vmatrix} 1 & 2 & 3 \\ 5 & 1 & 2 \\ -1 & 0 & 4 \end{vmatrix} = -222.$$

Problems

2.2.1 If A and B are the 2×2 matrices

$$\begin{bmatrix} a_{11} & a_{12} \\ a_{21} & a_{22} \end{bmatrix} \quad \text{and} \quad \begin{bmatrix} b_{11} & b_{12} \\ b_{21} & b_{22} \end{bmatrix},$$

verify by explicit calculation that $\det(AB) = \det A \det B$.

2.2.2 Show that

$$e_{pqr}a_{2p}a_{2q}a_{3r} = 0,$$

where a_{ij} is the i–j entry of a 3×3 matrix A. Hence show that, if A' is the matrix obtained from A by adding a multiple λ of the second row to the first row,

$$\det A' = \det A,$$

i.e. the value of the determinant is unchanged.

2.3 MATRIX INVERSION AND DIAGONALIZATION

2.3.1 Matrix inverses

We have seen that the 3×3 unit matrix I may be considered as equivalent to unity for the purposes of matrix multiplication. One might ask whether a process of matrix division is possible, whereby, given a 3×3 matrix A, solutions in the form of a 3×3 matrix B could be found to equations of the form

$$AB = I, \tag{2.23}$$

so that B could in some sense be regarded as the reciprocal of A. It turns out that, although it is not meaningful to talk about a reciprocal I/A, it *is* possible to define (except for a certain category of matrices to be explained below) an **inverse** matrix, denoted by A^{-1}, satisfying (2.23). Such a matrix B ($B = A^{-1}$) will have elements b_{ij} satisfying

$$a_{ip}b_{pj} = \delta_{ij}, \tag{2.24}$$

by (2.2), using the summation convention.

Now, from equation (2.19), we may deduce that

$$e_{ljk}(e_{pqr}a_{ip}a_{jq}a_{kr}) = (\det A)e_{ljk}e_{ijk} = 2(\det A)\delta_{il},$$

by problem 1.3.1, and this may be rewritten as

$$a_{ip}(e_{jlk}e_{pqr}a_{lq}a_{kr}) = 2(\det A)\delta_{ij}, \tag{2.25}$$

writing j for l and l for j.

Comparing (2.25) with (2.24), we see that a solution for B is certainly given, provided that $\det A \neq 0$, by

$$b_{pj} = \frac{1}{2 \det A} e_{jlk}e_{pqr}a_{lq}a_{kr}.$$

Now evaluation of the expression $\frac{1}{2}e_{jlk}e_{pqr}a_{lq}a_{kr}$ for all possible values of the free suffixes j and p produces precisely the set of cofactors A_{jp} of the matrix A as previously defined. For example, putting $j = p = 1$,

$$\frac{1}{2}e_{1lk}e_{1qr}a_{lq}a_{kr}$$
$$= \frac{1}{2}(e_{123}e_{123}a_{22}a_{33} + e_{132}e_{132}a_{33}a_{22} + e_{123}e_{132}a_{23}a_{32} + e_{132}e_{123}a_{32}a_{23})$$
$$= \frac{1}{2}(a_{22}a_{33} + a_{33}a_{22} - a_{23}a_{32} - a_{32}a_{23}) = a_{22}a_{33} - a_{23}a_{32}$$
$$= \begin{vmatrix} a_{22} & a_{23} \\ a_{32} & a_{33} \end{vmatrix} = A_{11},$$

and similarly for other elements. Thus

$$A_{ip} = \tfrac{1}{2} e_{ijk} e_{pqr} a_{jq} a_{kr}, \tag{2.26}$$

and

$$b_{pj} = \frac{1}{\det A} A_{jp}.$$

The transpose of the matrix whose elements are the cofactors A_{ij} is called the **adjoint** matrix of A, and is denoted by adj A. The inverse matrix B has the same elements as the adjoint, but divided by the determinant. Thus, provided that $\det A \neq 0$, we have the formula

$$A^{-1} = \frac{1}{\det A} (\text{adj } A). \tag{2.27}$$

For example,

$$\begin{bmatrix} 1 & 9 & 8 \\ 4 & 5 & 2 \\ 7 & 3 & 6 \end{bmatrix}^{-1} = -\frac{1}{250} \begin{bmatrix} 24 & -10 & -23 \\ -30 & -50 & 60 \\ -22 & 30 & -31 \end{bmatrix}^{\text{T}} = \begin{bmatrix} -0.096 & 0.12 & 0.088 \\ 0.04 & 0.2 & -0.12 \\ 0.092 & -0.24 & 0.124 \end{bmatrix}.$$

A matrix whose determinant is non-zero is called **non-singular**. **Singular** matrices (those whose determinant is zero) cannot have an inverse, since (2.23) cannot be satisfied. To see this, take determinants of both sides, using (2.21) and the observation that $\det I = 1$. The result is

$$\det A \det B = 0 \det B = \det I = 1,$$

which cannot be true for any matrix B.

For non-singular matrices, applying (2.21) to the equation

$$AA^{-1} = I \tag{2.28}$$

produces the result

$$\det(AA^{-1}) = \det A \det A^{-1} = \det I = 1.$$

Thus

$$\det(A^{-1}) = \frac{1}{\det A}. \tag{2.29}$$

It follows that, if A is non-singular, so is A^{-1}.

Given that A^{-1} exists and satisfies (2.28), matrix multiplication on the left by A^{-1} gives

$$A^{-1}(AA^{-1}) = A^{-1}I = A^{-1}$$

and hence $(A^{-1}A)A^{-1} = A^{-1}$ by associativity. Now we know that A^{-1} is also non-singular and has an inverse $(A^{-1})^{-1}$. Multiplying on the right by this matrix gives

$$(A^{-1}A)[A^{-1}(A^{-1})^{-1}] = (A^{-1})(A^{-1})^{-1} = I.$$

Hence

$$(A^{-1}A)I = A^{-1}A = I.$$

Thus we have shown that, if A^{-1} exists,
$$A^{-1}A = AA^{-1} = I, \qquad (2.30)$$
and, clearly,
$$(A^{-1})^{-1} = A. \qquad (2.31)$$

Knowledge of the inverse A^{-1} of a non-singular 3×3 matrix A leads to the unique solution in the form of a 3×1 column vector $\mathbf{x}$ of the equation
$$A\mathbf{x} = \mathbf{b}, \qquad (2.32)$$
where $\mathbf{b}$ is a given 3×1 column vector. Multiplication on the left by A^{-1} and use of (2.30) gives
$$\mathbf{x} = A^{-1}\mathbf{b}.$$
If $\mathbf{b}$ is the zero vector $\mathbf{0}$, (2.32) has the unique solution $\mathbf{x} = A^{-1}\mathbf{0} = \mathbf{0}$ when A is non-singular. For the equation $A\mathbf{x} = \mathbf{0}$ to have solutions $\mathbf{x}$ not equal to $\mathbf{0}$, it is necessary for $\mathbf{A}$ to be singular, i.e. $\det A = 0$.

Problems

2.3.1 If A is the 2×2 matrix
$$\begin{bmatrix} a_{11} & a_{12} \\ a_{21} & a_{22} \end{bmatrix},$$
verify that
$$\begin{bmatrix} a_{11} & a_{12} \\ a_{21} & a_{22} \end{bmatrix} \begin{bmatrix} a_{22} & -a_{12} \\ -a_{21} & a_{11} \end{bmatrix} = \det A \begin{bmatrix} 1 & 0 \\ 0 & 1 \end{bmatrix},$$
and hence that, if $\det A \neq 0$,
$$A^{-1} = \frac{1}{\det A} \begin{bmatrix} a_{22} & -a_{12} \\ -a_{21} & a_{11} \end{bmatrix}.$$

2.3.2 Using equation (2.27) to evaluate $(A^{-1})^{-1}$, show that
$$\text{adj}(A^{-1}) = \frac{1}{\det A} A.$$

2.3.3 Use equation (2.27) to solve the set of simultaneous equations
$$\left.\begin{aligned} x_1 - x_2 + x_3 &= 4, \\ 2x_1 + x_2 - 3x_3 &= -3, \\ -3x_1 + 2x_2 + x_3 &= -6. \end{aligned}\right\}$$

2.3.4 Verify that the matrix of coefficients in the set of simultaneous equations
$$\left.\begin{aligned} x_1 + x_2 + x_3 &= 0, \\ 4x_1 + x_2 + 3x_3 &= 0, \\ 2x_1 - 7x_2 - x_3 &= 0 \end{aligned}\right\}$$
is singular, and find the solutions $[x_1 \; x_2 \; x_3]^T \neq 0$.

2.3.2 Orthogonal matrices

Matrices R satisfying (2.8) are such that

$$R^{-1} = R^T. \qquad (2.33)$$

This is the defining property of **orthogonal** matrices. Using (2.30), we may deduce (2.9) from (2.8). This means that the set of equations (1.28) could have been deduced from the set (1.26) by matrix methods.

If we take determinants of both sides of (2.8), using (2.21) and (2.22), we obtain

$$\det R \det(R^T) = (\det R)^2 = 1.$$

Hence

$$\det R = \pm 1. \qquad (2.34)$$

Orthogonal matrices satisfying $\det R = +1$ are called **proper orthogonal**. In terms of transformations between rectangular cartesian systems of reference, proper orthogonal matrices correspond to the set of direction cosines relating two systems in which one system can be simply rotated (without reflection) into the position of the other system. For, when two systems are coincident, the matrix R of direction cosines reduces to the identity matrix, so that $\det R = 1$. If we then consider a rotation taking place progressively away from the first system towards the second system, the direction cosines will change in a continuous manner, and so also will the determinant of the matrix R. However, since the only values that $\det R$ can take are $+1$ and -1, $\det R$ must remain fixed at $+1$. The value -1 can arise only if we consider a discontinuous change involving a transformation from a right-handed set of axes to a left-handed set, or vice versa.

As examples, the matrix of coefficients in (1.29)

$$\begin{bmatrix} \cos\theta & \sin\theta & 0 \\ -\sin\theta & \cos 0 & 0 \\ 0 & 0 & 1 \end{bmatrix}$$

is proper orthogonal. Reversing the direction of the $\mathbf{i}_3'$ axis in Fig. 1.15, however, would give a left-handed system $\{\mathbf{i}_1', \mathbf{i}_2', \mathbf{i}_3'\}$, and the corresponding matrix R,

$$\begin{bmatrix} \cos\theta & \sin\theta & 0 \\ -\sin\theta & \cos\theta & 0 \\ 0 & 0 & -1 \end{bmatrix},$$

has determinant equal to -1.

2.3.3 Diagonalization of symmetric matrices

Given a 3×3 symmetric matrix A, consider the matrix equation

$$A\mathbf{x} = \lambda \mathbf{x}, \qquad (2.35)$$

where $\mathbf{x}$ is an initially unknown column vector and λ is an initially unknown scalar. Solutions $\mathbf{x}$ not equal to the zero vector are called **eigenvectors** of A, and the corresponding λ are called **eigenvalues** (proper values, or characteristic values).

Equation (2.35) may be expressed as
$$(A - \lambda I)\mathbf{x} = 0,$$
and this equation has solutions not identically zero only if
$$\det(A - \lambda I) = 0, \tag{2.36}$$
i.e. if $A - \lambda I$ is singular.

Now (2.36) gives a cubic equation (the **characteristic** equation) in λ, and this must be solved to give the three possible eigenvalues $\lambda^{(1)}$, $\lambda^{(2)}$, $\lambda^{(3)}$ of A. A cubic with real coefficients must have either three real roots *or* one real root and two complex conjugate roots. However, for *symmetric* matrices A, it may be shown that the roots are necessarily all real.

For, taking the complex conjugate of both sides of (2.35) gives
$$A\bar{\mathbf{x}} = \bar{\lambda}\bar{\mathbf{x}}, \tag{2.37}$$
where $\bar{\lambda}$ is the complex conjugate of λ (supposing λ to be possibly complex) and $\bar{\mathbf{x}}$ is the 3×1 column vector whose elements are the complex conjugates of the elements of $\mathbf{x}$. Equation (2.37) says that, if λ is a complex eigenvalue of A with (complex) eigenvector $\mathbf{x}$, then $\bar{\lambda}$ is a complex eigenvalue with eigenvector $\bar{\mathbf{x}}$. However, a little matrix manipulation shows that

$$\begin{aligned}
\bar{\mathbf{x}}^T A \mathbf{x} &= \bar{\mathbf{x}}^T(\lambda \mathbf{x}) = \lambda(\bar{\mathbf{x}}^T \mathbf{x}) \\
&= (\bar{\mathbf{x}}^T A \mathbf{x})^T && \text{(transposing a } 1 \times 1 \text{ matrix)} \\
&= \mathbf{x}^T A^T \bar{\mathbf{x}} = \mathbf{x}^T A \bar{\mathbf{x}} && \text{since } A \text{ is symmetric} \\
&= \mathbf{x}^T(\bar{\lambda}\bar{\mathbf{x}}) = \bar{\lambda}(\mathbf{x}^T \bar{\mathbf{x}}) = \bar{\lambda}(\bar{\mathbf{x}}^T \mathbf{x}).
\end{aligned}$$

Moreover, $\bar{\mathbf{x}}^T \mathbf{x}$, obtained by multiplying the entries of $\mathbf{x}$ by their complex conjugates in $\bar{\mathbf{x}}$ and summing, is certainly a real and positive quantity, since $\mathbf{x}$ cannot be zero, and the product of a complex number with its conjugate is equal to square of its modulus.

We deduce that, cancelling $(\bar{\mathbf{x}}^T \mathbf{x})$, $\lambda = \bar{\lambda}$, which proves that λ is real.

Suppose that the three (real) eigenvectors are $\mathbf{x}^{(1)}$, $\mathbf{x}^{(2)}$ and $\mathbf{x}^{(3)}$, corresponding to $\lambda^{(1)}$, $\lambda^{(2)}$ and $\lambda^{(3)}$, respectively. Clearly, by (2.35) these eigenvectors are undefined up to a scalar multiple. Let us take them all to be unit vectors. This means that we can impose the conditions $(\mathbf{x}^{(1)})^T \mathbf{x}^{(1)} = (\mathbf{x}^{(2)})^T \mathbf{x}^{(2)} = (\mathbf{x}^{(3)})^T \mathbf{x}^{(3)} = 1$.

Our second important proof demonstrates that, if the eigenvalues are all distinct, the eigenvectors are mutually orthogonal. The proof follows similar lines to the one above. Consider $\mathbf{x}^{(1)}$ and $\mathbf{x}^{(2)}$ without loss of generality. We have
$$A\mathbf{x}^{(1)} = \lambda^{(1)}\mathbf{x}^{(1)}$$
and
$$A\mathbf{x}^{(2)} = \lambda^{(2)}\mathbf{x}^{(2)}.$$
Hence
$$\begin{aligned}
(\mathbf{x}^{(2)})^T A \mathbf{x}^{(1)} &= \lambda^{(1)}(\mathbf{x}^{(2)})^T \mathbf{x}^{(1)} \\
&= [(\mathbf{x}^{(2)})^T A \mathbf{x}^{(1)}]^T \\
&= (\mathbf{x}^{(1)})^T A^T \mathbf{x}^{(2)} = (\mathbf{x}^{(1)})^T A \mathbf{x}^{(2)} && \text{since } A \text{ is symmetric} \\
&= (\mathbf{x}^{(1)})^T \lambda^{(2)} \mathbf{x}^{(2)} = \lambda^{(2)}(\mathbf{x}^{(1)})^T \mathbf{x}^{(2)} \\
&= \lambda^{(2)}(\mathbf{x}^{(2)})^T \mathbf{x}^{(1)}
\end{aligned}$$

(transposing a number of 1×1 quantities). Now, if $\lambda^{(1)} \neq \lambda^{(2)}$, the equality

$$\lambda^{(1)}[(\mathbf{x}^{(2)})^T\mathbf{x}^{(1)}] = \lambda^{(2)}[(\mathbf{x}^{(2)})^T\mathbf{x}^{(1)}]$$

can hold only if $(\mathbf{x}^{(2)})^T\mathbf{x}^{(1)} = 0$. We interpret this result as meaning that the scalar product of the vectors whose components are $\mathbf{x}^{(1)}$ and $\mathbf{x}^{(2)}$ with respect to some set of rectangular cartesian axes is zero. This completes the proof.

If any of the eigenvalues are the same, we have a greater range of eigenvectors at our disposal, and it may be shown that we can always choose a set of eigenvectors having the orthogonality property. This may now be written

$$(\mathbf{x}^{(i)})^T\mathbf{x}^{(j)} = \delta_{ij}, \qquad (2.38)$$

since the eigenvectors have been 'normalized' as unit vectors.

Suppose we now write the three eigenvectors as the columns of a 3×3 matrix $[\mathbf{x}^{(1)} \ \mathbf{x}^{(2)} \ \mathbf{x}^{(3)}]$. Then from (2.38) we may see that this matrix satisfies an equation similar to (2.8) and hence must be orthogonal. If its determinant turns out to be -1 rather than $+1$, we can make the matrix proper orthogonal by varying the order of the eigenvalues, e.g. taking $\lambda^{(2)}$ as $\lambda^{(1)}$ and vice versa, with corresponding change in order of $\mathbf{x}^{(1)}$ and $\mathbf{x}^{(2)}$. This has the effect of interchanging two columns of the matrix, thus changing the sign of the determinant, while leaving (2.38) unaltered. From the remarks at the end of section 2.2.1, of course, the determinant is equal to the scalar triple product $(\mathbf{x}^{(1)} \times \mathbf{x}^{(2)}) \cdot \mathbf{x}^{(3)}$.

Suppose that

$$\mathbf{x}^{(1)} = \begin{bmatrix} l_{11} \\ l_{12} \\ l_{13} \end{bmatrix}, \quad \mathbf{x}^{(2)} = \begin{bmatrix} l_{21} \\ l_{22} \\ l_{23} \end{bmatrix}, \quad \mathbf{x}^{(3)} = \begin{bmatrix} l_{31} \\ l_{32} \\ l_{33} \end{bmatrix}.$$

Then

$$[\mathbf{x}^{(1)} \ \mathbf{x}^{(2)} \ \mathbf{x}^{(3)}] = \begin{bmatrix} l_{11} & l_{21} & l_{31} \\ l_{12} & l_{22} & l_{32} \\ l_{13} & l_{23} & l_{33} \end{bmatrix},$$

and it is convenient to call this R^T, where

$$R = \begin{bmatrix} l_{11} & l_{12} & l_{13} \\ l_{21} & l_{22} & l_{23} \\ l_{31} & l_{32} & l_{33} \end{bmatrix}.$$

We may regard l_{ip} as the direction cosines ($p = 1, 2, 3$) defining the direction of $\mathbf{x}^{(i)}$, and R as the matrix of direction cosines.

The three solutions to the eigenvector equation (2.35) may now be assembled together in a matrix equation

$$A[\mathbf{x}^{(1)} \ \mathbf{x}^{(2)} \ \mathbf{x}^{(3)}] = [\lambda^{(1)}\mathbf{x}^{(1)} \ \lambda^{(2)}\mathbf{x}^{(2)} \ \lambda^{(3)}\mathbf{x}^{(3)}],$$

which may be written as

$$AR^T = R^T \begin{bmatrix} \lambda^{(1)} & 0 & 0 \\ 0 & \lambda^{(2)} & 0 \\ 0 & 0 & \lambda^{(3)} \end{bmatrix}.$$

Matrix inversion and diagonalization

It follows that

$$RAR^T = RR^T \begin{bmatrix} \lambda^{(1)} & 0 & 0 \\ 0 & \lambda^{(2)} & 0 \\ 0 & 0 & \lambda^{(3)} \end{bmatrix} = I \begin{bmatrix} \lambda^{(1)} & 0 & 0 \\ 0 & \lambda^{(2)} & 0 \\ 0 & 0 & \lambda^{(3)} \end{bmatrix} = \begin{bmatrix} \lambda^{(1)} & 0 & 0 \\ 0 & \lambda^{(2)} & 0 \\ 0 & 0 & \lambda^{(3)} \end{bmatrix}.$$

(2.39)

Thus we have demonstrated that for any symmetric matrix A there exists a proper orthogonal matrix R such that RAR^T is a diagonal matrix.

Example 1

Let

$$A = \begin{bmatrix} 1 & -2 & 1 \\ -2 & 1 & 1 \\ 1 & 1 & 0 \end{bmatrix}.$$

Then

$$A - \lambda I = \begin{bmatrix} 1-\lambda & -2 & 1 \\ -2 & 1-\lambda & 1 \\ 1 & 1 & -\lambda \end{bmatrix},$$

and the characteristic equation is

$$\begin{vmatrix} 1-\lambda & -2 & 1 \\ -2 & 1-\lambda & 1 \\ 1 & 1 & -\lambda \end{vmatrix} = 0,$$

which reduces to

$$(-2-\lambda)(1-\lambda)(3-\lambda) = 0.$$

Hence the eigenvalues are $\lambda^{(1)} = -2$, $\lambda^{(2)} = 1$, $\lambda^{(3)} = 3$, the order being arbitrary chosen. Substituting the first eigenvalue $\lambda^{(1)}$ into the eigenvalue equation (2.35) gives

$$\begin{bmatrix} 3 & -2 & 1 \\ -2 & 3 & 1 \\ 1 & 1 & 2 \end{bmatrix} \mathbf{x} = \mathbf{0},$$

with solutions which are scalar multiples of

$$\begin{bmatrix} 1 \\ 1 \\ -1 \end{bmatrix}.$$

Similarly $\lambda^{(2)}$ and $\lambda^{(3)}$ give eigenvectors which are scalar multiples of

$$\begin{bmatrix} 1 \\ 1 \\ 2 \end{bmatrix} \quad \text{and} \quad \begin{bmatrix} 1 \\ -1 \\ 0 \end{bmatrix}, \text{ respectively.}$$

The 'normalization' procedure gives a set of unit eigenvectors

$$\mathbf{x}^{(1)} = \begin{bmatrix} \frac{1}{\sqrt{3}} \\ \frac{1}{\sqrt{3}} \\ -\frac{1}{\sqrt{3}} \end{bmatrix}, \quad \mathbf{x}^{(2)} = \begin{bmatrix} \frac{1}{\sqrt{6}} \\ \frac{1}{\sqrt{6}} \\ \frac{2}{\sqrt{6}} \end{bmatrix}, \quad \mathbf{x}^{(3)} = \begin{bmatrix} \frac{1}{\sqrt{2}} \\ -\frac{1}{\sqrt{2}} \\ 0 \end{bmatrix},$$

and this defines an orthogonal matrix

$$\begin{bmatrix} \frac{1}{\sqrt{3}} & \frac{1}{\sqrt{6}} & \frac{1}{\sqrt{2}} \\ \frac{1}{\sqrt{3}} & \frac{1}{\sqrt{6}} & -\frac{1}{\sqrt{2}} \\ -\frac{1}{\sqrt{3}} & \frac{2}{\sqrt{6}} & 0 \end{bmatrix}.$$

The determinant of this matrix turns out to be $+1$. Thus it is proper orthogonal, and there is no need to interchange the labels (1) and (2) on the first two eigenvectors and eigenvalues, as suggested above.

$$R = \begin{bmatrix} \frac{1}{\sqrt{3}} & \frac{1}{\sqrt{3}} & -\frac{1}{\sqrt{3}} \\ \frac{1}{\sqrt{6}} & \frac{1}{\sqrt{6}} & \frac{2}{\sqrt{6}} \\ \frac{1}{\sqrt{2}} & -\frac{1}{\sqrt{2}} & 0 \end{bmatrix},$$

and we know that RAR^T is diagonal and equal to

$$\begin{bmatrix} -2 & 0 & 0 \\ 0 & 1 & 0 \\ 0 & 0 & 3 \end{bmatrix}.$$

Example 2

If repeated roots occur in the characteristic equation, the procedure is similar. For example, let

$$A = \begin{bmatrix} 3 & -1 & 0 \\ -1 & 3 & 0 \\ 0 & 0 & 2 \end{bmatrix}.$$

Matrix inversion and diagonalization

The characteristic equation is

$$\det(A - \lambda I) = \begin{vmatrix} 3-\lambda & -1 & 0 \\ -1 & 3-\lambda & 0 \\ 0 & 0 & 2-\lambda \end{vmatrix} = (2-\lambda)^2(4-\lambda) = 0.$$

and the eigenvalues are

$$\lambda^{(1)} = 4, \quad \lambda^{(2)} = \lambda^{(3)} = 2.$$

The eigenvalue $\lambda^{(1)}$ has a corresponding eigenvector which is a scalar multiple of $\begin{bmatrix} 1 \\ -1 \\ 0 \end{bmatrix}$, and normalization gives

$$\mathbf{x}^{(1)} = \begin{bmatrix} \dfrac{1}{\sqrt{2}} \\ -\dfrac{1}{\sqrt{2}} \\ 0 \end{bmatrix}.$$

However, $\lambda^{(2)}$ gives rise to the eigenvector equation

$$\begin{bmatrix} 1 & -1 & 0 \\ -1 & 1 & 0 \\ 0 & 0 & 0 \end{bmatrix} \mathbf{x} = \mathbf{0},$$

with solutions $\mathbf{x} = \begin{bmatrix} \alpha \\ \alpha \\ \beta \end{bmatrix}$ for any values of α and β. These solutions define a whole plane of possible eigenvectors, and it is sufficient for us to choose two mutually orthogonal unit vectors in this plane. The simplest choices are to take $\alpha = 1/\sqrt{2}, \beta = 0$ for $\mathbf{x}^{(2)}$ and $\alpha = 0, \beta = 1$ for $\mathbf{x}^{(3)}$. Then we have an orthogonal matrix of eigenvectors

$$\begin{bmatrix} \dfrac{1}{\sqrt{2}} & \dfrac{1}{\sqrt{2}} & 0 \\ -\dfrac{1}{\sqrt{2}} & \dfrac{1}{\sqrt{2}} & 0 \\ 0 & 0 & 1 \end{bmatrix}.$$

The determinant again turns out to be $+1$. Hence, with

$$R = \begin{bmatrix} \dfrac{1}{\sqrt{2}} & -\dfrac{1}{\sqrt{2}} & 0 \\ \dfrac{1}{\sqrt{2}} & \dfrac{1}{\sqrt{2}} & 0 \\ 0 & 0 & 1 \end{bmatrix},$$

we have

$$RAR^T = \begin{bmatrix} 4 & 0 & 0 \\ 0 & 2 & 0 \\ 0 & 0 & 2 \end{bmatrix}.$$

Physical applications of the diagonalization procedure in the context of tensor analysis are described in section 3.2.6.

3

Tensor analysis

3.1 INTRODUCTION

There are a number of approaches to the theory of tensors. One is via the algebraic theory of linear operators. A second-order tensor, for example, is equivalent to a linear operator which 'acts' on vectors and produces vectors. It may be regarded as a 'black box' with vectors as input and vectors as output. There are a great many physical problems for which this view is useful, and a number of them will be discussed later. In these examples, where the vectors belong to a three dimensional euclidean space, it is possible to represent a second-order tensor as a 3×3 matrix of components, and the linear operation in terms of matrix multiplication. The 3×3 matrix T of components is multiplied into a 3×1 column vector of vector components **u** and produces a 3×1 column vector of components **v**. This is represented as the matrix equation

$$T\mathbf{u} = \mathbf{v}. \tag{3.1}$$

Here the column vectors are the components of the given vectors with respect to a given (or chosen) reference system of co-ordinates. The 3×3 matrix consists of the components of the tensor with respect to the same system. For a different reference system the components of **u**, **v** and the second-order tensor **T** will all change, in general, and (3.1) would be represented by matrices and column vectors with different numerical values. However, we may adopt the view that (3.1) has an invariant ('co-ordinate-free') validity, irrespective of the reference system chosen. This validity is generally a consequence of the physical characteristics of the problem being considered. The choice of reference system cannot affect the underlying physical laws in a given situation. Tensor analysis as developed here is a means of giving mathematical substance to this underlying physical idea.

For a given reference system of curvilinear co-ordinates a vector **u** will have contravariant components (u^1, u^2, u^3). With respect to a new reference system the components may become $(\bar{u}^1, \bar{u}^2, \bar{u}^3)$, and there is a set of transformation equations by which the new components can be calculated. The vector **u** itself is essentially unchanged, if we continue to think of it as a physical quantity with a certain magnitude and direction. Nevertheless, the components of **u** with respect to a given reference system, together with the transformation equations, are of fundamental importance to the theory, and we focus on these aspects in this chapter in order to reformulate the definition of a vector. We then extend the new definition to a hierarchy of quantities, which are **generalized tensors of order (or rank)** n. (A vector is in fact a tensor of order one.) Our definition sees a vector as a correspondence between general co-ordinate systems and sets of three ordered numbers. In other words, a vector is the set of all its possible component representations with respect to all possible reference systems.

Tensors of order two in three-dimensional euclidean space have nine components, just as a 3×3 matrix has nine elements. These components change according to a particular transformation law when the reference system is changed. The tensor itself may be regarded as the set of all its possible *matrix* representations with respect to all possible reference systems. We may also wish to regard a tensor, like a vector, as having an objective physical reality, although it is generally not possible to see it in terms as simple as those of magnitude and direction, as for vectors.

Our approach, then, is to emphasize primarily the transformation laws which relate the various component representations of tensors. We should also mention that, with respect to general reference systems, tensors may have contravariant and/or covariant components, and that the transformation laws will depend on the particular types of components under consideration. The linear operator properties of tensors will be demonstrated *after* presentation of the transformation laws.

For many purposes it is perfectly adequate to consider only rectangular cartesian reference systems. In courses on tensors, students are commonly introduced to *cartesian* tensors (for which only rectangular cartesian systems of co-ordinates are considered), before being exposed, if at all, to the generalized variety. Experience in handling the former should, in principle, help to make the latter more approachable, and we certainly recommend that the reader should have a thorough grasp of section 3.2 here before proceeding to section 3.3. Indeed, one aspect of tensor analysis on which we place particular emphasis in this book is the technique of proceeding from equations based on rectangular cartesian reference systems to equations which apply to arbitrary reference systems. In Chapter 6 we take for granted certain mathematical equations of physics expressed in cartesian tensor form and show how they may be converted to equations valid with respect to any reference system whatsoever. It should be borne in mind that the generalized theory *incorporates* the theory of cartesian tensors as a special case.

3.2 CARTESIAN TENSORS

3.2.1 Vectors as tensors of order one

Suppose we have a rectangular cartesian reference system, with an orthonormal basis $(\mathbf{i}_1, \mathbf{i}_2, \mathbf{i}_3)$. To specify a vector **u**, three real numbers u_1, u_2 and u_3 are required, i.e. the components of **u** in the directions of the basis vectors, given by the scalar products

$$u_j = \mathbf{u} \cdot \mathbf{i}_j, \qquad j = 1, 2, 3.$$

In chapter 1 we saw that, if we consider a different orthonormal basis $\{\mathbf{i}_1', \mathbf{i}_2', \mathbf{i}_3'\}$, the same vector **u** has components u_1', u_2', u_3' with respect to this basis, where the relation between the components on the two different bases is equation (1.32), i.e. $u_j' = l_{jp} u_p$; here l_{jp} represents the cosine of the angle between $\mathbf{i}_j'$ and $\mathbf{i}_p$. We now use the transformation law (1.32) to *define* a **cartesian tensor of order one** (written CT1, for short), which we may regard as equivalent to a vector.

A CT1 is a correspondence between orthonormal bases and sets of three real numbers which transform according to (1.32). To any given orthonormal basis $\{\mathbf{i}_1, \mathbf{i}_2, \mathbf{i}_3\}$ there corresponds a set of numbers $\{u_1, u_2, u_3\}$; to any other basis $\{\mathbf{i}_1', \mathbf{i}_2', \mathbf{i}_3'\}$ there corresponds another set $\{u_1', u_2', u_3'\}$, where

$$u_j' = l_{jp} u_p \quad \text{and} \quad l_{jp} = \mathbf{i}_j' \cdot \mathbf{i}_p. \tag{3.2}$$

This transformation law guarantees that the components always determine the same vector.

Thus, if we know the components of a vector in one rectangular cartesian system, we can obtain the components of the vector in any other rectangular cartesian system. Since there is an infinite choice of systems we cannot write down all the possible sets of components of any particular CT1, with the exception of the particular case where the components with respect to some basis are all zero. We can see that the transformation law then ensures that the components remain $\{0, 0, 0\}$ with respect to any other basis whatsoever. This collection of components, i.e. $\{0, 0, 0\}$ with respect to any basis, corresponds to the zero vector **0**.

The symbol **u** may stand for a vector quantity and also for a column vector of components

$$\begin{bmatrix} u_1 \\ u_2 \\ u_3 \end{bmatrix}$$

of the vector with respect to some orthonormal basis. The meaning of the symbol should be clear from the context. The transformation law for vectors (1.32) can be expressed in terms of column vectors and matrices as

$$\mathbf{u}' = R\mathbf{u}, \tag{3.3}$$

where the column vector **u**′ contains the components of the same vector (denoted by **u**) with respect to a different orthonormal basis, and R is the matrix of direction cosines.

We can use the definition (3.2) to show that the gradient of a scalar quantity, grad Φ, as defined by (1.44), is a CT1. We recall that for a given set of rectangular cartesian axes $Oy_1 y_2 y_3$, a point in space has position vector $\mathbf{r} = y_1 \mathbf{i}_1 + y_2 \mathbf{i}_2 + y_3 \mathbf{i}_3$, and a scalar quantity Φ at that point has partial derivatives $\partial \Phi / \partial y_i$, which constitute the components of grad Φ. To test whether or not $\partial \Phi / \partial y_i$ is a CT1, we must discover how the components change when we take a new set of axes $Oy_1' y_2' y_3'$, in the directions of the unit vectors $\{\mathbf{i}_1', \mathbf{i}_2', \mathbf{i}_3'\}$. (It is not necessary to consider a different origin O).

Now the position vector **r** is also expressible as $\mathbf{r} = y_1' \mathbf{i}_1' + y_2' \mathbf{i}_2' + y_3' \mathbf{i}_3'$, and the

cartesian co-ordinates of **r** transform like vector components, i.e.
$$y_i' = l_{ip} y_p, \qquad (3.4)$$
where the direction cosines l_{ip} fix the relative positions of the two sets of axes. The gradient has components $\partial \Phi / \partial y_i'$ in terms of the new co-ordinates. Now, by the chain rule for partial derivatives,

$$\frac{\partial \Phi}{\partial y_i'} = \frac{\partial \Phi}{\partial y_1} \frac{\partial y_1}{\partial y_i'} + \frac{\partial \Phi}{\partial y_2} \frac{\partial y_2}{\partial y_i'} + \frac{\partial \Phi}{\partial y_3} \frac{\partial y_3}{\partial y_i'}$$

$$= \frac{\partial \Phi}{\partial y_p} \frac{\partial y_p}{\partial y_i'} \qquad \text{using the summation convention.}$$

Inverting (3.4) to express y_p in terms of the y_i', as in (1.31), gives
$$y_p = l_{jp} y_j' \qquad (3.5)$$

Hence
$$\frac{\partial y_p}{\partial y_i'} = l_{jp} \frac{\partial y_j'}{\partial y_i'}$$

since the l_{jp} are constants (making sure, moreover, that no dummy suffix appears more than twice in the same expression), i.e.

$$\frac{\partial y_p}{\partial y_i'} = l_{jp} \delta_{ij} = l_{ip}. \qquad (3.6)$$

Thus
$$\frac{\partial \Phi}{\partial y_i'} = \frac{\partial \Phi}{\partial y_p} l_{ip},$$

i.e.
$$(\text{grad } \Phi)_i' = l_{ip} (\text{grad } \Phi)_p,$$

demonstrating that grad Φ transforms according to (3.2) and must be a CT1.

The upshot of this result is that grad Φ is a genuine vector and that we are justified in representing it by an arrow with magnitude and direction.

Problems

3.2.1 If u_i and v_i are CT1s, show that both $u_i + v_i$ and λu_i, where λ is a scalar, are CT1s. (Prove this in two ways, showing that both the suffix form (3.2) and the matrix form (3.3) of the transformation law are satisified.)

3.2.2 (a) If u_i is a CT1, show using (3.2) that
$$u_i' u_i' = u_j u_j,$$
with the equivalent matrix result from (3.3)
$$(\mathbf{u}')^T \mathbf{u}' = \mathbf{u}^T \mathbf{u}.$$

(This shows that the square of the magnitude of **u**, and thus the magnitude of **u** itself, is an **invariant**, i.e. it remains unchanged by cartesian transformations.)

(b) If v_i is also a CT1, show similarly that the scalar product $u_i v_i$ is an invariant.

3.2.2 Second-order tensors

A cartesian tensor of order two (CT2) is defined similarly as a set of components which transform according to a certain law when the basis vectors change. In three dimensions a CT2 has nine components, and we represent these components by a symbol such as T_{ij}, where the free suffixes i and j can take any value from 1 to 3. It is often convenient to write the components in a 3×3 matrix array

$$\begin{bmatrix} T_{11} & T_{12} & T_{13} \\ T_{21} & T_{22} & T_{23} \\ T_{31} & T_{32} & T_{33} \end{bmatrix}. \qquad (3.7)$$

These are the components of the tensor when referred to some orthonormal basis $\{\mathbf{i}_1, \mathbf{i}_2, \mathbf{i}_3\}$. When we choose a different basis $\{\mathbf{i}_1', \mathbf{i}_2', \mathbf{i}_3'\}$, the components (of the same tensor) become T_{ij}', say, where the new components are given by the transformation law

$$T_{ij}' = l_{ip} l_{jq} T_{pq}, \qquad (3.8)$$

and l_{ip}, as usual, represent the direction cosines of the new basis with respect to the old one.

Note that in this formulation of the law the free suffixes i and j associated with the 'new' components T_{ij}' occupy the first of the two possible positions of the suffixes in the expressions l_{ip}, l_{jq}, but the repeated (summed) suffixes p and q associated with the 'old' components T_{pq} occupy the second position. Any other letters may be used in the same positions, e.g. (3.8) may be written $T_{kl}' = l_{kr} l_{ls} T_{rs}$ or $T_{pq}' = l_{pi} l_{qj} T_{ij}$.

Written out in full, the transformation law is

$$\left. \begin{aligned}
T_{11}' &= l_{11}^2 T_{11} + l_{11} l_{12} T_{12} + l_{11} l_{13} T_{13} + l_{12} l_{11} T_{21} + l_{12}^2 T_{22} \\
&\quad + l_{12} l_{13} T_{23} + l_{13} l_{11} T_{31} + l_{13} l_{12} T_{32} + l_{13}^2 T_{33}, \\
T_{12}' &= l_{11} l_{21} T_{11} + l_{11} l_{22} T_{12} + l_{11} l_{23} T_{13} + l_{12} l_{21} T_{21} + l_{12} l_{22} T_{22} \\
&\quad + l_{12} l_{23} T_{23} + l_{13} l_{21} T_{31} + l_{13} l_{22} T_{32} + l_{13} l_{23} T_{33}, \\
T_{13}' &= l_{11} l_{31} T_{11} + l_{11} l_{32} T_{12} + l_{11} l_{33} T_{13} + l_{12} l_{31} T_{21} + l_{12} l_{32} T_{22} \\
&\quad + l_{12} l_{33} T_{23} + l_{13} l_{31} T_{31} + l_{13} l_{32} T_{32} + l_{13} l_{33} T_{33}, \\
T_{21}' &= l_{21} l_{11} T_{11} + l_{21} l_{12} T_{12} + l_{21} l_{13} T_{13} + l_{22} l_{11} T_{21} + l_{22} l_{12} T_{22} \\
&\quad + l_{22} l_{13} T_{23} + l_{23} l_{11} T_{31} + l_{23} l_{12} T_{32} + l_{23} l_{13} T_{33}, \\
T_{22}' &= l_{21}^2 T_{11} + l_{21} l_{22} T_{12} + l_{21} l_{23} T_{13} + l_{22} l_{21} T_{21} + l_{22}^2 T_{22} \\
&\quad + l_{22} l_{23} T_{23} + l_{23} l_{21} T_{31} + l_{23} l_{22} T_{32} + l_{23}^2 T_{33}, \\
T_{23}' &= l_{21} l_{31} T_{11} + l_{21} l_{32} T_{12} + l_{21} l_{33} T_{13} + l_{22} l_{31} T_{21} + l_{22} l_{32} T_{22} \\
&\quad + l_{22} l_{33} T_{23} + l_{23} l_{31} T_{31} + l_{23} l_{32} T_{32} + l_{23} l_{33} T_{33}, \\
T_{31}' &= l_{31} l_{11} T_{11} + l_{31} l_{12} T_{12} + l_{31} l_{13} T_{13} + l_{32} l_{11} T_{21} + l_{32} l_{12} T_{22} \\
&\quad + l_{32} l_{13} T_{23} + l_{33} l_{11} T_{31} + l_{33} l_{12} T_{32} + l_{33} l_{13} T_{33}, \\
T_{32}' &= l_{31} l_{21} T_{11} + l_{31} l_{22} T_{12} + l_{31} l_{23} T_{13} + l_{32} l_{21} T_{21} + l_{32} l_{22} T_{22} \\
&\quad + l_{32} l_{23} T_{23} + l_{33} l_{21} T_{31} + l_{33} l_{22} T_{32} + l_{33} l_{23} T_{33}, \\
\text{and} \quad T_{33}' &= l_{31}^2 T_{11} + l_{31} l_{32} T_{12} + l_{31} l_{33} T_{13} + l_{32} l_{31} T_{21} + l_{32}^2 T_{22} \\
&\quad + l_{32} l_{33} T_{23} + l_{33} l_{31} T_{31} + l_{33} l_{32} T_{32} + l_{33}^2 T_{33}.
\end{aligned} \right\} \qquad (3.9)$$

The sheer space occupied by these equations emphasizes the compactness of (3.8) and the utility of the suffix notation and summation convention, although it may be necessary in actual problems to write out equations in full.

The transformation law may also be written in an equivalent matrix form. We shall use the bold-face symbol **T** as a co-ordinate-free representation of some CT2 (which we can now think of as the set of *all* its possible component representations) and the symbol T to stand for the matrix array (3.7) of the components of **T** with respect to a *particular* orthonormal basis. The transformation law becomes

$$T' = RTR^T. \qquad (3.10)$$

This follows in a straightforward manner after

(i) writing (3.8) as $T_{ij}' = l_{ip} T_{pq} l_{jq}$ with repeated suffixes, as far as possible, following their partners, and
(ii) recognizing that l_{jq} is the q–j entry of the transposed matrix R^T, as discussed in Chapter 2.

The matrix form of the law is not generally speaking, applicable to higher-order tensors, which we define later; the suffix form remains basic.

Example 1

A simple example of a CT2 is supplied by the so-called **dyadic product** of two vectors (two first-order tensors). If these are **u** and **v**, with components u_i, v_i with respect to some basis, we form the set of products of components

$$T_{ij} = u_i v_j,$$

forming the matrix array

$$T = \begin{bmatrix} u_1 v_1 & u_1 v_2 & u_1 v_3 \\ u_2 v_1 & u_2 v_2 & u_2 v_3 \\ u_3 v_1 & u_3 v_2 & u_3 v_3 \end{bmatrix}.$$

We now show that these are the components of a CT2 by showing that the transformation law (3.8) is satisfied when we choose a different basis. The vector components become

$$u_i' = l_{ip} u_p \quad \text{and} \quad v_j' = l_{jq} v_q$$

by (1.32), and the components of the dyadic product are

$$\begin{aligned} T_{ij}' = u_i' v_j' &= (l_{ip} u_p)(l_{jq} v_q) \\ &= l_{ip} l_{jq} u_p v_q \\ &= l_{ip} l_{jq} T_{pq}, \end{aligned}$$

thus satisfying (3.8).

Alternatively, we can write the matrix of components of the dyadic product as

$$T = \mathbf{u}\mathbf{v}^T,$$

where **u** is the 3×1 column vector and $\mathbf{v}^T$ the 1×3 row vector of the components of the two vectors with respect to the original basis. Then, since

$$\mathbf{u}' = R\mathbf{u} \quad \text{and} \quad \mathbf{v}'^{\mathrm{T}} = \mathbf{v}^{\mathrm{T}} R^{\mathrm{T}},$$

by (3.3), we obtain new components

$$\begin{aligned} T' &= \mathbf{u}'\mathbf{v}'^{\mathrm{T}} = (R\mathbf{u})(\mathbf{v}^{\mathrm{T}} R^{\mathrm{T}}) \\ &= R(\mathbf{u}\mathbf{v}^{\mathrm{T}})R^{\mathrm{T}} = RTR^{\mathrm{T}}, \end{aligned}$$

as required by (3.10). The dyadic product of $\mathbf{u}$ and $\mathbf{v}$ will be denoted by $\mathbf{u} \otimes \mathbf{v}$.

The transformation law (3.8) expresses each 'new' component T_{ij}' of the tensor $\mathbf{T}$ as a linear combination of the 'old' components T_{ij}. This linearity means that we can add tensors and perform scalar multiplication on them in a natural way. Thus, if $\mathbf{S}$ and $\mathbf{T}$ are CT2s, with components S_{ij} and T_{ij} with respect to some basis, then the corresponding sums of components $S_{ij} + T_{ij}$ form the components of a CT2 (written $\mathbf{S} + \mathbf{T}$), since we can show that the transformation law is satisfied. Moreover, if λ is a scalar, multiplication of all the components of $\mathbf{T}$ by λ produces a set of components λT_{ij} which satisfy the transformation law and may consequently be regarded as the components of a CT2, which we may write as $\lambda \mathbf{T}$.

Another way of generating a different tensor from a given second-order tensor $\mathbf{T}$ is to take the transpose of the matrix array of components. Transposing both sides of the matrix transformation law (3.10) gives

$$\begin{aligned} (T')^{\mathrm{T}} &= (RTR^{\mathrm{T}})^{\mathrm{T}} = (R^{\mathrm{T}})^{\mathrm{T}} T^{\mathrm{T}} R^{\mathrm{T}} \\ &= RT^{\mathrm{T}} R^{\mathrm{T}}, \end{aligned} \tag{3.11}$$

and this shows that applying the transformation law to the matrix array T^{T} yields the transpose of the matrix array that we obtain by transforming T itself. Thus the set of all transposed matrix arrays forms a CT2, which we write $\mathbf{T}^{\mathrm{T}}$. As an example, $(\mathbf{u} \otimes \mathbf{v})^{\mathrm{T}} = \mathbf{v} \otimes \mathbf{u}$.

If the matrix of components of $\mathbf{T}$ with respect to some basis is symmetric, it follows from (3.11) that the matrix of components with respect to any basis is symmetric, i.e. the property of symmetry is preserved by the transformation law. We then call $\mathbf{T}$ a **symmetric** CT2, and we have $\mathbf{T}^{\mathrm{T}} = \mathbf{T}$.

Similarly, the matrix of components of a CT2 $\mathbf{S}$ with respect to some basis may be skew symmetric, with $S^{\mathrm{T}} = -S$, or $S_{ji} = -S_{ij}$, and this property is also preserved by the transformation law when we consider a different basis. The tensor $\mathbf{S}$ is then called **skew symmetric**.

A result which makes use of all the simple algebraic operations introduced here is the following: any CT2 can be expressed as the sum of a symmetric and skew symmetric tensor. As a proof, we merely write down the identity

$$\mathbf{T} = \tfrac{1}{2}(\mathbf{T} + \mathbf{T}^{\mathrm{T}}) + \tfrac{1}{2}(\mathbf{T} - \mathbf{T}^{\mathrm{T}}) \tag{3.12}$$

and observe that the matrix array $\tfrac{1}{2}(T + T^{\mathrm{T}})$ is necessarily symmetric and $\tfrac{1}{2}(T - T^{\mathrm{T}})$ skew symmetric.

Example 2

Another example of a CT2 is provided by considering the components of the

Kronecker delta δ_{ij}, with matrix

$$\begin{bmatrix} 1 & 0 & 0 \\ 0 & 1 & 0 \\ 0 & 0 & 1 \end{bmatrix}.$$

Supposing this to be the matrix of components of a CT2 **T** with respect to some basis, and applying the transformation law with $T_{pq} = \delta_{pq}$, we obtain

$$l_{ip}l_{jq}\delta_{pq} = l_{ip}l_{jp} \quad \text{because of the substitution properties of } \delta_{ij}$$
$$= \delta_{ij} \quad \text{using the properties of direction cosines.}$$

Hence we obtain precisely the same set of components that we started with. This demonstrates that we can regard δ_{ij} as the components of a CT2 whose components are

$$\begin{bmatrix} 1 & 0 & 0 \\ 0 & 1 & 0 \\ 0 & 0 & 1 \end{bmatrix}$$

with respect to *any* orthonormal basis whatsoever. It is clearly a symmetric tensor. A cartesian tensor whose components remain the same under arbitrary co-ordinate transformations is called **isotropic**.

From a CT2 **T** we can generate further second-order tensors, e.g. the nine components

$$T_{ik}T_{jk} \tag{3.13}$$

where i and j are free suffixes and k is a repeated suffix. To see this, consider the corresponding components $T_{ik}'T_{jk}'$ with respect to a 'new' basis, and apply the transformation law

$$T_{ik}'T_{jk}' = (l_{ip}l_{kq}T_{pq})(l_{jr}l_{ks}T_{rs}).$$

Note that we are careful here to substitute valid expressions

$$T_{ik}' = l_{ip}l_{kq}T_{pq} \quad \text{and} \quad T_{ik}' = l_{jr}l_{ks}T_{rs}$$

with free suffixes agreeing (in the correct places) on both sides, and that it has been necessary when using both expressions together to introduce new repeated suffixes r and s in order to avoid any repeated suffix occurring more than twice in one expression.

Rearranging the order of terms,

$$T_{ik}'T_{jk}' = l_{ip}l_{jr}(l_{kq}l_{ks})T_{pq}T_{rs}$$
$$= l_{ip}l_{jr}\delta_{qs}T_{pq}T_{rs} \quad \text{by (1.27)}$$
$$= l_{ip}l_{jr}T_{ps}T_{rs} \quad \text{since } \delta_{qs}T_{pq} = T_{ps}.$$

This shows that the components $T_{ik}T_{jk}$ transform like a CT2. In fact we could obtain a slightly closer correspondence with (3.8) by writing q in place of r, giving

$$T_{ik}'T_{jk}' = l_{ip}l_{jq}(T_{ps}T_{qs}).$$

Alternatively, using matrix methods, we can represent the expression (3.13) by the

Sec. 3.2] **Cartesian tensors** 71

matrix TT^T. If T is a CT2, with $T' = RTR^T$, we obtain

$$\begin{aligned} T'T'^T &= (RTR^T)(RTR^T)^T = (RTR^T)(RT^T R^T) \\ &= RT(R^T R)T^T R^T = RT^T IT^T R^T \quad \text{since } R \text{ is orthogonal} \\ &= R(TT^T)R^T, \end{aligned}$$

showing that TT^T transforms according to (3.10).

Example 3

Given a vector field $\mathbf{v}(\mathbf{r})$, such as the velocity field in a fluid in motion, the partial derivatives $\partial v_i/\partial y_j$ of the components of $\mathbf{v}$ with respect to the rectangular cartesian co-ordinates of the position vector $\mathbf{r}$ form the components of a CT2. To show this, we consider the corresponding components $\partial v_i'/\partial y_j'$ with respect to a new system of axes. Now

$$\frac{\partial v_i'}{\partial y_j'} = \frac{\partial (l_{ip} v_p)}{\partial y_q} \frac{\partial y_q}{\partial y_j'}$$

by the chain rule for partial differentiation

$$= l_{ip} \frac{\partial v_p}{\partial y_q} l_{jq} \quad \text{by (3.6)}$$

$$= l_{ip} l_{jq} \frac{\partial v_p}{\partial y_q}.$$

Hence $\partial v_i/\partial y_j$ transforms according to the transformation law (3.8) for a CT2.

Problems

3.2.3 If $\mathbf{S}$ and $\mathbf{T}$ are CT2s, use the matrix transformation law (3.10) to show that $\mathbf{S} + 2\mathbf{T}$ is also a CT2.

3.2.4 If $\mathbf{T}$ is a CT2 with components

$$\begin{bmatrix} 1 & 1 & 1 \\ 0 & 1 & 1 \\ 0 & 0 & 1 \end{bmatrix}$$

with respect to the basis $(\mathbf{i}_1, \mathbf{i}_2, \mathbf{i}_3)$, find the components of $\mathbf{T}$ with respect to the basis $\{\mathbf{i}_1', \mathbf{i}_2', \mathbf{i}_3'\}$, where $\mathbf{i}_1' = -\mathbf{i}_1$, $\mathbf{i}_2' = -\mathbf{i}_2$, $\mathbf{i}_3' = \mathbf{i}_3$.

3.2.5 What are the changes in axes corresponding to the proper orthogonal matrices of direction cosines

(i) $\begin{bmatrix} -1 & 0 & 0 \\ 0 & -1 & 0 \\ 0 & 0 & 1 \end{bmatrix}$ and

(ii) $\begin{bmatrix} 0 & 1 & 0 \\ -1 & 0 & 0 \\ 0 & 0 & 1 \end{bmatrix}$?

Write out equations (3.9) for each of these transformations. Deduce that, if **T** is an isotropic CT2, $T_{13} = T_{31} = T_{23} = T_{32} = 0$ and $T_{11} = T_{22}$. (By similar transformations we can also show that $T_{12} = T_{21} = 0$ and $T_{22} = T_{33}$. Thus the *only* isotropic CT2 is a multiple of δ_{ij}.)

3.2.6 Show using both suffix and matrix methods that, if T_{ij} is a CT2, then so are

(i) $T_{ik}T_{kj}$ and
(ii) $T_{ik}T_{kl}T_{lj}$.

3.2.7 Write out in full the set of equations

$$S_{ij} = 2\mu E_{ij} + \lambda \delta_{ij} E_{kk}.$$

If E_{ij} is a CT2 and μ, λ are scalars, show that S_{ij} is a CT2. Show that

(i) $S_{kk} = (2\mu + 3\lambda)E_{kk}$,

(ii) $E_{ij} = \frac{1}{2\mu} S_{ij} - \frac{\lambda}{2\mu(2\mu + 3\lambda)} \delta_{ij} S_{kk}$ and

(iii) $S_{ij}E_{ij} = 2\mu E_{ij}E_{ij} + \lambda E_{kk}^2$.

3.2.3 The two-dimensional case

Formulas for the transformation of vector and tensor components in two dimensions are often required by engineers in practical situations, such as in the analysis of plane strain and plane stress problems in solid mechanics. Given an orthonormal basis $\{\mathbf{i}_1, \mathbf{i}_2\}$ in two dimensions, a vector **u** has components $\{u_1, u_2\}$ and a second-order tensor **T** has a 2×2 matrix of components

$$\begin{bmatrix} T_{11} & T_{12} \\ T_{21} & T_{22} \end{bmatrix}.$$

Alternatively, we can remain in three dimensions, with basis $\{\mathbf{i}_1, \mathbf{i}_2, \mathbf{i}_3\}$, but consider only changes of basis in which $\mathbf{i}_3$ remains fixed and ignore the possibly non-zero u_3 component of **u** and the components $T_{13}, T_{23}, T_{31}, T_{32}$ and T_{33} of **T**. For rotations of the vectors $\mathbf{i}_1$ and $\mathbf{i}_2$ through an angle θ about $\mathbf{i}_3$, the required 2×2 rotation matrix is given by

$$R = \begin{bmatrix} \cos\theta & \sin\theta \\ -\sin\theta & \cos\theta \end{bmatrix},$$

and the transformation law (3.10) yields

$$\begin{bmatrix} T'_{11} & T'_{12} \\ T'_{21} & T'_{22} \end{bmatrix} = \begin{bmatrix} \cos\theta & \sin\theta \\ -\sin\theta & \cos\theta \end{bmatrix} \begin{bmatrix} T_{11} & T_{12} \\ T_{21} & T_{22} \end{bmatrix} \begin{bmatrix} \cos\theta & -\sin\theta \\ \sin\theta & \cos\theta \end{bmatrix},$$

i.e.

$$T' = \begin{bmatrix} T_{11}\cos^2\theta + T_{22}\sin^2\theta & -T_{11}\sin\theta\cos\theta + T_{22}\sin\theta\cos\theta \\ + T_{12}\sin\theta\cos\theta + T_{21}\sin\theta\cos\theta & + T_{12}\cos^2\theta - T_{21}\sin^2\theta \\ -T_{11}\sin\theta\cos\theta + T_{22}\sin\theta\cos\theta & T_{11}\sin^2\theta + T_{22}\cos^2\theta \\ -T_{12}\sin^2\theta + T_{21}\cos^2\theta & -T_{12}\sin\theta\cos\theta - T_{21}\sin\theta\cos\theta \end{bmatrix}$$

Sec. 3.2]   **Cartesian tensors**   73

$$= \begin{bmatrix} \frac{1}{2}(T_{11}+T_{22})+\frac{1}{2}(T_{11}-T_{22})\cos(2\theta) & \frac{1}{2}(T_{22}-T_{11})\sin(2\theta) \\ +\frac{1}{2}(T_{12}+T_{21})\sin(2\theta) & +\frac{1}{2}(T_{12}-T_{21})+\frac{1}{2}(T_{12}+T_{21})\cos(2\theta) \\ \frac{1}{2}(T_{22}-T_{11})\sin(2\theta) & \frac{1}{2}(T_{11}+T_{22})-\frac{1}{2}(T_{11}-T_{22})\cos(2\theta) \\ +\frac{1}{2}(T_{21}-T_{12})+\frac{1}{2}(T_{12}+T_{21})\cos(2\theta) & -\frac{1}{2}(T_{12}+T_{21})\sin(2\theta) \end{bmatrix}$$
(3.14)

Note that $T_{11}' + T_{22}' = T_{11} + T_{22}$, i.e. the sum of the diagonal elements of each 2×2 matrix of components is *invariant*. In other words, it is unaffected by the change of basis. Another invariant would seem to be $T_{12}' - T_{21}' = T_{12} - T_{21}$. However, although the magnitude of this term is certainly invariant, its sign would change if we changed the basis by relabelling $\mathbf{i}_1$ as $\mathbf{i}_2$ and $\mathbf{i}_2$ as $\mathbf{i}_1$. Such a transformation cannot be accomplished by a rigid rotation in two dimensions (about $\mathbf{i}_3$) but could correspond to a rigid rotation of the base vectors in three dimensions if $\mathbf{i}_3$ reverses direction, i.e. $\mathbf{i}_1' = \mathbf{i}_2, \mathbf{i}_2' = \mathbf{i}_1, \mathbf{i}_3' = -\mathbf{i}_3$. Squaring this quantity, however, gives a genuine invariant $(T_{12} - T_{21})^2$. Another invariant in two dimensions may be discovered by taking determinants of both sides of the matrix transformation law, bearing in mind that $\det R = 1$. We obtain $\det T' = \det T$, i.e.

$$T_{11}'T_{22}' - T_{12}'T_{21}' = T_{11}T_{22} - T_{12}T_{21}.$$

which may also be verified directly from (3.14).

For symmetric tensors, $T_{12} = T_{21}$, and the transformation becomes

$$T' = \begin{bmatrix} \frac{1}{2}(T_{11}+T_{22})+\frac{1}{2}(T_{11}-T_{22})\cos(2\theta) & \frac{1}{2}(T_{22}-T_{11})\sin(2\theta) \\ +T_{12}\sin(2\theta) & +T_{12}\cos(2\theta) \\ \frac{1}{2}(T_{22}-T_{11})\sin(2\theta) & \frac{1}{2}(T_{11}+T_{22})+\frac{1}{2}(T_{22}-T_{11})\cos(2\theta) \\ +T_{12}\cos(2\theta) & -T_{12}\sin(2\theta) \end{bmatrix}.$$
(3.15)

We note the fact that T' is also symmetric. Moreover, it is now always possible to choose a particular rotation angle θ so that

$$\tfrac{1}{2}(T_{22} - T_{11})\sin(2\theta) + T_{12}\cos(2\theta) = 0,$$

i.e.

$$\tan(2\theta) = \frac{T_{12}}{T_{11} - T_{22}},$$
(3.16)

thus making $T_{12}' = T_{21}' = 0$. When the off-diagonal matrix components of the tensor **T** vanish, the tensor **T** is said to be **diagonalized**. We consider the three-dimensional equivalent of diagonalization in section 3.2.6.

Problems

3.2.8 From (3.14) verify that $T_{11}'T_{22}' - T_{12}'T_{21}' = T_{11}T_{22} - T_{12}T_{21}$.

3.2.4 Invariants

Invariants are equivalent in concept to scalars; they are quantities which do not depend on any particular choice of basis or co-ordinate system. The components T_{ij} of a CT2 **T** are not scalars, since they change when the reference system changes.

We can easily generate invariants of **T**, however, as certain functions of the components.

For example, the sum of the diagonal elements in the matrix array of components, $T_{11} + T_{22} + T_{33}$, i.e. the **trace** Tr T of the matrix, is an invariant. This follows by writing the sum as T_{ii} and putting $i = j$ in the transformation law (3.8). Thus

$$T_{ii}' = l_{ip}l_{iq}T_{pq}$$
$$= \delta_{pq}T_{pq} \quad \text{by the properties of direction cosines}$$
$$= T_{pp} = T_{ii}.$$

This means that

$$T_{11}' + T_{22}' + T_{33}' = T_{11} + T_{22} + T_{33},$$

or
$$\text{Tr } T' = \text{Tr } T. \tag{3.17}$$

This invariant has been obtained from T_{ij} by making the two suffixes identical, with implied summation over the remaining suffix (T_{ii} or T_{jj}), a procedure which is called **contraction**. It may be used to generate another invariant from (3.13). Putting $i = j$ in this expression gives $T_{jk}'T_{jk}'$ (or $T_{ik}T_{ik}$), and

$$T_{jk}'T_{jk}' = (l_{jp}l_{kq}T_{pq})(l_{jr}l_{ks}T_{rs})$$
$$= (l_{jp}l_{jr})(l_{kq}l_{ks})T_{pq}T_{rs}$$
$$= \delta_{pr}\delta_{qs}T_{pq}T_{rs}$$
$$= T_{pq}T_{pq} \quad \text{as required.}$$

Written out in full, this invariant is

$$T_{11}^2 + T_{12}^2 + T_{13}^2 + T_{21}^2 + T_{22}^2 + T_{23}^2 + T_{31}^2 + T_{32}^2 + T_{33}^2.$$

In two dimensions it reduces to

$$T_{11}^2 + T_{12}^2 + T_{21}^2 + T_{22}^2$$

which can be expressed in terms of the invariants previously noted in section 3.2.3 as

$$(T_{11} + T_{22})^2 + (T_{12} - T_{21})^2 - 2(T_{11}T_{22} - T_{12}T_{21}).$$

Applied to the second-order tensor $\partial v_i/\partial y_j$, where $\mathbf{v}(y_1, y_2, y_3)$ is a vector field, contraction yields $\partial v_i/\partial y_i = \text{div } \mathbf{v}$; thus we have demonstrated that div **v** is a genuine scalar.

A further invariant of a CT2 is the determinant of its matrix of components, det T. This may be demonstrated immediately, as mentioned in the last section, by taking the determinant of both sides of the matrix version (3.10) of the transformation law to obtain

$$\det T' = \det(RTR^T) = \det R \det T \det(R^T)$$
$$= \det T. \tag{3.18}$$

since $\det R = \det(R^T) = 1$. It follows that the property of non-singularity (and of singularity) of the matrix of components of a CT2 is preserved under tensor transformations.

The term 'invariant' is also used to describe properties of tensors, such as the linear operator property of second-order tensors to be discussed in the next section, which do not depend on any particular choice of co-ordinate system. This concept of invariance, as applied to fundamental physical principles,, is central in the historical development of tensor analysis.

Problems

3.2.9 Show that contraction applied to the dyadic product $\mathbf{u} \otimes \mathbf{v}$ produces the scalar product $\mathbf{u} \cdot \mathbf{v}$.

3.2.10 Show that, if T is the matrix array of components of a CT2 $\mathbf{T}$, the invariants $T_{pq}T_{pq}$ and $T_{pq}T_{qp}$ are given by $\mathrm{Tr}(TT^\mathrm{T})$ and $\mathrm{Tr}(T^2)$, respectively.

3.2.11 Show that, if T_{ij} is a CT2, $T_{ij}T_{jk}T_{ki}$ is an invariant, and equal to $\mathrm{Tr}(T^3)$.

3.2.5 Second-order tensors and linear operators

If $\mathbf{T}$ is a CT2 and $\mathbf{u}$ is a vector (CT1), then the expression

$$T_{ij}u_j$$

has three components (corresponding to the free suffix i), and we can show that it satisfies the transformation rules for a vector.

Using (3.2) and (3.8), we have

$$\begin{aligned}
T_{ij}'u_j' &= (l_{ip}l_{jq}T_{pq})(l_{jr}u_r) \\
&= l_{ip}(l_{jq}l_{jr})T_{pq}u_r \\
&= l_{ip}\delta_{qr}T_{pq}u_r \\
&= l_{ip}(T_{pr}u_r) \qquad \text{as required.}
\end{aligned}$$

If we call the resulting vector $\mathbf{v}$, we may write

$$\mathbf{v} = \mathbf{T}\mathbf{u}.$$

Thus $\mathbf{T}$ acts on the vector $\mathbf{u}$ and produces the vector $\mathbf{v}$. The requirements for a *linear operator* are satisfied; namely,

(i) if λ is a scalar, $\lambda\mathbf{u}$ is a vector, and $\mathbf{T}(\lambda\mathbf{u}) = \lambda\mathbf{v}$;
(ii) if $\mathbf{T}\mathbf{u}_1 = \mathbf{v}_1$ and $\mathbf{T}\mathbf{u}_2 = \mathbf{v}_2$, then $\mathbf{T}(\mathbf{u}_1 + \mathbf{u}_2) = \mathbf{v}_1 + \mathbf{v}_2$.

Turning the above result around, we obtain the **quotient theorem**; this may be stated as follows.

Suppose $\mathbf{T}$ is an entity with nine components T_{ij} with respect to any given orthonormal basis. Then, if $T_{ij}u_j$, for *arbitrary* vectors $\mathbf{u}$, are the components ($i = 1, 2, 3$) of a vector, $\mathbf{T}$ must be a CT2.

In proof, suppose that $T_{ij}u_j = v_i$, where $\mathbf{u}$ is an arbitrary vector and $\mathbf{v}$ is a vector. With respect to a different basis, we shall have $T_{ij}'u_j' = v_i'$, where $u_j' = l_{jp}u_p$ and $v_i' = l_{iq}v_q$. Hence $T_{ij}'(l_{jp}u_p) = l_{iq}v_q = l_{iq}(T_{qp}u_p)$, and thus $(T_{ij}'l_{jp} - l_{iq}T_{qp})u_p = 0$. If this is to hold for *arbitrary* vectors $\mathbf{u}$, it follows, e.g. by considering $\mathbf{u} = (1, 0, 0), (0, 1, 0), (0, 0, 1)$ in turn,

that
$$T_{ij}'l_{jp} - l_{iq}T_{qp} = 0 \qquad \text{for } p = 1, 2, 3.$$
So p can be treated as a free suffix, and
$$l_{jp}T_{ij}' = l_{iq}T_{qp}.$$
Multiplying both sides by l_{kp}, implying summation over p,
$$l_{kp}l_{jp}T_{ij}' = \delta_{kj}T_{ij}' = T_{ik}'$$
$$= l_{kp}(l_{iq}T_{qp}) = l_{iq}l_{kp}T_{qp},$$
or, adjusting suffixes,
$$T_{ij}' = l_{ip}l_{jq}T_{pq} \qquad \text{as required.}$$

The quotient theorem supplies a useful test for many second-order tensors arising in physical problems. In other approaches to tensors, the operator property is taken to be the defining property of second-order tensors, and the transformation law is deduced from it.

Example 1

The dyadic product of two vectors **u** and **v**, written as $\mathbf{u} \otimes \mathbf{v}$, is sometimes defined as a linear operator. Its action on vectors **w** may be written as
$$(\mathbf{u} \otimes \mathbf{v})\mathbf{w} = (\mathbf{v} \cdot \mathbf{w})\mathbf{u}, \qquad (3.19)$$
i.e. the result is a vector which, explicitly, is the vector **u** multiplied by the scalar product of **v** and **w**.

In suffix terms, we have the equivalent version
$$(u_i v_j)w_j = (v_j w_j)u_i.$$
We could deduce from the quotient theorem that the dyadic product is a second-order tensor.

Example 2

If we consider a rigid body rotating with one point O fixed, the velocity **v** of a point P of the body with position vector **r** (taking O as the origin) is given by the vector product $\boldsymbol{\omega} \times \mathbf{r}$, where $\boldsymbol{\omega}$ is the instantaneous spin vector of the body, whose magnitude is the angular velocity of the body and whose direction is the axis of spin (the sense being given by a right-handed screw convention). The angular momentum (or *moment of momentum*) of a particle of mass m at P is given by
$$\mathbf{h} = \mathbf{r} \times m\mathbf{v} = m\mathbf{r} \times (\boldsymbol{\omega} \times \mathbf{r}).$$
Summing over all the particles of the body gives the total angular momentum vector of the body:
$$\mathbf{H} = \sum m\mathbf{r} \times (\boldsymbol{\omega} \times \mathbf{r}).$$
$$= \sum m[(\mathbf{r} \cdot \mathbf{r})\boldsymbol{\omega} - (\mathbf{r} \cdot \boldsymbol{\omega})\mathbf{r}] \qquad \text{by (1.21).}$$

If we choose rectangular cartesian axes $Oy_1y_2y_3$ with origin O, the relationship between the components H_i and ω_i of **H** and **ω** may be expressed as

$$H_1 = \sum m[(y_1^2 + y_2^2 + y_3^2)\omega_1 - (y_1\omega_1 + y_2\omega_2 + y_3\omega_3)y_1]$$
$$= \sum m[(y_2^2 + y_3^2)\omega_1 - y_1y_2\omega_2 - y_1y_3\omega_3]$$
$$H_2 = \sum m[(y_3^2 + y_1^2)\omega_2 - y_2y_1\omega_1 - y_2y_3\omega_3]$$

and
$$H_3 = \sum m[(y_1^2 + y_2^2)\omega_3 - y_3y_1\omega_1 - y_3y_2\omega_2],$$

where (y_1, y_2, y_3) are the co-ordinates of the general point P.

These equations may be assembled into the matrix form

$$\begin{bmatrix} H_1 \\ H_2 \\ H_3 \end{bmatrix} = \begin{bmatrix} \mathscr{I}_{11} & \mathscr{I}_{12} & \mathscr{I}_{13} \\ \mathscr{I}_{21} & \mathscr{I}_{22} & \mathscr{I}_{23} \\ \mathscr{I}_{31} & \mathscr{I}_{32} & \mathscr{I}_{33} \end{bmatrix} \begin{bmatrix} \omega_1 \\ \omega_2 \\ \omega_3 \end{bmatrix},$$

where

$$\mathscr{I}_{11} = \sum m(y_2^2 + y_3^2), \quad \mathscr{I}_{22} = \sum m(y_3^2 + y_1^2), \quad \mathscr{I}_{33} = \sum m(y_1^2 + y_2^2),$$
$$\mathscr{I}_{12} = \mathscr{I}_{21} = -\sum my_1y_2,$$
$$\mathscr{I}_{23} = \mathscr{I}_{32} = -\sum my_2y_3$$

and $\mathscr{I}_{31} = \mathscr{I}_{13} = -\sum my_3y_1.$

In matrix terms,

$$H = \mathscr{I}\omega,$$

where $\mathscr{I}$ is the (symmetric) 3×3 inertia matrix at O with respect to the axes $Oy_1y_2y_3$.
In suffixes,

$$H_i = \mathscr{I}_{ij}\omega_j,$$

and this equation is valid, with appropriate components, with respect to any set of axes.

The inertia matrix acts on arbitrary spin vectors **ω** and produces angular momentum vectors **H**. It follows from the quotient theorem that $\mathscr{I}_{ij}$ are the components of a CT2 $\mathscr{I}$, which we may call the **inertia tensor** of the body at O. The components may be expressed explicitly as

$$\mathscr{I}_{ij} = \sum m(y_k y_k \delta_{ij} - y_i y_j).$$

The diagonal elements $\mathscr{I}_{11}, \mathscr{I}_{22}, \mathscr{I}_{33}$ of the inertia matrix may be identified as the moments of inertia of the rigid body about the axes Oy_1, Oy_2, Oy_3, respectively.

Example 3

Consider a rigid rotation through an angle α about an axis parallel to the unit vector **n** and passing through the origin O (Fig. 3.1). A point P with position vector **r** relative to O moves in a circle, centre N (the projection of P on the axis), radius $r \sin \beta$, where $r = |\mathbf{r}|$ and $\beta = \angle$ PON, to the point P* with position vector **r***. We may consider **r*** as the sum of vectors along ON, NM and MP*, where M is the projection of P* on NP, and it may be verified by checking magnitude and directions that these vectors

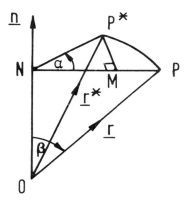

Fig. 3.1.

are $(\mathbf{n}\cdot\mathbf{r})\mathbf{n}$, $[\mathbf{r}-(\mathbf{n}\cdot\mathbf{r})\mathbf{n}]\cos\alpha$, and $(\mathbf{n}\times\mathbf{r})\sin\alpha$, respectively. For example, the magnitude of MP* = NP* $\sin\alpha$ = NP $\sin\alpha$ = $r\sin\beta\sin\alpha$ = $|\mathbf{n}\times\mathbf{r}|\sin\alpha$, and $\mathbf{n}\times\mathbf{r}$ also gives the correct direction of MP*. Hence

$$\mathbf{r}^* = (\mathbf{n}\cdot\mathbf{r})\mathbf{n} + [\mathbf{r}-(\mathbf{n}\cdot\mathbf{r})\mathbf{n}]\cos\alpha + (\mathbf{n}\times\mathbf{r})\sin\alpha$$
$$= \mathbf{r}\cos\alpha + (\mathbf{n}\cdot\mathbf{r})\mathbf{n}(1-\cos\alpha) + (\mathbf{n}\times\mathbf{r})\sin\alpha.$$

In component form, with respect to an orthonormal basis at O, this vector equation becomes

$$y_i^* = y_i\cos\alpha + (n_k y_k)n_i(1-\cos\alpha) + e_{ijk}n_j y_k \sin\alpha$$
$$\text{by (1.18) and (1.19)}$$
$$= Q_{ik} y_k$$

where

$$Q_{ik} = \delta_{ik}\cos\alpha + n_i n_k(1-\cos\alpha) + e_{ijk}n_j \sin\alpha. \qquad (3.20)$$

We see that these are the components of a linear operator $\mathbf{Q}$ which sends vectors $\mathbf{r}$ to vectors $\mathbf{r}^*$, i.e.

$$\mathbf{r}^* = \mathbf{Q}\mathbf{r}. \qquad (3.21)$$

By the quotient theorem, $\mathbf{Q}$ must be a CT2 and is called the **rotation tensor**.

The inverse operator to $\mathbf{Q}$ corresponds to a restoring rotation through the negative angle $-\alpha$ about the same axis (or through α about the oppositely directed axis $-\mathbf{n}$), i.e.

$$Q_{ik}^{-1} = \delta_{ik}\cos(-\alpha) + n_i n_k[1-\cos(-\alpha)] + e_{ijk}n_j \sin(-\alpha)$$
$$= \delta_{ik}\cos\alpha + n_i n_k(1-\cos\alpha) - e_{ijk}n_j \sin\alpha,$$

but note that this is identical with Q_{ki}, by the symmetric properties of δ_{ik} and $n_i n_k$ and the skew symmetric property of e_{ijk}. Hence, in matrix terms,

$$Q^{-1} = Q^T,$$

showing that the matrix array Q_{ik} is *orthogonal*.

We may note the formal similarity between (3.21) and (3.3). In the latter case, however, the vector remains unchanged while the co-ordinate system is rotated, and

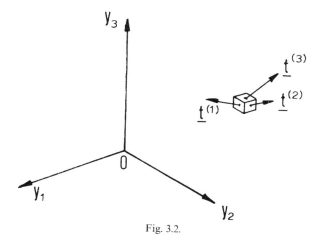

Fig. 3.2.

the orthogonal matrix R does not represent a tensor, whereas in (3.21) the co-ordinate system remains unchanged while the vector **r** is rotated by the linear operator **Q**.

Example 4

The forces acting within a continuous body (solid or fluid) subjected to external forces may be represented by a stress matrix S_{ij}. If we imagine a small cube of material centred on some point within the body at a certain instant, with edges parallel to given rectangular cartesian axes Oy_1, Oy_2, Oy_3, each face of the cube is being subjected to a certain force per unit area because of contact with the material immediately adjacent to it and not belonging to the cube. This force per unit area may be represented by a *vector*, called the **traction**.

The traction vectors $\mathbf{t}^{(1)}, \mathbf{t}^{(2)}$ and $\mathbf{t}^{(3)}$ acting on the faces of a typical cube whose unit outward normals are parallel to the positive y_1, y_2 and y_3 axes, respectively, are illustrated in Fig. 3.2. The tractions on the opposite faces will be $-\mathbf{t}^{(1)}, -\mathbf{t}^{(2)}$ and $-\mathbf{t}^{(3)}$, i.e. equal and opposite, if the cube is sufficiently small, as a consequence of Newton's third law of motion. If the cartesian components of $\mathbf{t}^{(1)}, \mathbf{t}^{(2)}$ and $\mathbf{t}^{(3)}$ are $(S_{11}, S_{21}, S_{31}), (S_{12}, S_{22}, S_{32})$ and (S_{13}, S_{23}, S_{33}), respectively, we can assemble the components into a matrix

$$(S_{ij}) = \begin{bmatrix} S_{11} & S_{12} & S_{13} \\ S_{21} & S_{22} & S_{23} \\ S_{31} & S_{32} & S_{33} \end{bmatrix}$$

whose columns are the traction components.

It was shown by Cauchy using the principles of mechanics that, if we consider a material surface within the body at some instant and an element of area on that surface at a point for which the traction components acting on the faces of a cube centred at the point are S_{ij}, then, if **n** is a unit normal to that element and **t** is the traction exerted on the element by the material immediately adjacent to it (and on the side of the element

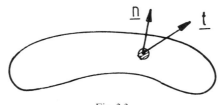

Fig. 3.3.

given by the direction of **n**) (Fig. 3.3), the components of **t** are given in terms of the components of **n** by the equation

$$t_i = S_{ij}n_j.$$

It follows from this result and the quotient theorem (slightly modified because **n** is an arbitrary *unit* vector) that S_{ij} are the components of a CT2.

Problems

3.2.12 Prove the following versions of the quotient theorem.
(a) If T_{ij} is an entity with nine components with respect to any cartesian coordinate system, show that, if $T_{ij}u_iv_j$ is an invariant for arbitrary vectors **u** and **v**, T_{ij} is a CT2.
(b) If T_{ij} is an entity with nine components with respect to any cartesian coordinate system, and $T_{ij} = T_{ji}$, show that, if $T_{ij}u_iu_j$ is an invariant for arbitrary vectors **u**, T_{ij} is a symmetric CT2.

3.2.13 If $\mathscr{I}$ is the inertia tensor of a rigid body at the origin O, show that, if **n** is a unit vector,

$$\mathscr{I}_{ij}n_in_j = \sum m[|\mathbf{r}|^2 - (\mathbf{r}\cdot\mathbf{n})^2],$$

with summation over the particles of the body, and interpret this expression as the moment of inertia of the body about an axis through O parallel to **n**.

3.2.14 A rigid body consists of eight equal particles of mass m rigidly connected by light rods in the shape of a cube of side a. Three edges lie along the axes Oy_1, Oy_2, Oy_3, and one particle is at the origin O (Fig. 3.4). Show that the

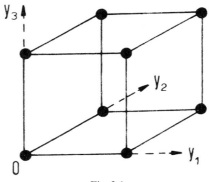

Fig. 3.4.

Sec. 3.2]   **Cartesian tensors**   81

inertia matrix of the body at O with respect to the given axes is

$$ma^2 \begin{bmatrix} 8 & -2 & -2 \\ -2 & 8 & -2 \\ -2 & -2 & 8 \end{bmatrix}.$$

3.2.15 From (3.20) show that

$$Q_{kk} = 1 + 2\cos\alpha$$

and $\quad e_{ijk}Q_{kj} = 2n_i \sin\alpha.$

Given that $\begin{bmatrix} \frac{2}{3} & -\frac{1}{3} & \frac{2}{3} \\ \frac{2}{3} & \frac{2}{3} & -\frac{1}{3} \\ -\frac{1}{3} & \frac{2}{3} & \frac{2}{3} \end{bmatrix}$ are the components of a rotation tensor, find

the angle α and the components n_i of the axis of rotation.

3.2.16 If a rigid-body rotation is carried out as in Example 3 such that α is a given function of time (with derivative $\dot\alpha$) and the rotation axis **n** is fixed, establish the following results:
 (a) $\dot Q_{ik} = \dot\alpha(-\delta_{ik}\sin\alpha + n_i n_k \sin\alpha + e_{ijk}n_j \cos\alpha)$;
 (b) $\mathbf{v} = \dot{\mathbf{Q}}\mathbf{r} = \dot{\mathbf{Q}}\mathbf{Q}^T\mathbf{r}^*$,
  where the velocity vector $\mathbf{v} = \dot{\mathbf{r}}^*$;
 (c) $\dot Q_{ik} Q_{jk} = \dot\alpha e_{ikj} n_k$;
 (d) $\mathbf{v} = \mathbf{\Omega}\mathbf{r}^*$, where the components of the spin tensor $\mathbf{\Omega}$ are given by $\Omega_{ij} = \dot\alpha e_{ikj} n_k$;
 (e) $\mathbf{v} = \boldsymbol{\omega} \times \mathbf{r}^*$, where the spin vector $\boldsymbol{\omega} = \dot\alpha\mathbf{n}$.

3.2.6 Diagonalization of a symmetric tensor

We have seen in Chapter 2 that for any symmetric 3×3 matrix T there is an orthogonal matrix R such that RTR^T is a diagonal matrix, the diagonal elements then being the eigenvalues of T, i.e. the roots λ of the (cubic) *characteristic* equation

$$\det(T - \lambda I) = 0.$$

In view of (3.10), this result may be directly interpreted as implying the following.

For any symmetric CT2 **T** there exists some orthonormal basis for which the matrix array of components of **T** is diagonal.

Given an arbitrary orthonormal basis, for which the matrix of components of **T** is T, we merely have to rotate the basis to a new basis whose direction cosines with respect to the first are given by the appropriate matrix R, and the new matrix T' of components will be diagonal. The new basis defines directions which are called the **principal axes** of the tensor **T**, and the diagonal entries are then the **principal values** (or **eigenvalues**) of **T**.

If we take the inertia tensor $\mathcal{I}$ in rigid-body dynamics as an example, the result shows that it is always possible to choose a set of mutually perpendicular axes at a fixed point O of the body such that the inertia matrix is diagonal. These axes are called the **principal axes of inertia** of the body at O, and the diagonal elements are called the **principal moments of inertia** of the body at O. If the principal moments of inertia are denoted A, B and C, the relationship between the components of the spin vector and the angular momentum vector with respect to the principal axes has the simple form

$$\begin{bmatrix} H_1 \\ H_2 \\ H_3 \end{bmatrix} = \begin{bmatrix} A & 0 & 0 \\ 0 & B & 0 \\ 0 & 0 & C \end{bmatrix} \begin{bmatrix} \omega_1 \\ \omega_2 \\ \omega_3 \end{bmatrix},$$

i.e.

$$H_1 = A\omega_1, \quad H_2 = B\omega_2, \quad H_3 = C\omega_3.$$

In two-dimensional problems, equation (3.16) explicitly gives the rotation angle required to diagonalize a CT2.

The characteristic equation of a symmetric CT2 **T** may be written out in full as

$$\det(T - \lambda I) = -\lambda^3 + I_1\lambda^2 - I_2\lambda + I_3 = 0. \tag{3.22}$$

If the roots (the eigenvalues) of this equation are τ_1, τ_2 and τ_3, the coefficients in the equation are, by simple algebra,

$$\left.\begin{aligned} I_1 &= \tau_1 + \tau_2 + \tau_3, \\ I_2 &= \tau_1\tau_2 + \tau_2\tau_3 + \tau_3\tau_1, \\ I_3 &= \tau_1\tau_2\tau_3. \end{aligned}\right\}$$

Now I_1 is the trace of the matrix representation

$$\begin{bmatrix} \tau_1 & 0 & 0 \\ 0 & \tau_2 & 0 \\ 0 & 0 & \tau_3 \end{bmatrix}$$

of **T** with respect to the principal axes of **T**, and we know from (3.17) that this is invariant; I_3 is also clearly the determinant of this matrix, and by (3.18) this is also an invariant. Finally, we can write

$$\begin{aligned} I_2 &= \tfrac{1}{2}(\tau_1 + \tau_2 + \tau_3)^2 - \tfrac{1}{2}(\tau_1^2 + \tau_2^2 + \tau_3^2) \\ &= \tfrac{1}{2}I_1^2 - \tfrac{1}{2}(\tau_1^2 + \tau_2^2 + \tau_3^2). \end{aligned}$$

The expression in parentheses is also an invariant. It is the trace of the matrix

$$\begin{bmatrix} \tau_1^2 & 0 & 0 \\ 0 & \tau_2^2 & 0 \\ 0 & 0 & \tau_3^2 \end{bmatrix},$$

which is just the matrix of components of T^2 with respect to the principal axes of T. Thus

$$\left.\begin{aligned} I_1 &= \operatorname{Tr} T = T_{kk}, \\ I_2 &= \tfrac{1}{2}(\operatorname{Tr} T)^2 - \tfrac{1}{2}\operatorname{Tr}(T^2) = \tfrac{1}{2}(T_{jj}T_{kk} - T_{jk}T_{jk}) \\ I_3 &= \det T. \end{aligned}\right\} \tag{3.23}$$

These are called the **principal invariants** of **T**.

Problems

3.2.17 Find the eigenvalues of the following symmetric tensors, whose components are given with respect to a right-handed orthonormal basis $\{i_1, i_2, i_3\}$. Express the eigenvectors as a right-handed orthonormal basis, and write down an orthogonal matrix which diagonalizes the tensor.

(a) $\begin{bmatrix} 6 & 0 & 0 \\ 0 & 1 & -1 \\ 0 & -1 & 1 \end{bmatrix}$;

(b) $\begin{bmatrix} 2 & 1 & 1 \\ 1 & 1 & 2 \\ 1 & 2 & 1 \end{bmatrix}$.

3.2.18 Find the principal moments of inertia at O of the rigid body in problem 3.2.14, and the directions of the principal axes of inertia at O.

3.2.7 Higher-order cartesian tensors

Having defined a CT1 and a CT2 in terms of components satisfying a particular transformation law, we now extend the definition to higher-order quantities. A cartesian tensor of order three (CT3) is an entity (denoted, say, by **U**) with 27 components U_{ijk} with respect to a given orthonormal basis, such that, if we choose a different orthonormal basis with direction cosines l_{ip} with respect to the first basis, the components of **U** become U_{ijk}', where

$$U_{ijk}' = l_{ip}l_{jq}l_{kr}U_{pqr} \qquad (3.24)$$

with summation over p, q and r.

Examples

It is easy to see that if **u**, **v** and **w** are vectors, the set of products of components $u_i v_j w_k$ forms a CT3. Another example would be the set of products $u_i T_{jk}$, where **u** is a CT1 and **T** a CT2.

We now demonstrate that the set of numbers e_{ijk} defined in (1.15) also represents a CT3, with the special property that it is **isotropic**, i.e. a CT3 whose components are the same with respect to any right-handed orthonormal basis. Applying the transformation law (3.24), we obtain

$$l_{ip}l_{jq}l_{kr}e_{pqr} = (\det R)e_{ijk} \qquad \text{by (2.19).}$$

If we confine our attention to proper orthogonal transformations (e.g. if we use only right-handed systems), we have $\det R = 1$, and hence

$$l_{ip}l_{jq}l_{kr}e_{pqr} = e_{ijk},$$

as required. We give e_{ijk} the name of **alternating tensor**.

A cartesian tensor of order n is an entity with 3^n components $U_{ij\cdots m}$ with respect to a given orthonormal basis, the components changing on change of basis according to the law

$$U_{ij\cdots m}' = l_{ip}l_{jq}\cdots l_{mt}U_{pq\cdots t}. \qquad (3.25)$$

Tensors of the same order may be added together, any tensor may be multiplied by a scalar, and components of tensors may be multiplied together to form higher-order tensors. For example, if T_{ij} is a CT2 and U_{ijk} is a CT3, the set of products $T_{ij}U_{klm}$ forms a cartesian tensor of order five (CT5). The operation of contraction will reduce the order by two, e.g. in the above example $T_{ij}U_{klj}$, $T_{ij}U_{jlm}$, $T_{ij}U_{kli}$, etc., are all obtained from $T_{ij}U_{klm}$ by this procedure, and each represents a CT3. (See problem 3.2.18.) Further contraction will produce a CT1. For example, $T_{ij}U_{kij}$ is doubly contracted from $T_{ij}U_{klm}$ and is a CT1. Explicitly,

$$\begin{aligned}T_{ij}'U_{kij}' &= (l_{ip}l_{jq}T_{pq})(l_{kr}l_{is}l_{jt}U_{rst})\\ &= (l_{ip}l_{is})(l_{jq}l_{jt})l_{kr}T_{pq}U_{rst}\\ &= \delta_{ps}\delta_{qt}l_{kr}T_{pq}U_{rst}\\ &= l_{kr}(T_{pq}U_{rpq}),\end{aligned}$$

showing that the appropriate rule (3.2) is satisfied.

It follows as a case in point that, if **u** and **v** are vectors, the combination of components $u_i v_j e_{kij}$ forms a CT1, since we now know that e_{kij} is a CT3 (and also $u_i v_j$ is a CT2). The vector $e_{kij}u_i v_j$ is just the vector product of **u** and **v**, and we can write $\mathbf{u}\times\mathbf{v}=e_{pqr}u_q v_r \mathbf{i}_p$, as in (1.19). Another example, if **u** is a vector *field*, is the set of components $e_{kij}\partial u_j/\partial y_i$ since we know $\partial u_j/\partial y_i$ is a CT2. This vector is curl **u**, and curl $\mathbf{u}=e_{pqr}(\partial u_r/\partial y_q)\mathbf{i}_p$, as in (1.48).

Various extensions of the quotient theorem can be formulated. In linear elasticity, for example, the second-order tensors of stress S_{ij} and strain E_{ij} are linearly related by the equation

$$S_{ij}=L_{ijkl}E_{kl},$$

where the 81 elastic moduli L_{ijkl} depend on the material involved. We may regard L_{ijkl} as acting on the arbitrary strain tensor E_{ij} and producing the stress tensor S_{ij}. A quotient theorem (slightly modified to take account of the fact that S_{ij} and E_{ij} are symmetric tensors) shows that L_{ijkl} must be the components of a cartesian tensor of order four (CT4).

Problems

3.2.19 Show that, if T_{ij} is a CT2 and U_{ijk} is a CT3, then $T_{ij}U_{klj}$ constitutes a CT3.

3.2.20 Show that, if T_{ij} is a second-order cartesian tensor *field* (the components depending on position in space), then the partial derivatives $\partial T_{ij}/\partial y_k$ constitute a CT3.

3.2.21 (A version of the quotient theorem) If U_{ijkl} is an entity with 81 components with respect to any given orthonormal basis, such that, if **T** is an arbitrary CT2, the set of nine components $U_{ijkl}T_{kl}$ always constitute a CT2; then U_{ijkl} is a CT4.

3.2.22 If a CT3 U_{ijk} is symmetric with respect to its first and last pair of indices, i.e. $U_{ijk}=U_{jik}=U_{ikj}$, and is skew symmetric with respect to its first and third indices, i.e. $U_{ijk}=-U_{kji}$, show that all its components must be zero.

3.2.8 Tensor products

The set of nine dyadic products $\mathbf{i}_j \otimes \mathbf{i}_k$ ($j, k = 1, 2, 3$) provides a useful way of representing second-order tensors. In fact, just as the set of three orthonormal vectors $\{\mathbf{i}_1, \mathbf{i}_2, \mathbf{i}_3\}$ forms a basis for vectors (CT1s), the set of dyadic products forms a **basis** for CT2s. This means that any CT2 $\mathbf{T}$ can be expressed as a linear combination (with summation over j and k) of dyadic products

$$\mathbf{T} = T_{jk} \mathbf{i}_j \otimes \mathbf{i}_k.$$

The coefficients T_{jk} are just the components of $\mathbf{T}$ with respect to the given basis. The action of $\mathbf{T}$, regarded as a linear operator, on a vector $\mathbf{u}$ may be expressed as

$$\begin{aligned}\mathbf{Tu} &= (T_{jk} \mathbf{i}_j \otimes \mathbf{i}_k)\mathbf{u} \\ &= T_{jk}(\mathbf{i}_k \cdot \mathbf{u})\mathbf{i}_j \qquad \text{by (3.19)} \\ &= T_{jk} u_k \mathbf{i}_j.\end{aligned}$$

This is a vector whose component in the $\mathbf{i}_j$ direction is $T_{jk} u_k$.

Example

The isotropic second-order tensor whose components are δ_{jk} with respect to any orthonormal basis whatsoever has the dyadic form $\delta_{jk} \mathbf{i}_j \otimes \mathbf{i}_k$, which is equivalent to $\mathbf{i}_j \otimes \mathbf{i}_j$ by the substitution property of δ_{jk}, i.e.

$$\mathbf{i}_1 \otimes \mathbf{i}_1 + \mathbf{i}_2 \otimes \mathbf{i}_2 + \mathbf{i}_3 \otimes \mathbf{i}_3. \tag{3.26}$$

It is natural to call this the **identity** operator $\mathbf{I}$, since its action on any vector $\mathbf{u}$ is given by

$$\mathbf{Iu} = (\mathbf{i}_j \otimes \mathbf{i}_j)\mathbf{u} = \mathbf{i}_j(\mathbf{i}_j \cdot \mathbf{u}) = \mathbf{i}_j u_j = \mathbf{u}.$$

The dyadic product representation of tensors may be extended to higher-order tensors. We may introduce the notation $\mathbf{u} \otimes \mathbf{v} \otimes \mathbf{w}$ for the third-order tensor obtained from the three vectors $\mathbf{u}$, $\mathbf{v}$ and $\mathbf{w}$ by taking products of components $u_i v_j w_k$ with respect to any orthonormal basis $\{\mathbf{i}_1, \mathbf{i}_2, \mathbf{i}_3\}$. The set of tensors $\mathbf{i}_p \otimes \mathbf{i}_q \otimes \mathbf{i}_r$ then forms a basis for the set of third-order tensors, which may typically be expressed as

$$\mathbf{U} = U_{pqr} \mathbf{i}_p \otimes \mathbf{i}_q \otimes \mathbf{i}_r.$$

A CT3 such as $\mathbf{u} \otimes \mathbf{v} \otimes \mathbf{w}$ may be regarded as a linear operator acting on vectors to produce CT2s, e.g.

$$(\mathbf{u} \otimes \mathbf{v} \otimes \mathbf{w})\mathbf{a} = (\mathbf{w} \cdot \mathbf{a})(\mathbf{u} \otimes \mathbf{v})$$

or, in component form,

$$(u_i v_j w_k) a_k = (w_k a_k)(u_i v_j).$$

Problems

3.2.23 (i) Find the cartesian components of the following second-order tensors with respect to the basis $\{\mathbf{i}_1, \mathbf{i}_2, \mathbf{i}_3\}$:

(a) $\mathbf{i}_1 \otimes \mathbf{i}_1$;
(b) $\mathbf{i}_1 \otimes \mathbf{i}_2$;
(c) $\mathbf{i}_2 \otimes \mathbf{i}_1$;
(d) $\mathbf{i}_1 \otimes \mathbf{i}_1 + 2\mathbf{i}_2 \otimes \mathbf{i}_2$.

Obtain the components of the vector that results when each tensor acts on the vector with cartesian components (u_1, u_2, u_3).

(ii) Obtain the components of the CT2 that results when the CT3

$$\mathbf{i}_1 \otimes \mathbf{i}_3 \otimes \mathbf{i}_1 + 2\mathbf{i}_2 \otimes \mathbf{i}_1 \otimes \mathbf{i}_2 + 3\mathbf{i}_3 \otimes \mathbf{i}_2 \otimes \mathbf{i}_3$$

acts on the vector with cartesian components (3, 4, 7).

3.3 GENERALIZED TENSORS

3.3.1 General co-ordinate transformations

The transformation equations (3.4) between different rectangular cartesian co-ordinate systems having the same origin are of a simple linear form, the coefficients being the appropriate direction cosines. The Jacobian J of the transformation is the determinant of the matrix of coefficients and we always have $J = 1$ if we confine our attention to right-handed systems. The transformation equations that relate two general curvilinear co-ordinate systems $\{x^1, x^2, x^3\}$ and $\{\bar{x}^1, \bar{x}^2, \bar{x}^3\}$ will be of *non-linear* form in general:

$$\bar{x}^i = \bar{x}^i(x^1, x^2, x^3). \tag{3.27}$$

However, the differentials $d\bar{x}^i$ and dx^i of the co-ordinates are related by the usual formula

$$d\bar{x}^i = \frac{\partial \bar{x}^i}{\partial x^1} dx^1 + \frac{\partial \bar{x}^i}{\partial x^2} dx^2 + \frac{\partial \bar{x}^i}{\partial x^3} dx^3 = \frac{\partial \bar{x}^i}{\partial x^j} dx^j. \tag{3.28}$$

The partial derivatives which constitute the coefficients in this formula take particular values at a given point of space, and so *locally* the relation between the differentials may be regarded as linear.

For example, the relation between rectangular cartesians and polar co-ordinates in two dimensions has the non-linear form

$$\left. \begin{array}{l} x = \rho \cos \phi, \\ y = \rho \sin \phi, \end{array} \right\}$$

but the relation between the differentials at a point is

$$\left. \begin{array}{l} dx = \dfrac{\partial x}{\partial \rho} d\rho + \dfrac{\partial x}{\partial \phi} d\phi = \cos \phi \, d\rho - \rho \sin \varphi \, d\phi \\ dy = \dfrac{\partial y}{\partial \rho} d\rho + \dfrac{\partial y}{\partial \phi} d\phi = \sin \phi \, d\rho + \rho \cos \phi \, d\phi, \end{array} \right\}$$

We can still define the Jacobian J of the transformation, although now it is the

Generalized tensors

determinant of the matrix of coefficients $\partial \bar{x}^i/\partial x^j$ in equation (3.28) governing the transformation of differentials. So, in general, J is not a constant but varies with position. In the above example,

$$J = \begin{vmatrix} \cos\phi & -\rho\sin\phi \\ \sin\phi & \rho\cos\phi \end{vmatrix} = \rho\cos^2\phi + \rho\sin^2\phi = \rho.$$

The transformation between differentials will be invertible locally if the matrix of coefficients is non-singular, i.e. if $J \neq 0$. Thus the transformation between rectangular cartesians and polars in two dimensions is singular at the origin, where $\rho = 0$.

For the three-dimensional example of the transformation (1.54) from spherical polar co-ordinates to rectangular cartesians, with $y_1 = \bar{x}^1, y_2 = \bar{x}^2, y_3 = \bar{x}^3, r = x^1, \theta = x^2, \phi = x^3$, we have

$$\frac{\partial \bar{x}^1}{\partial x^1} = \sin x^2 \cos x^3, \quad \frac{\partial \bar{x}^1}{\partial x^2} = x^1 \cos x^2 \cos x^3, \quad \frac{\partial \bar{x}^1}{\partial x^3} = -x^1 \sin x^2 \sin x^3$$

$$\frac{\partial \bar{x}^2}{\partial x^1} = \sin x^2 \sin x^3, \quad \frac{\partial \bar{x}^2}{\partial x^2} = x^1 \cos x^2 \sin x^3, \quad \frac{\partial \bar{x}^2}{\partial x^3} = x^1 \sin x^2 \cos x^3$$

and

$$\frac{\partial \bar{x}^3}{\partial x^1} = \cos x^2, \quad \frac{\partial \bar{x}^3}{\partial x^2} = -x^1 \sin x^2, \quad \frac{\partial \bar{x}^3}{\partial x^3} = 0.$$

Hence

$$J = \begin{vmatrix} \frac{\partial \bar{x}^1}{\partial x^1} & \frac{\partial \bar{x}^1}{\partial x^2} & \frac{\partial \bar{x}^1}{\partial x^3} \\ \frac{\partial \bar{x}^2}{\partial x^1} & \frac{\partial \bar{x}^2}{\partial x^2} & \frac{\partial \bar{x}^2}{\partial x^3} \\ \frac{\partial \bar{x}^3}{\partial x^1} & \frac{\partial \bar{x}^3}{\partial x^2} & \frac{\partial \bar{x}^3}{\partial x^3} \end{vmatrix} = \begin{vmatrix} \sin x^2 \cos x^3 & x^1 \cos x^2 \cos x^3 & -x^1 \sin x^2 \sin x^3 \\ \sin x^2 \sin x^3 & x^1 \cos x^2 \sin x^3 & x^1 \sin x^2 \cos x^3 \\ \cos x^2 & -x^1 \sin x^2 & 0 \end{vmatrix}$$

$$= (x^1)^2 \sin x^2 = r^2 \sin\theta,$$

which shows that $J \neq 0$ except on the axis given by $\theta = 0$ or π.

At points where $J \neq 0$ the transformation (3.27) itself may be inverted to express x^i in terms of $\bar{x}^1$, $\bar{x}^2$ and $\bar{x}^3$, i.e.

$$x^i = x^i(\bar{x}^1, \bar{x}^2, \bar{x}^3). \tag{3.29}$$

In the above example we have

$$x^1 = \sqrt{(\bar{x}^1)^2 + (\bar{x}^2)^2 + (\bar{x}^3)^2},$$

$$x^2 = \cos^{-1}\left(\frac{\bar{x}^3}{\sqrt{(\bar{x}^1)^2 + (\bar{x}^2)^2 + (\bar{x}^3)^2}}\right)$$

and

$$x^3 = \tan^{-1}\left(\frac{\bar{x}^2}{\bar{x}^1}\right).$$

The corresponding inverse relationship between the differentials may be obtained by inverting the matrix of coefficients in equation (3.28). It follows that the Jacobian of the transformation (3.29) is the determinant of the inverse of the matrix of partial derivatives $\partial \bar{x}^i/\partial x^j$. By the property of determinants, this is equal to the reciprocal of the determinant of the matrix $(\partial \bar{x}^i/\partial x^j)$, i.e. equal to $1/J$.

This result follows alternatively from considering the chain rule for partial derivatives in the form

$$\frac{\partial \bar{x}^i}{\partial x^j}\frac{\partial x^j}{\partial \bar{x}^k} = \frac{\partial \bar{x}^i}{\partial \bar{x}^k} = \delta^i_k. \tag{3.30}$$

Taking determinants of the matrices on both sides gives

$$\det\left(\frac{\partial \bar{x}^i}{\partial x^j}\frac{\partial x^j}{\partial \bar{x}^k}\right) = \det\left(\frac{\partial \bar{x}^i}{\partial x^j}\right)\det\left(\frac{\partial x^j}{\partial \bar{x}^k}\right) = \det(\delta^i_k) = 1,$$

and hence

$$\det\left(\frac{\partial x^j}{\partial \bar{x}^k}\right) = 1 \Big/ \det\left(\frac{\partial \bar{x}^i}{\partial x^j}\right) = \frac{1}{J}.$$

(Recall that the determinant of a matrix product is equal to the product of the determinants and that the matrix array corresponding to δ^i_k is just the identity matrix.)

Suppose we consider two successive transformations, the first given by

$$\bar{x}^i = \bar{x}^i(x^1, x^2, x^3)$$

with Jacobian J_1 at some point in space with co-ordinates (x^1, x^2, x^3), and the second given by

$$\hat{x}^i = \hat{x}^i(\bar{x}^1, \bar{x}^2, \bar{x}^3),$$

where $\{\hat{x}^1, \hat{x}^2, \hat{x}^3\}$ are a new set of generalized co-ordinates, the Jacobian being J_2 at the same point. Then we may deduce from the chain rule

$$\frac{\partial \hat{x}^i}{\partial x^k} = \frac{\partial \hat{x}^i}{\partial \bar{x}^j}\frac{\partial \bar{x}^j}{\partial x^k} \tag{3.31}$$

that the Jacobian J_3 of the composite transformation

$$\hat{x}^i = \hat{x}^i(x^1, x^2, x^3)$$

is given by

$$J_3 = \det\left(\frac{\partial \hat{x}^i}{\partial x^k}\right) = \det\left(\frac{\partial \hat{x}^i}{\partial \bar{x}^j}\right)\det\left(\frac{\partial \bar{x}^i}{\partial x^k}\right) = J_2 J_1, \tag{3.32}$$

i.e. it is equal to the product of the Jacobians. Thus, if J_1 and J_2 are both non-zero, so too is J_3.

It is convenient to denote the matrix whose i–j entry is $\partial \bar{x}^i/\partial x^j$ by A and the matrix whose i–j entry is $\partial x^j/\partial \bar{x}^i$ by B. Then equation (3.30) may be written as the matrix equation

$$AB^T = I. \tag{3.33}$$

Thus $B^T = A^{-1}$ and we also have

$$B^T A = BA^T = A^T B = I. \tag{3.34}$$

Suppose that in addition to the curvilinear co-ordinates x^i and $\bar{x}^i$ we have a set of rectangular cartesian co-ordinates $\{y^1, y^2, y^3\}$ (the indices now written as superscripts for consistency with the other co-ordinates). Assume that there exist transformations

$$x^i = x^i(y^1, y^2, y^3)$$

and

$$\bar{x}^i = \bar{x}^i(y^1, y^2, y^3),$$

with corresponding matrices of partial derivatives $\partial x^i/\partial y^j$ and $\partial \bar{x}^i/\partial y^j$ denoted by M and $\bar{M}$, respectively, and that there are inverse transformations with partial derivatives $\partial y^j/\partial x^i$ and $\partial y^j/\partial \bar{x}^i$ which are the i–j entries of matrices L and $\bar{L}$, respectively. (This notation is consistent with that used in section 1.5.3.) Then from the chain rule (3.31) we may deduce that

$$\bar{M} = AM. \tag{3.35}$$

Since we also have $LM^T = \bar{L}\bar{M}^T = I$, it follows by matrix manipulation (or by further use of the chain rule) that

$$\bar{L} = BL. \tag{3.36}$$

If we introduce corresponding Jacobians J^* and $\bar{J}^*$, where $J^* = \det M$ and $\bar{J}^* = \det \bar{M}$, then we also have

$$\bar{J}^* = JJ^*, \tag{3.37}$$

taking determinants of (3.35), since $J = \det A$.

From (1.71) and (1.73) we may deduce useful matrix expressions for g_{ij} and g^{ij}. If we let G be the matrix with components g_{ij}, we have, by (1.71),

$$G = LL^T, \tag{3.38}$$

and by (1.73) the matrix with components g^{ij} must be MM^T. This is just the inverse of G.

Similar expressions hold for the corresponding matrices $\bar{G}$ and its inverse in the system $\{\bar{x}^1, \bar{x}^2, \bar{x}^3\}$.

If we put $\det G = g$, and $\det \bar{G} = \bar{g}$, we have

$$g = \det(LL^T) = (\det L)^2 = (\det M)^{-2} = (J^*)^{-2}, \tag{3.39}$$

and, similarly,

$$\bar{g} = (\bar{J}^*)^{-2}.$$

Hence, from (3.37),

$$J = \frac{\bar{J}^*}{J^*} = \sqrt{\frac{g}{\bar{g}}}. \tag{3.40}$$

Note that g and $\bar{g}$ are always positive, but that J will not necessarily be positive (for completely general systems), unless we confine our attention to (say) right-handed systems, i.e. systems for which the scalar triple product of the base vectors $\mathbf{g}_1 \cdot (\mathbf{g}_2 \times \mathbf{g}_3)$ is always positive.

3.3.2 Transformation of base vectors and vector components

Given a set of curvilinear co-ordinates $\{x^1, x^2, x^3\}$ we can define a natural set of base vectors at any point in space as in section 1.7 by setting $\mathbf{g}_i = \partial \mathbf{r}/\partial x^i$, $i = 1, 2, 3$. For a different set of co-ordinates $\{\bar{x}^1, \bar{x}^2, \bar{x}^3\}$ the base vectors will be $\bar{\mathbf{g}}_i = \partial \mathbf{r}/\partial \bar{x}^i$, $i = 1, 2, 3$. The relationship between the two different sets of base vectors may be obtained from the transformation equation (3.27). Again, the chain rule for partial derivatives plays an important part. We have

$$\frac{\partial \mathbf{r}}{\partial x^j} = \frac{\partial \mathbf{r}}{\partial \bar{x}^k} \frac{\partial \bar{x}^k}{\partial x^j},$$

(with summation over the suffix k). Hence

$$\mathbf{g}_j = \bar{\mathbf{g}}_k \frac{\partial \bar{x}^k}{\partial x^j},$$

i.e.

$$\mathbf{g}_j = \frac{\partial \bar{x}^k}{\partial x^j} \bar{\mathbf{g}}_k. \tag{3.41}$$

Similarly, another version of the chain rule is

$$\frac{\partial \mathbf{r}}{\partial \bar{x}^j} = \frac{\partial \mathbf{r}}{\partial x^k} \frac{\partial x^k}{\partial \bar{x}^j},$$

which leads to

$$\bar{\mathbf{g}}_j = \frac{\partial x^k}{\partial \bar{x}^j} \mathbf{g}_k. \tag{3.42}$$

Thus we can express each base vector in one set of co-ordinates as a linear combination of the base vectors in the other set, the coefficients being the appropriate partial derivatives.

Similar relationships exist between the corresponding reciprocal base vectors. Taking the scalar products of both sides of (3.41) with $\bar{\mathbf{g}}^l$ gives

$$\mathbf{g}_j \cdot \bar{\mathbf{g}}^l = \frac{\partial \bar{x}^k}{\partial x^j} \bar{\mathbf{g}}_k \cdot \bar{\mathbf{g}}^l = \frac{\partial \bar{x}^k}{\partial x_j} \delta_k^l = \frac{\partial \bar{x}^l}{\partial x_j}, \tag{3.43}$$

and, operating similarly with (3.42),

$$\bar{\mathbf{g}}_j \cdot \mathbf{g}^l = \frac{\partial x^l}{\partial \bar{x}^j}. \tag{3.44}$$

Now, writing $\mathbf{g}^l$ as a linear combination of $\bar{\mathbf{g}}^1, \bar{\mathbf{g}}^2, \bar{\mathbf{g}}^3$, i.e.

$$\mathbf{g}^l = \alpha_k^l \bar{\mathbf{g}}^k,$$

for some α_k^l, we have

$$\mathbf{g}^l \cdot \bar{\mathbf{g}}_j = \alpha_k^l \bar{\mathbf{g}}^k \cdot \bar{\mathbf{g}}_j = \alpha_k^l \delta_j^k = \alpha_j^l$$

$$= \frac{\partial x^l}{\partial \bar{x}_j} \quad \text{from (3.44),}$$

thus determining $\alpha_k^l = \partial x^l / \partial \bar{x}^k$.

Sec. 3.3]   **Generalized tensors**   91

Hence

$$g^l = \frac{\partial x^l}{\partial \bar{x}^k} \bar{g}^k. \tag{3.45}$$

Similarly, we can obtain the relationship

$$\bar{g}^l = \frac{\partial \bar{x}^l}{\partial x^j} g^j. \tag{3.46}$$

We have seen that an arbitrary vector **u** has two different sets of components, depending on whether it is expressed as a linear combination of the base vectors or the reciprocal base vectors, i.e.

$$\mathbf{u} = u^i \mathbf{g}_i = u_i \mathbf{g}^i,$$

where u^i and u_i are the contravariant and covariant components of **u**, respectively. Each set of components of **u** will transform differently under the co-ordinate transformation (3.27), and we now proceed to establish the appropriate transformation rules.

For contravariant components, we have

$$\mathbf{u} = u^i \mathbf{g}_i = \bar{u}^i \bar{\mathbf{g}}_i,$$

and

$$u^i \mathbf{g}_i = u^i \frac{\partial \bar{x}^k}{\partial x^i} \bar{\mathbf{g}}_k \qquad \text{by (3.41)}.$$

Hence

$$u^i \frac{\partial \bar{x}^k}{\partial x^i} \bar{\mathbf{g}}_k = \bar{u}^i \bar{\mathbf{g}}_i = \bar{u}^k \bar{\mathbf{g}}_k.$$

Comparing coefficients of $\bar{\mathbf{g}}_k$ on each side, we obtain

$$\bar{u}^k = u^i \frac{\partial \bar{x}^k}{\partial x^i},$$

or, changing i to j and k to i,

$$\bar{u}^i = \frac{\partial \bar{x}^i}{\partial x^j} u^j. \tag{3.47}$$

It is useful to be familiar with this transformation rule and to be able to remember it. If we think of the $\bar{x}^i$ as the 'new' co-ordinates (and the x^i as the 'old' ones), then for *contravariant* components (for which the index i appears as a superscript) the new co-ordinates appear *on the top* in the coefficients $\partial \bar{x}^i / \partial x^j$.

For covariant components, the same vector is expressible as

$$\mathbf{u} = u_j \mathbf{g}^j = \bar{u}_k \bar{\mathbf{g}}^k.$$

Taking scalar products with $\bar{\mathbf{g}}_i$ gives

$$u_j \mathbf{g}^j \cdot \bar{\mathbf{g}}_i = u_j \frac{\partial x^j}{\partial \bar{x}^i} \qquad \text{by (3.44)}$$

and

$$\bar{u}_k \bar{\mathbf{g}}^k \cdot \bar{\mathbf{g}}_i = \bar{u}_k \delta^k_i = \bar{u}_i.$$

Thus we have the transformation rule

$$\bar{u}_i = \frac{\partial x^j}{\partial \bar{x}^i} u_j. \tag{3.48}$$

Note carefully the difference between this rule and (3.47). For *covariant* components (for which the index i occurs as a *subscript*) the 'new' co-ordinates appear *on the bottom* in the coefficients $\partial x^j / \partial \bar{x}^i$.

It is interesting to compare the transformation rules (3.47) and (3.48) with the rule (3.2) for CT1s. With rectangular cartesian systems having the same origin we have

$$y_i' = l_{ij} y_j \quad \text{and} \quad y_j = l_{ij} y_i'$$

as in (3.4) and (3.5), and it follows that

$$\frac{\partial y_i'}{\partial y_j} = \frac{\partial y_j}{\partial y_i'} = l_{ij}.$$

Thus in this case there is no difference between the coefficients in the transformation rules (3.47) and (3.48), as is to be expected, since in rectangular cartesian systems there is no distinction between covariant and contravariant components.

Note that the differentials dx^i and $d\bar{x}^i$ are naturally related by the chain rule (3.28). The similarity between this equation and (3.47) shows that the differentials behave like *contravariant* components of vectors, the vectors in fact being the differentials of the position vector $d\mathbf{r} = dx^i \, \partial \mathbf{r}/\partial x^i = dx^i \, \mathbf{g}_i$. This is the reason why the index for the co-ordinates $\{x^1, x^2, x^3\}$ is written in the upper position rather than the lower one, although it is important to keep in mind that it is the *differentials* $\{dx^1, dx^2, dx^3\}$ which behave like contravariant components and not the co-ordinates $\{x^1, x^2, x^3\}$ themselves. While the *cartesian* co-ordinates are the cartesian components of the position vector, the curvilinear co-ordinates $\{x^1, x^2, x^3\}$ are *not* contravariant components of that vector.

A naturally occurring set of quantities which transform according to the *covariant* rule (3.48) is the set of partial derivatives $\{\partial \Phi/\partial x^1, \partial \Phi/\partial x^2, \partial \Phi/\partial x^3\}$ of a scalar field Φ. The chain rule again shows immediately that

$$\frac{\partial \Phi}{\partial \bar{x}_i} = \frac{\partial x^j}{\partial \bar{x}^i} \frac{\partial \Phi}{\partial x^j},$$

as required. Thus $\partial \Phi / \partial x^i$ are the covariant components of a vector which is the same gradient vector grad Φ that we have met previously, i.e.

$$\text{grad } \Phi = \frac{\partial \Phi}{\partial x^i} \mathbf{g}^i.$$

The transformation rules may be expressed in matrix form using the matrices introduced in the last section. A contravariant vector has a set of components represented by the 3×1 column vector $\mathbf{u}^c$ with respect to the co-ordinates $\mathbf{x}^i$, and the transformation rule (3.47) becomes

$$\bar{\mathbf{u}}^c = A \mathbf{u}^c. \tag{3.49}$$

Similarly, a covariant vector is represented by a 3×1 column vector $\mathbf{u}_c$, and transforms according to (3.48), which becomes

$$\bar{u}_c = B u_c. \tag{3.50}$$

3.3.3 Definition of generalized tensors of orders one and two

We adopt the same approach to the definition of generalized tensors that we followed for cartesian tensors, i.e.

(1) we derive the transformation laws for vector components, using vectors as a familiar mathematical concept,
(2) we use the transformation laws to redefine a vector as a tensor of order one, and
(3) we extend the transformation laws to define higher-order tensor quantities.

The situation is somewhat more complicated in the case of generalized tensors because we must deal with both covariant and contravariant components.

We have established the rule (3.47) for transformation of contravariant vector components and the rule (3.48) for transformation of covariant vector components. The next step is to define a **contravariant tensor of order one** (a contravariant vector) to be a correspondence between sets of curvilinear co-ordinates $\{x^1, x^2, x^3\}$ and sets of three numbers $\{u^1, u^2, u^3\}$ which transform according to (3.47). A **covariant tensor of order one** (a covariant vector) is defined in the same way except that the transformation rule is (3.48) and the set of three components is denoted $\{u_1, u_2, u_3\}$.

The set $\{u^i\}$ and the set $\{u_i\}$ are said to be **associated** if they satisfy

$$u_i = g_{ij} u^j \tag{3.51}$$

as in (1.67). The contravariant and covariant components are then guaranteed to represent the same vector.

Let us check that, if two sets of tensor components are associated with respect to one choice of curvilinear co-ordinates (x^1, x^2, x^3), then they remain associated with respect to any other choice $(\bar{x}^1, \bar{x}^2, \bar{x}^3)$. By the transformation laws we have

$$\bar{u}_i = \frac{\partial x^j}{\partial \bar{x}^i} u_j = \frac{\partial x^j}{\partial \bar{x}^i} g_{jk} u^k \qquad \text{by (3.51)}$$

$$= \frac{\partial x^j}{\partial \bar{x}^i} \frac{\partial y_l}{\partial x^j} \frac{\partial y_l}{\partial x^k} u^k \qquad \text{by (1.71)}$$

(for some choice of cartesian system $\{y_1, y_2, y_3\}$)

$$= \frac{\partial y_l}{\partial \bar{x}^i} \frac{\partial y_l}{\partial x^k} u^k \qquad \text{by the chain rule}$$

$$= \frac{\partial y_l}{\partial \bar{x}^i} \left(\frac{\partial y_l}{\partial \bar{x}^j} \frac{\partial \bar{x}^j}{\partial x^k} \right) u^k \qquad \text{by the chain rule}$$

$$= \bar{g}_{ij} \frac{\partial \bar{x}^j}{\partial x^k} u^k$$

$$= \bar{g}_{ij} \bar{u}^j \qquad \text{by the transformation rule (3.47)}$$

as required.

Our definition can now be extended to higher-order tensors. A **contravariant second-order tensor** is a correspondence between sets of generalized co-ordinates $\{x^1, x^2, x^3\}$ and sets of nine components T^{ij} which transform according to the rule

$$\bar{T}^{ij} = \frac{\partial \bar{x}^i}{\partial x^k} \frac{\partial \bar{x}^j}{\partial x^l} T^{kl}. \tag{3.52}$$

That is, nine numbers T^{ij} are associated with a particular point P in space with curvilinear co-ordinates (x^1, x^2, x^3) and, if the same point has co-ordinates $(\bar{x}^1, \bar{x}^2, \bar{x}^3)$ with respect to some other co-ordinate system, the nine components become $\bar{T}^{ij}$ as given by (3.52), the coefficients in the summation on the RHS being given by the various partial derivatives of the transformation (3.27) evaluated at P. We refrain from writing out these equations in full, as in (3.9) and (3.14).

A simple example of a contravariant second-order tensor is provided by the product of components $u^i v^j$ of two contravariant vectors, since we can immediately verify the transformation property

$$\bar{u}^i \bar{v}^j = \left(\frac{\partial \bar{x}^i}{\partial x^k} u^k \right) \left(\frac{\partial \bar{x}^j}{\partial x^l} v^l \right) = \frac{\partial \bar{x}^i}{\partial x^k} \frac{\partial \bar{x}^j}{\partial x^l} u^k v^l. \tag{3.53}$$

A **covariant second-order tensor** may be defined in precisely the same way except that we write the components T_{ij} with respect to a particular co-ordinate system with both indices as subscripts and the corresponding transformation law becomes

$$\bar{T}_{ij} = \frac{\partial x^k}{\partial \bar{x}^i} \frac{\partial x^l}{\partial \bar{x}^j} T_{kl}. \tag{3.54}$$

Note that these transformation laws are not difficult to remember if we observe that 'new' co-ordinates associated with contravariant or covariant components must occur on the top or bottom, respectively, of the appropriate partial derivative. Note also the consistency of these equations with respect to free and repeated indices; *summation is carried out only over repeated indices occurring once in an upper position and once in a lower position*. In a partial derivative term such as $\partial \bar{x}^i / \partial x^j$, the i is regarded as occupying an upper and the j a lower position.

We may also define **mixed second-order tensors** $T^i_{\cdot j}$ and $T_i^{\cdot j}$ containing one contravariant and one covariant component. The dot before the j indicates that in each case i is the first index and j the second. The transformation laws are

$$\bar{T}^i_{\cdot j} = \frac{\partial \bar{x}^i}{\partial x^k} \frac{\partial x^l}{\partial \bar{x}^j} T^k_{\cdot l}, \tag{3.55}$$

and

$$\bar{T}_i^{\cdot j} = \frac{\partial x^k}{\partial \bar{x}^i} \frac{\partial \bar{x}^j}{\partial x^l} T_k^{\cdot l}. \tag{3.56}$$

In these laws, then, we are careful to distinguish between the cases where the covariant or the contravariant component of the mixed tensor occurs first.

In some circumstances, however, it may be unnecessary to make this distinction,

Sec. 3.3]   **Generalized tensors**   95

since the transformation law (3.56) may be written, interchanging i and j, as

$$\bar{T}^{\cdot i}_j = \frac{\partial x^k}{\partial \bar{x}^j}\frac{\partial \bar{x}^i}{\partial x^l} T^{\cdot l}_k = \frac{\partial \bar{x}^i}{\partial x^l}\frac{\partial x^k}{\partial \bar{x}^j} T^{\cdot l}_k$$

$$= \frac{\partial \bar{x}^i}{\partial x^k}\frac{\partial x^l}{\partial \bar{x}^j} T^{\cdot k}_l,$$

interchanging k and l also. This form of the rule is identical with (3.55) except that the order of the indices in the tensor has been changed. So both mixed tensors transform under essentially the same rule. We may wish to preserve the distinction between $T^i_{\cdot j}$ and $T_i^{\cdot j}$, however, in the context of *associated* components of the same tensor.

Example 1

The Kronecker delta δ^i_j (with one covariant and one contravariant component, the order being irrelevant) may be regarded as a mixed second-order tensor. Its components remain unchanged when we change from one general co-ordinate system to another.

To see this, apply the transformation rule (3.55). We obtain

$$\frac{\partial \bar{x}^i}{\partial x^k}\frac{\partial x^l}{\partial \bar{x}^j}\delta^k_l = \frac{\partial \bar{x}^i}{\partial x^k}\frac{\partial x^k}{\partial \bar{x}^j} \qquad \text{using the substitution property of } \delta^k_l$$

$$= \frac{\partial \bar{x}^i}{\partial \bar{x}^j} \qquad\qquad\qquad \text{by the chain rule}$$

$$= \delta^i_j \qquad \text{since the } \bar{x}^i \text{ are an independent set of co-ordinates.}$$

Example 2

The nine quantities g_{ij} constitute a covariant second-order tensor. Applying the transformation law (3.54) gives

$$\frac{\partial x^k}{\partial \bar{x}^i}\frac{\partial x^l}{\partial \bar{x}^j} g_{kl} = \frac{\partial x^k}{\partial \bar{x}^i}\frac{\partial x^l}{\partial \bar{x}^j} \mathbf{g}_k \cdot \mathbf{g}_l$$

$$= \left(\frac{\partial x^k}{\partial \bar{x}^i}\mathbf{g}_k\right)\cdot\left(\frac{\partial x^l}{\partial \bar{x}^j}\mathbf{g}_l\right)$$

$$= \bar{\mathbf{g}}_i \cdot \bar{\mathbf{g}}_j \qquad\qquad\qquad\qquad\qquad\qquad \text{(by 3.42)}$$

$$= \bar{g}_{ij}.$$

This tensor is called the **metric tensor**, since by (1.68) it governs the measure of distance in a given co-ordinate system. The transformation law applied to the metric tensor thus generates the new metric tensor for the new co-ordinate system.

It may sometimes be convenient to represent the transformation laws for second-order tensors in matrix form. Using the same symbol T for the different matrix arrays of contravariant, covariant and mixed components, equations (3.52), (3.54), (3.55) and (3.56) may be written

$$\bar{T} = ATA^T, \qquad (3.57)$$

$$\bar{T} = BTB^T, \qquad (3.58)$$

$$\bar{T} = ATB^T \qquad (3.59)$$

and $\qquad \bar{T} = BTA^T. \qquad (3.60)$

We can immediately deduce that, if S and T represent contravariant tensors and λ is a scalar we can generate further contravariant tensors by taking the sum $S + T$ of corresponding components, and by multiplying components throughout by λ to produce λS or λT. This follows by direct application of (3.57), using the fundamental linear properties of the transformation law, i.e.

$$\bar{T} = ATA^T \quad \text{and} \quad \bar{S} = ASA^T.$$

Hence
$$\bar{S} + \bar{T} = A(S + T)A^T,$$

and $\qquad \lambda \bar{S} = A(\lambda S)A^T,$

showing that $S + T$ and λS transform in the correct way. Similar results hold for covariant and mixed components, although note that we cannot add a contravariant tensor to a covariant tensor to produce a tensor.

Mixed tensors may be formed from a covariant tensor S_{ij} and a contravariant tensor T^{ij} by taking sums of products $S_{ik}T^{kj}$. To show that these components transform in the required way, we start with

$$\bar{S} = BSB^T \quad \text{and} \quad \bar{T} = ATA^T,$$

and consider the corresponding matrix product ST, which satisfies the transformation law

$$\bar{S}\bar{T} = (BSB^T)(ATA^T) = BS(B^T A)TA^T$$
$$= BSITA^T \qquad \text{by (3.34)}$$
$$= B(ST)A^T \qquad \text{as required,}$$

since (3.60) is satisfied.

Compare this result with a similar procedure starting with contravariant tensors S^{ij} and T^{ij}. The matrix product ST generates components $S^{ik}T^{kj}$, but now these do *not* transform as a contravariant tensor, since

$$\bar{S}\bar{T} = (ASA^T)(ATA^T) = ASA^T ATA^T,$$

and we cannot satisfy the law (3.57).

Given a contravariant tensor T^{ij} with matrix array T we can also generate another contravariant tensor with components T^{ji} and matrix T^T. The transpose of (3.57)

$$\bar{T}^T = (ATA^T)^T = AT^T A^T$$

shows that T^{ji} behaves in the correct manner.

Problems

3.3.1. If u_i is a covariant vector and v^i is a contravariant vector, show that $u_i v^j$ is a mixed second-order tensor.

3.3.2. (a) Show that the set of nine partial derivatives $\partial^2 \Phi/(\partial x^i \partial x^j)$ of a scalar field Φ transforms according to the rule

$$\frac{\partial^2 \Phi}{\partial \bar{x}^i \partial \bar{x}^j} = \frac{\partial x^k}{\partial \bar{x}^i} \frac{\partial x^l}{\partial \bar{x}^j} \frac{\partial^2 \Phi}{\partial x^k \partial x^l} + \frac{\partial^2 x^l}{\partial \bar{x}^i \partial \bar{x}^j} \frac{\partial \Phi}{\partial x^l},$$

and therefore does not transform as a second-order tensor.

(b) Find the transformation rule for the set of partial derivatives $\partial v^i/\partial x^j$ of a contravariant vector field v^i, showing that it does not transform as a second-order tensor.

3.3.3. If T is the 3×3 matrix of rectangular cartesian components of a CT2 **T** with respect to co-ordinates (y^1, y^2, y^3), show that the matrices of associated components of **T** with respect to a set of curvilinear co-ordinates $\{x^1, x^2, x^3\}$, are given by the matrix products MTM^T (contravariant components), LTL^T (covariant components), and MTL^T and LTM^T (mixed components), where L and M are the matrices defined in section 3.3.1.

3.3.4. Assuming that the matrix T in equation (3.57) is non-singular, deduce that

$$(\bar{T})^{-1} = BT^{-1}B^T.$$

Hence show that, if T represents the matrix of components of a contravariant tensor, with $\det T \neq 0$, then the inverse matrix T^{-1} represents the components of a covariant tensor.

3.3.4 Associated components of tensors

We saw in section 3.3.2 that quantities such as the differentials dx^i of the co-ordinates x^i and the partial derivatives $\partial \Phi/\partial x^i$ of a scalar field Φ fall naturally within the definitions of contravariant and covariant tensors of order one, respectively. A vector at a point in space, however, will normally be regarded as having an existence independent of any co-ordinate system, although its components may be specified in either contravariant or covariant form, once a set of curvilinear co-ordinates has been chosen. The components will be 'associated' according to (3.51).

A similar view may be taken of second-order tensors. We may wish to regard them as linear operators, independent of any co-ordinate system, which act on vectors to produce vectors. A simple example is the dyadic product $\mathbf{u} \otimes \mathbf{v}$ of two vectors. We have already seen the contravariant form $u^i v^j$ of this tensor in (3.53). Not surprisingly, its (associated) covariant and mixed components are $u_i v_j$, $u^i v_j$ and $u_i v^j$. The action of this tensor on arbitrary vectors $\mathbf{w}$ is given by

$$(\mathbf{u} \otimes \mathbf{v})\mathbf{w} = \mathbf{u}(\mathbf{v} \cdot \mathbf{w}). \tag{3.61}$$

It is possible to write an explicit form for the tensor using components if we introduce the various dyadic products $\mathbf{g}_i \otimes \mathbf{g}_j$, $\mathbf{g}^i \otimes \mathbf{g}^j$, $\mathbf{g}_i \otimes \mathbf{g}^j$ and $\mathbf{g}^i \otimes \mathbf{g}_j$ of the base vectors and reciprocal base vectors.

In fact we have

$$\mathbf{u}\otimes\mathbf{v} = u^i v^j \mathbf{g}_i \otimes \mathbf{g}_j = u_i v_j \mathbf{g}^i \otimes \mathbf{g}^j = u^i v_j \mathbf{g}_i \otimes \mathbf{g}^j = u_i v^j \mathbf{g}^i \otimes \mathbf{g}_j. \tag{3.62}$$

It is easy to check that each component form has the required operator property (3.61). For example,

$$\begin{aligned}(u^i v^j \mathbf{g}_i \otimes \mathbf{g}_j)\mathbf{w} &= u^i v^j \mathbf{g}_i (\mathbf{g}_j \cdot \mathbf{w}) \\ &= u^i v^j \mathbf{g}_i w_j \qquad &\text{from (1.61)} \\ &= (v^j w_j) u^i \mathbf{g}_i \\ &= (\mathbf{v}\cdot\mathbf{w})\mathbf{u} \qquad &\text{from (1.60).}\end{aligned}$$

A general second-order tensor **T**, viewed as a linear operator, can be expressed as a *linear combination* of dyadic products:

$$\mathbf{T} = T^{ij}\mathbf{g}_i \otimes \mathbf{g}_j = T_{ij}\mathbf{g}^i \otimes \mathbf{g}^j = T^i_{\cdot j}\mathbf{g}_i \otimes \mathbf{g}^j = T_i^{\cdot j}\mathbf{g}^i \otimes \mathbf{g}_j. \tag{3.63}$$

Here the contravariant components T^{ij}, covariant components T_{ij} and mixed components $T^i_{\cdot j}$ and $T_i^{\cdot j}$, are all components of the same entity, the tensor **T**, and are said to be **associated**. If we adopted a different set of curvilinear co-ordinates $\bar{x}^i$, we would obtain different components $\bar{T}^{ij}, \bar{T}_{ij}, \bar{T}^i_{\cdot j}$ and $\bar{T}_i^{\cdot j}$, but the basic entity **T** would be unchanged. The new components would still be associated (see problem 3.3.6).

Relations between the associated components of second-order tensors, similar to those for first-order tensors (3.51), are not difficult to obtain. For example, consider the identity

$$T^{ij}\mathbf{g}_i \otimes \mathbf{g}_j = T^i_{\cdot j}\mathbf{g}_i \otimes \mathbf{g}^j.$$

Operating on each side with the vector $\mathbf{g}_k$ gives

$$(T^{ij}\mathbf{g}_i \otimes \mathbf{g}_j)\mathbf{g}_k = T^{ij}\mathbf{g}_i(\mathbf{g}_j \cdot \mathbf{g}_k) = T^{ij}\mathbf{g}_i g_{jk} \qquad \text{by (1.65)}$$

and

$$(T^i_{\cdot j}\mathbf{g}_i \otimes \mathbf{g}^j)\mathbf{g}_k = T^i_{\cdot j}\mathbf{g}_i(\mathbf{g}^j \cdot \mathbf{g}_k) = T^i_{\cdot j}\mathbf{g}_i \delta^j_k$$

i.e.

$$\mathbf{T}\mathbf{g}_k = T^i_{\cdot k}\mathbf{g}_i.$$

Comparing coefficients of $\mathbf{g}_i$ gives the equation

$$T^{ij}g_{jk} = T^i_{\cdot k}. \tag{3.64}$$

The action of g_{jk} on T^{ij} (with summation over j) is thus seen to be to *lower* the repeated index (and to replace it with k).

Similarly we have

$$g_{ik}T^{ij} = T_k^{\cdot j}$$

and

$$g_{ik}g_{jl}T^{ij} = T_{kl}. \tag{3.65}$$

In the latter case both indices are lowered.

We may also deduce in the same way (or by using (1.74)) that

$$g^{jk}T_{ij} = T_i^{\cdot k}.$$

Thus g^{jk} acting on T_{ij} has the effect of *raising* the repeated index (and replacing it with k).

It turns out that the above examples δ^i_j of a mixed tensor and g_{ij} of a covariant tensor are in fact associated, since applying g_{ik} to δ^i_j, thus lowering the index i and replacing it with k, yields

$$g_{ik}\delta^i_j = g_{kj},$$

by the substitution property of δ^i_j. They are associated components of the tensor (here, a linear operator)

$$g_{ij}\mathbf{g}^i \otimes \mathbf{g}^j = \delta^i_j \mathbf{g}_i \otimes \mathbf{g}^j = \mathbf{g}_i \otimes \mathbf{g}^i$$
$$= \mathbf{g}_1 \otimes \mathbf{g}^1 + \mathbf{g}_2 \otimes \mathbf{g}^2 + \mathbf{g}_3 \otimes \mathbf{g}^3.$$

This tensor is just the *identity* second-order tensor $\mathbf{I}$, and in fact is precisely the same operator which features in the expression (3.26).

Its action on an arbitrary vector $\mathbf{u}$ produces $\mathbf{u}$, as may be seen by considering

$$(\mathbf{g}_i \otimes \mathbf{g}^i)\mathbf{u} = \mathbf{g}_i(\mathbf{g}^i \cdot \mathbf{u}) = \mathbf{g}_i u^i = \mathbf{u}.$$

The associated contravariant components are just g^{ij}; so we have

$$\mathbf{I} = g^{ij}\mathbf{g}_i \otimes \mathbf{g}_j = g_{ij}\mathbf{g}^i \otimes \mathbf{g}^j = \mathbf{g}_i \otimes \mathbf{g}^i. \tag{3.66}$$

Still regarding a tensor $\mathbf{T}$ as a linear operator, we may deduce from (3.63) the results of operating with $\mathbf{T}$ on the base vectors:

$$\mathbf{Tg}_j = T_{ij}\mathbf{g}^i = T^i_{\cdot j}\mathbf{g}_i,$$
$$\mathbf{Tg}^j = T^{ij}\mathbf{g}_i = T^{\cdot j}_i \mathbf{g}^i.$$

Taking scalar products of these equations with the appropriate base vectors gives the following results:

$$\left.\begin{array}{l} T_{ij} = \mathbf{g}_i \cdot \mathbf{Tg}_j, \\ T^{ij} = \mathbf{g}^i \cdot \mathbf{Tg}^j, \\ T^{\cdot j}_i = \mathbf{g}_i \cdot \mathbf{Tg}^j, \\ T^i_{\cdot j} = \mathbf{g}^i \cdot \mathbf{Tg}_j. \end{array}\right\} \tag{3.67}$$

and

Now we can verify that the transformation laws remain satisfied. For example,

$$\bar{T}_{ij} = \bar{\mathbf{g}}_i \cdot \mathbf{T}\bar{\mathbf{g}}_j = \left(\frac{\partial x^k}{\partial \bar{x}^i}\mathbf{g}_k\right) \cdot \mathbf{T}\left(\frac{\partial x^l}{\partial \bar{x}^j}\mathbf{g}_l\right) \quad \text{by (3.42)}$$

$$= \frac{\partial x^k}{\partial \bar{x}^i}\frac{\partial x^l}{\partial \bar{x}^j}\mathbf{g}_k \cdot \mathbf{Tg}_l.$$

Here we are able to take the coefficients $\partial x^l/\partial \bar{x}^j$ outside the parentheses, since $\mathbf{T}$ is a linear operator. Hence

$$\bar{T}_{ij} = \frac{\partial x^k}{\partial \bar{x}^i}\frac{\partial x^l}{\partial \bar{x}^j} T_{kl},$$

as required, using the first equation in (3.67).

The tensor $\mathbf{T}^\mathrm{T}$, which we may call the **transpose** of the tensor $\mathbf{T}$, has the linear operator representation

$$\mathbf{T}^\mathrm{T} = T^{ji}\mathbf{g}_i \otimes \mathbf{g}_j = T^{ij}\mathbf{g}_j \otimes \mathbf{g}_i.$$

Now consideration of the components of the dyadic product $\mathbf{u} \otimes \mathbf{v}$ shows that

$$(\mathbf{u} \otimes \mathbf{v})^T = \mathbf{v} \otimes \mathbf{u},$$

and it follows that $\mathbf{T}^T$ also has the representations

$$\mathbf{T}^T = (T^i_{\cdot j}\mathbf{g}_i \otimes \mathbf{g}^j)^T = T^i_{\cdot j}\mathbf{g}^j \otimes \mathbf{g}_i = T^j_{\cdot i}\mathbf{g}^i \otimes \mathbf{g}_j$$
$$= (T^{\cdot j}_i \mathbf{g}^i \otimes \mathbf{g}_j)^T = T^{\cdot j}_i \mathbf{g}_j \otimes \mathbf{g}^i = T^{\cdot i}_j \mathbf{g}_i \otimes \mathbf{g}^j. \qquad (3.68)$$

Now, for any vectors $\mathbf{u}$ and $\mathbf{v}$,

$$\mathbf{u} \cdot \mathbf{T}^T \mathbf{v} = \mathbf{u} \cdot (T^{ij}\mathbf{g}_j \otimes \mathbf{g}_i)\mathbf{v} = \mathbf{u} \cdot T^{ij}\mathbf{g}_j(\mathbf{g}_i \cdot \mathbf{v})$$
$$= \mathbf{u} \cdot T^{ij}\mathbf{g}_j v_i = T^{ij}v_i(\mathbf{u} \cdot \mathbf{g}_j) = T^{ij}v_i u_j$$
$$= v_i T^{ij} u_j = \mathbf{v} \cdot (\mathbf{T}\mathbf{u}).$$

In the theory of linear transformations in vector spaces, the identity

$$\mathbf{u} \cdot \mathbf{T}^* \mathbf{v} = \mathbf{T}\mathbf{u} \cdot \mathbf{v} \qquad (3.69)$$

for arbitrary vectors $\mathbf{u}$ and $\mathbf{v}$ defines the **adjoint** operator $\mathbf{T}^*$ of a linear operator $\mathbf{T}$. The adjoint is here equivalent to the transpose.

Problems

3.3.5. From equations (3.64) and (1.73), deduce that associated components T^{ij} and $T^i_{\cdot j}$ of a second-order tensor $\mathbf{T}$ satisfy

$$T^{ij} = g^{jk} T^i_{\cdot k}.$$

3.3.6. If T_{ij} is a covariant second-order tensor with associated mixed components $T^k_{\cdot l}$ with respect to co-ordinates $\{x^1, x^2, x^3\}$, so that $T_{ij} = g_{ik} T^k_{\cdot j}$, show that, if the corresponding components with respect to co-ordinates $\{\bar{x}^1, \bar{x}^2, \bar{x}^3\}$ are $\bar{T}_{ij}$ and $\bar{T}^k_{\cdot l}$, obtained from the appropriate transformation laws (3.54) and (3.55), then these components are still associated, i.e.

$$\bar{T}_{ij} = \bar{g}_{ik} \bar{T}^k_{\cdot j}.$$

3.3.7. In a co-ordinate system with metric tensor

$$g_{ij} = \begin{bmatrix} 6 & 1 & 3 \\ 1 & 2 & -1 \\ 3 & -1 & 6 \end{bmatrix}$$

at a given point of space the action of a second-order tensor $\mathbf{T}$ on a vector $\mathbf{v}$ with covariant components (v_1, v_2, v_3) is to produce the vector $\mathbf{u}$ with contravariant components $(v_1 + v_2, 2v_2, v_3)$. Show that the contravariant components T^{ij}, the mixed components $T^i_{\cdot j}$ and $T^{\cdot j}_i$, and the covariant components T_{ij} of $\mathbf{T}$ are given respectively by the matrices

$$\begin{bmatrix} 1 & 1 & 0 \\ 0 & 2 & 0 \\ 0 & 0 & 1 \end{bmatrix}, \begin{bmatrix} 7 & 3 & 2 \\ 2 & 4 & -2 \\ 3 & -1 & 6 \end{bmatrix}, \begin{bmatrix} 6 & 8 & 3 \\ 1 & 5 & -1 \\ 3 & 1 & 6 \end{bmatrix}, \begin{bmatrix} 53 & 19 & 28 \\ 8 & 12 & -8 \\ 37 & -1 & 44 \end{bmatrix}.$$

3.3.8. If the action of the second-order tensor $\mathbf{T}$ on a vector $\mathbf{v}$ with contravariant components (v^1, v^2, v^3) is to produce the vector $\mathbf{u}$ with covariant components

$(v^1 + 3v^2, -2v^3, -v^2 - 4v^3)$, show that the action of the tensor $\mathbf{T}^T$ on $\mathbf{v}$ is to produce a vector with covariant components $(v^1, 3v^1 - v^3, -2v^2 - 4v^3)$.

3.3.5 Higher-order tensors

Tensors of higher order than two may be defined in terms of the appropriate transformation law, taking account of contravariant and covariant components. The transformation laws are the natural extension of those already met in (3.47), (3.48), (3.52), (3.54), (3.55) and (3.56). For example, a **contravariant tensor** U^{ijk} **of order three** transforms according to the rule

$$\bar{U}^{ijk} = \frac{\partial \bar{x}^i}{\partial x^l} \frac{\partial \bar{x}^j}{\partial x^m} \frac{\partial \bar{x}^k}{\partial x^n} U^{lmn}. \tag{3.70}$$

Simple examples of such a quantity are

(1) the set of products of contravariant components $u^i v^j w^k$ of three vectors $\mathbf{u}, \mathbf{v}$ and $\mathbf{w}$ and
(2) the set of products of contravariant components $u^i T^{jk}$ of a vector $\mathbf{u}$ and a tensor $\mathbf{T}$ of order two.

We can also define a **covariant tensor** U_{ijk} **of order three**, which transforms according to the rule

$$\bar{U}_{ijk} = \frac{\partial x^l}{\partial \bar{x}^i} \frac{\partial x^m}{\partial \bar{x}^j} \frac{\partial x^n}{\partial \bar{x}^k} U_{lmn}, \tag{3.71}$$

and various mixed tensors, such as $U^i{}_{jk}$, for which the transformation rule is

$$\bar{U}^i{}_{jk} = \frac{\partial \bar{x}^i}{\partial x^l} \frac{\partial x^m}{\partial \bar{x}^j} \frac{\partial x^n}{\partial \bar{x}^k} U^l{}_{mn}. \tag{3.72}$$

These third-order tensors would be **associated** if their components were related by the process of raising or lowering indices, i.e. if

$$U^{ijk} = g^{jl} g^{km} U^i{}_{lm} \quad \text{and} \quad U_{ijk} = g_{il} U^l{}_{jk}.$$

In this case all the components would be generated from the same third-order entity $\mathbf{U}$ which could be expressed as a linear combination of tensor products

$$\mathbf{U} = U^{ijk} \mathbf{g}_i \otimes \mathbf{g}_j \otimes \mathbf{g}_k = U_{ijk} \mathbf{g}^i \otimes \mathbf{g}^j \otimes \mathbf{g}^k = U^i{}_{jk} \mathbf{g}_i \otimes \mathbf{g}^j \otimes \mathbf{g}^k,$$

where each tensor product (of three vectors) may be regarded as a linear operator acting on vectors to produce second-order tensors, in accordance with the rule

$$(\mathbf{u} \otimes \mathbf{v} \otimes \mathbf{w}) = (\mathbf{w} \cdot \mathbf{z}) \mathbf{u} \otimes \mathbf{v}$$

for the action of the tensor product $\mathbf{u} \otimes \mathbf{v} \otimes \mathbf{w}$ of three vectors on the vector $\mathbf{z}$.

Let us investigate the effect of applying the transformation (3.70) to the alternating symbol e_{ijk}, given that we know already that e_{ijk} may be regarded as a CT3. Writing the indices as superscripts, we obtain

$$\frac{\partial \bar{x}^i}{\partial x^l} \frac{\partial \bar{x}^j}{\partial x^m} \frac{\partial \bar{x}^k}{\partial x^n} e^{lmn} = (\det A) e^{ijk} = J e^{ijk}, \tag{3.73}$$

using (2.19).

The factor J on the RHS shows that e^{ijk} does *not* transform like a contravariant third-order tensor. Similarly, writing the indices as subscripts, we see that e_{ijk} does not transform like a covariant tensor either, since

$$\frac{\partial x^l}{\partial \bar{x}^i} \frac{\partial x^m}{\partial \bar{x}^j} \frac{\partial x^n}{\partial \bar{x}^k} e_{ijk} = (\det B) e_{lmn}$$

$$= (\det A)^{-1} e_{lmn} = J^{-1} e_{lmn}. \qquad (3.74)$$

However, if we refer back to equation (3.40), we may see how to generate a genuine tensor by considering an appropriate multiple of the alternating symbol. We have

$$\sqrt{\frac{g}{\bar{g}}} e^{ijk} = \frac{\partial \bar{x}^i}{\partial x^l} \frac{\partial \bar{x}^j}{\partial x^m} \frac{\partial \bar{x}^k}{\partial x^n} e^{lmn}.$$

Thus, writing

$$\bar{\varepsilon}^{ijk} = (\bar{g})^{-1/2} e^{ijk} \quad \text{and} \quad \varepsilon^{lmn} = g^{-1/2} e^{lmn},$$

we have

$$\bar{\varepsilon}^{ijk} = \frac{\partial \bar{x}^i}{\partial x^l} \frac{\partial \bar{x}^j}{\partial x^m} \frac{\partial \bar{x}^k}{\partial x^n} \varepsilon^{lmn}.$$

Similarly, (3.74) may be re-written as

$$\bar{\varepsilon}_{ijk} = \frac{\partial x^l}{\partial \bar{x}^i} \frac{\partial x^m}{\partial \bar{x}^j} \frac{\partial x^n}{\partial \bar{x}^k} \varepsilon_{lmn},$$

where

$$\bar{\varepsilon}_{ijk} = \bar{g}^{1/2} e_{ijk} \quad \text{and} \quad \varepsilon_{lmn} = g^{1/2} e_{lmn}.$$

This shows that the quantities defined by

$$\varepsilon^{ijk} = g^{-1/2} e^{ijk} \quad \text{and} \quad \varepsilon_{ijk} = g^{1/2} e_{ijk} \qquad (3.75)$$

are contravariant and covariant third-order tensors, respectively. Moreover, they are *associated* components, since, for example, applying the lowering of indices to ε^{ijk} gives

$$g_{il} g_{jm} g_{kn} \varepsilon^{ijk} = g_{il} g_{jm} g_{kn} (g^{-1/2} e^{ijk}) = g^{-1/2} (g_{il} g_{jm} g_{kn} e^{ijk})$$

$$= g^{-1/2} (g e_{lmn}) \qquad \text{by (2.20)}$$

$$= g^{1/2} e_{lmn} = \varepsilon_{lmn}.$$

We may call the corresponding third-order tensor

$$\varepsilon^{ijk} \mathbf{g}_i \otimes \mathbf{g}_j \otimes \mathbf{g}_k.$$

the alternating tensor.

If we confine our attention to rectangular cartesian co-ordinate systems, of course, we have $g = \bar{g} = J = 1$, and the components (contravariant, covariant or mixed) of the alternating tensor reduce to those of the alternating symbol e_{ijk}. In fact, we could have generated equations (3.75) by applying the transformation equations (3.70) and (3.71) directly to the cartesian tensor components e_{ijk} (with respect to a right-handed rectangular cartesian system (y^1, y^2, y^3)). Thus, with (x^1, x^2, x^3) as the 'new' curvilinear

Sec. 3.3] **Generalized tensors** 103

co-ordinates and the cartesians (y^1, y^2, y^3) as the 'old' ones, we obtain from (3.70) the covariant components

$$\frac{\partial x^i}{\partial y^l} \frac{\partial x^j}{\partial y^m} \frac{\partial x^k}{\partial y^n} e_{lmn} = e_{ijk}(\det M) \qquad \text{by (2.19)}$$

$$= J^* e_{ijk} = g^{-1/2} e_{ijk} \qquad \text{by (3.39)},$$

and from (3.71) the contravariant components

$$\frac{\partial y^l}{\partial x^i} \frac{\partial y^m}{\partial x^j} \frac{\partial y^n}{\partial x^k} e_{lmn} = e_{ijk} \det L$$

$$= g^{1/2} e_{ijk}.$$

It is not difficult to see that the permutation identity (1.17) becomes

$$\varepsilon_{ijk}\varepsilon^{ilm} = \delta^l_j \delta^m_k - \delta^m_j \delta^l_k \qquad (3.76)$$

when expressed in terms of generalized tensor quantities.

A tensor of order $M + N$ with M contravariant components and N covariant components satisfies the transformation rule (supposing that the precise positions of the indices out of the $M + N$ possible positions are not of immediate concern here)

$$\bar{U}^{ij\cdots k}_{pq\cdots r} = \frac{\partial \bar{x}^i}{\partial x^l} \frac{\partial \bar{x}^j}{\partial x^m} \cdots \frac{\partial \bar{x}^k}{\partial x^n} \frac{\partial x^s}{\partial \bar{x}^p} \frac{\partial x^t}{\partial \bar{x}^q} \cdots \frac{\partial x^u}{\partial \bar{x}^r} U^{lm\cdots n}_{st\cdots u}, \qquad (3.77)$$

where $i, j, \ldots, k$, and $l, m, \ldots, n$ each represent M contravariant components, and $p, q, \ldots, r$ and $s, t, \ldots, u$, represent N covariant components. The 'new' $\bar{x}^i$ co-ordinates occur on the *top* of the partial derivatives for terms corresponding to *contravariant* components and on the *bottom* for *covariant* components.

3.3.6 Contraction, invariants and the quotient theorem

We have seen that contraction is the operation of making two indices identical in a tensor expression, implying summation according to the summation convention. Inspection of the general rule (3.77) will clarify why, when working with general tensors, the operation preserves tensorial character only when one index is a subscript and the other a superscript. Putting i and p identical, for example, we obtain (replacing p by i for the sake of argument)

$$\bar{U}^{ij\cdots k}_{iq\cdots r} = \frac{\partial \bar{x}^i}{\partial x^l} \frac{\partial \bar{x}^j}{\partial x^m} \cdots \frac{\partial \bar{x}^k}{\partial x^n} \frac{\partial x^s}{\partial \bar{x}^i} \frac{\partial x^t}{\partial \bar{x}^q} \cdots \frac{\partial x^u}{\partial \bar{x}^r} U^{lm\cdots n}_{st\cdots u}$$

$$= \left(\frac{\partial \bar{x}^i}{\partial x^l} \frac{\partial x^s}{\partial \bar{x}^i}\right) \frac{\partial \bar{x}^j}{\partial x^m} \cdots \frac{\partial \bar{x}^k}{\partial x^n} \frac{\partial x^t}{\partial \bar{x}^q} \cdots \frac{\partial x^u}{\partial \bar{x}^r} U^{lm\cdots n}_{st\cdots u}$$

changing the order of the partial derivative terms

$$= \left(\frac{\partial x^s}{\partial x^l}\right) \frac{\partial \bar{x}^j}{\partial x^m} \cdots \frac{\partial \bar{x}^k}{\partial x^n} \frac{\partial x^t}{\partial \bar{x}^q} \cdots \frac{\partial x^u}{\partial \bar{x}^r} U^{lm\cdots n}_{st\cdots u}$$

by the chain rule for partial differentiation

$$= \delta_l^s \frac{\partial \bar{x}^j}{\partial x^m} \cdots \frac{\partial \bar{x}^k}{\partial x^n} \frac{\partial x^t}{\partial \bar{x}^q} \cdots \frac{\partial x^u}{\partial \bar{x}^r} U_{st\cdots u}^{lm\cdots n}$$

since the x^i are independent co-ordinates

$$= \frac{\partial \bar{x}^j}{\partial x^m} \cdots \frac{\partial \bar{x}^k}{\partial x^n} \frac{\partial x^t}{\partial \bar{x}^q} \cdots \frac{\partial x^u}{\partial \bar{x}^r} U_{lt\cdots u}^{lm\cdots n} \qquad (3.78)$$

by the substitution operator property.

This proof shows that $U_{iq\cdots r}^{ij\cdots k}$, with $M + N - 2$ free indices $j, \ldots, k, q, \ldots, r$, is a tensor with $M - 1$ contravariant components and $N - 1$ covariant components. The proof hinges on the use of the chain rule, and a similar proof would apply for contraction on any of the upper and lower indices. The operation reduces the order of the tensor by two.

Contraction of any two of the upper indices (or two of the lower indices), however, would not allow the same proof to go through. The consequence is that expressions such as $U_{pq\cdots r}^{ii\cdots k}$, with $M + N - 2$ free indices, lose their tensor character, so that, although $U_{pq\cdots r}^{ij\cdots k}$ is a tensor of order $M + N$, $U_{pq\cdots r}^{ii\cdots k}$ (putting $j = i$) is *not* a tensor of order $M + N - 2$.

Thus, if T^{ij} is a contravariant second-order tensor, the contracted expression T^{ii} is *not* an invariant quantity but, for a mixed second-order tensor $T^i_{.j}$, the expression $T^i_{.i}$ *is* an invariant. We may call invariants **tensors of order zero** to widen the validity of the result that contraction (when properly applied) reduces the order of a tensor by two. Tensors of order zero will commonly represent physical scalar quantities, whose magnitudes are independent of the co-ordinate system employed. The transformation rule for a tensor Φ of order zero is thus simply

$$\bar{\Phi} = \Phi. \qquad (3.79)$$

If T^{ij} and $T^i_{.j}$ are associated components, then the invariant $T^i_{.i}$ is expressible in terms of the contravariant components T^{ij} as $g_{ij}T^{ij}$, which follows from contraction of the identity $T^i_{.k} = g_{kj}T^{ij}$. The expression $g_{ij}T^{ij}$ may be regarded as obtainable from the mixed fourth-order tensor $g_{ij}T^{kl}$ by two contractions.

Given two vectors **u** and **v**, the above considerations imply that their scalar product must be given by

$$\mathbf{u} \cdot \mathbf{v} = u^i v_i = u_i v^i$$
$$= g_{ij} u^i v^j = g^{ij} u_i v_j, \qquad (3.80)$$

in terms of contravariant and covariant components of **u** and **v**, as previously given in (1.64). How can we be sure that these expressions agree with the one previously given in (1.18) for rectangular cartesian co-ordinates? The following two tests are sufficient and have wider implications.

(a) If we are dealing with a single (scalar) quantity, we must ensure that it is a genuine invariant. This will be the case if it is obtained from genuine tensor quantities (in this case the vectors **u** and **v**) by a valid process of contraction (e.g. $u^i v_i$ obtained from $u^i v_j$, but *not* $u_i v_i$ obtained from $u_i v_j$).

(b) We must ensure that our expression reduces to the known cartesian form when the co-ordinates are rectangular cartesians. Since rectangular cartesians are just a particular case of the general co-ordinate systems considered here, the transformation law (3.79) ensure that our expression is an invariant for all possible co-ordinate transformations.

In the case of the scalar product, it is clear that test (b) is satisfied, since $u^i v_i$ reduces to $u_i v_i$ for cartesian systems.

The natural extension to general co-ordinate systems of the definition of the vector product (1.19) is the vector with contravariant components

$$(\mathbf{u} \times \mathbf{v})^i = \varepsilon^{ijk} u_j v_k, \tag{3.81}$$

which is the doubly contracted form of the mixed fifth-order tensor $\varepsilon^{ijk} u_l v_m$, where ε^{ijk} is the contravariant alternating tensor. The tests to apply to check that this is the required vector are similar to those above.

(a) Test that the expression is a genuine tensor of the required order.
(b) Check that it reduces to the known cartesian form.

Here we know that ε^{ijk} reduces to the alternating symbol e_{ijk} for rectangular cartesian systems, and hence (3.81) reduces to (1.19); so test (b) is satisfied. Moreover, we know that ε^{ijk} is a tensor, and hence $\varepsilon^{ijk} u_j v_k$ is a contravariant vector. (Note that $\varepsilon^{ijk} u^j v^k$ or $\varepsilon^{ijk} u_j v^k$ would not transform as contravariant vectors.) Thus test (a) is also satisfied, and (3.81) must be the correct form.

It is also straightforward to write down the associated covariant vector,

$$(\mathbf{u} \times \mathbf{v})_i = \varepsilon_{ijk} u^j v^k. \tag{3.82}$$

Equivalently,

$$\mathbf{u} \times \mathbf{v} = \varepsilon^{ijk} u_j v_k \mathbf{g}_i = \varepsilon_{ijk} u^j v^k \mathbf{g}^i.$$

Tests for tensor character are also supplied by various forms of the quotient theorem, but careful note must be taken of contravariant and covariant components. We give two examples here, the first extending the form of the theorem given in section 3.2.5 to generalized tensors.

Example 1

T^{ij} is a contravariant second-order tensor if the expression $T^{ij} u_j$, for arbitrary covariant vectors u_j, is contravariant vector.

Proof

Suppose $T^{ij} u_j = v^i$, where u_j is a covariant and v^i a contravariant vector. The implicit assumption behind the slightly oversimplified statement of the theorem which we have given here is that this identity holds in one particular co-ordinate system x^i and that a similar identity

$$\bar{T}^{ij} \bar{u}_j = \bar{v}^i$$

holds in another system $\bar{x}^i$, where the second-order quantity $\bar{T}^{ij}$ has 'new' components

$\bar{T}^{ij}$. By the transformation properties of *vectors*,

$$\bar{T}^{ij}\frac{\partial x^k}{\partial \bar{x}^j}u_k = \frac{\partial \bar{x}^i}{\partial x^l}v^l = \frac{\partial \bar{x}^i}{\partial x^l}(T^{lk}u_k),$$

i.e.

$$\left(\frac{\partial x^k}{\partial \bar{x}^j}\bar{T}^{ij}\right)u_k = \left(\frac{\partial \bar{x}^i}{\partial x^l}T^{lk}\right)u_k.$$

However, since u_k is arbitrary, we see that (e.g. by substituting the components (1, 0, 0), (0, 1, 0), (0, 0, 1), in turn for u_k)

$$\frac{\partial x^k}{\partial \bar{x}^j}\bar{T}^{ij} = \frac{\partial \bar{x}^i}{\partial x^l}T^{lk} \qquad \text{for each } k,$$

effectively cancelling u_k from both sides of the above equation. Thus k now becomes a free suffix on each side.

Multiplying both sides by $\partial \bar{x}^m / \partial x^k$ and using (3.30),

$$\frac{\partial \bar{x}^m}{\partial x^k}\frac{\partial x^k}{\partial \bar{x}^j}\bar{T}^{ij} = \delta_j^m \bar{T}^{ij} = \bar{T}^{im}$$

$$= \frac{\partial \bar{x}^m}{\partial x^k}\frac{\partial \bar{x}^i}{\partial x^l}T^{lk},$$

or, relabelling the indices,

$$\bar{T}^{ij} = \frac{\partial \bar{x}^i}{\partial x^k}\frac{\partial \bar{x}^j}{\partial x^l}T^{kl},$$

which shows that T^{ij} transforms according to (3.52), as required.

Example 2

U_{ijk} is a covariant third-order tensor if the expression $U_{ijk}T^{jk}$, for arbitrary contravariant second-order tensors T^{ij}, represents a covariant vector.

Proof

The proof follows similar lines to that for Example 1.

Problems

3.3.9 Prove the following identities:
  (a) $\varepsilon_{ijm}\varepsilon^{kjm} = 2\delta_i^k$;
  (b) $\varepsilon_{ijk}\varepsilon^{ijk} = 6$;
  (c) $\mathbf{g}_i \times \mathbf{g}_j = \varepsilon_{ijk}\mathbf{g}^k$.

3.3.10 Show that the angle θ between two vectors $\mathbf{u}$ and $\mathbf{v}$ is given by the expressions

$$\cos\theta = \frac{u_i v^i}{\sqrt{u_j u^j}\sqrt{v_k v^k}} = \frac{g_{ij}u^i v^j}{\sqrt{g_{kl}u^k u^l}\sqrt{g_{mn}v^m v^n}}.$$

3.3.11 (Forms of the quotient theorem)
- (a) Prove Example 2 above.
- (b) Show that u^i is a contravariant vector if, for arbitrary covariant vectors v_i, the expression $u^i v_i$ is an invariant.
- (c) Show that T_{ij} is a covariant second-order tensor if, for arbitrary contravariant vectors u^k, the expression $T_{ij} u^j$ is a covariant vector.

3.3.12 If $T^i{}_{.j}$ and $T_{.j}^i$ are associated mixed components of a tensor $\mathbf{T}$, show that
- (a) $T^i{}_{.i} = T_{.i}^i$ and
- (b) $\det(T^i{}_{.j}) = \det(T_{.j}^i)$.

3.3.7 Symmetry and skew symmetry

A tensor of order at least two is *symmetric* with respect to two of its indices if its components are unchanged in value when those indices are interchanged. A contravariant second-order tensor T^{ij}, for example, is symmetric if $T^{ij} = T^{ji}$. In this case there is only one pair of indices. For a third-order tensor such as U^{ijk}, there are three possible pairs of indices, and thus three possible symmetries:

(i) $U^{ijk} = U^{jik}$;
(ii) $U^{ijk} = U^{ikj}$;
(iii) $U^{ijk} = U^{kji}$.

The definition (1.65) of the metric tensor g_{ij} immediately shows that it is symmetric.

Symmetries with respect to two contravariant or two covariant indices are preserved under the transformation law for tensors. For example, if $T^{ij} = T^{ji}$, then in another co-ordinate system the components satisfy

$$\bar{T}^{ij} = \frac{\partial \bar{x}^i}{\partial x^k} \frac{\partial \bar{x}^j}{\partial x^l} T^{kl} = \frac{\partial \bar{x}^i}{\partial x^k} \frac{\partial \bar{x}^j}{\partial x^l} T^{lk}$$

$$= \frac{\partial \bar{x}^j}{\partial x^l} \frac{\partial \bar{x}^i}{\partial x^k} T^{lk} \quad \text{changing the order of partial derivative terms}$$

$$= \frac{\partial \bar{x}^j}{\partial x^k} \frac{\partial \bar{x}^i}{\partial x^l} T^{kl} \quad \text{writing } k \text{ for } l \text{ and } l \text{ for } k$$

$$= \bar{T}^{ji} \quad \text{by (3.52).}$$

Alternatively, the matrix form (3.57) of the transformation law gives the slightly more direct proof:

$$\bar{T}^{\mathrm{T}} = (A T A^{\mathrm{T}})^{\mathrm{T}} = A T^{\mathrm{T}} A^{\mathrm{T}} = A T A^{\mathrm{T}} = \bar{T}.$$

For covariant components, we have, similarly, from (3.58),

$$\bar{T}^{\mathrm{T}} = (B T B^{\mathrm{T}}) = B T^{\mathrm{T}} B^{\mathrm{T}} = B T B^{\mathrm{T}} = \bar{T}$$

if T is symmetric.

However, symmetries with respect to mixed contravariant and covariant indices, such as $T^i{}_{.j} = T^j{}_{.i}$, are of no interest, because they are not preserved under the transformation law, as may be seen immediately from (3.59) and (3.60). Even though the matrix array T in (3.59) may be symmetric, there is no reason in general why the array $\bar{T}$ with respect to another co-ordinate system should also be symmetric.

Skew symmetry with respect to two indices implies that the values of components change sign but not magnitude on interchange of the indices. The contravariant components ε^{ijk} and covariant components ε_{ijk} of the alternating tensor are skew symmetric with respect to any pair of indices, since $\varepsilon^{ijk} = -\varepsilon^{jik}$, etc. Just as for symmetry, the property of skew symmetry is preserved under transformations of co-ordinates when pairs of contravariant or pairs of covariant components are involved, but not for a mixed pair of contravariant and covariant components.

A skew-symmetric contravariant second-order tensor T^{ij} has matrix array

$$\begin{bmatrix} 0 & -T^{21} & T^{13} \\ T^{21} & 0 & -T^{32} \\ -T^{13} & T^{32} & 0 \end{bmatrix}$$

and so has only three independent components. These components are closely related to the three components of a covariant vector w_i defined by

$$w_i = \tfrac{1}{2}\varepsilon_{ijk}T^{kj}. \tag{3.83}$$

Note that for rectangular cartesian systems the components of the vector **w** reduce to

$$w_i = \tfrac{1}{2}e_{ijk}T_{kj},$$

i.e.

$$w_1 = \tfrac{1}{2}(T_{32} - T_{23}) = \tfrac{1}{2}(T_{32} + T_{32}) = T_{32},$$

$$w_2 = T_{13}, \quad w_3 = T_{21},$$

so that the matrix array of T_{ij} is

$$\begin{bmatrix} 0 & -w_3 & w_2 \\ w_3 & 0 & -w_1 \\ -w_2 & w_1 & 0 \end{bmatrix}.$$

From (3.81) we find that, for any vector **a**,

$$(\mathbf{w} \times \mathbf{a})^i = \varepsilon^{ijk}w_j a_k$$
$$= \tfrac{1}{2}\varepsilon^{ijk}\varepsilon_{jlm}T^{ml}a_k$$
$$= \tfrac{1}{2}\varepsilon^{jki}\varepsilon_{jlm}T^{ml}a_k \qquad \text{permuting indices cyclically,}$$
$$= \tfrac{1}{2}(\delta^k_l\delta^i_m - \delta^k_m\delta^i_l)T^{ml}a_k \qquad \text{by (3.76)}$$
$$= \tfrac{1}{2}(T^{ik} - T^{ki})a_k = T^{ik}a_k = (\mathbf{Ta})^i$$

i.e.

$$\mathbf{Ta} = \mathbf{w} \times \mathbf{a} \qquad \text{for any vector } \mathbf{a}.$$

Thus the effect of any skew-symmetric tensor **T** acting on a vector is equivalent to taking the vector product of that vector with the corresponding vector **w**.

3.3.8 Diagonalization

Suppose we have a second-order tensor **T** defined as in (3.63) as a linear operator acting on vectors to produce vectors. Its action on certain vectors **X** may be to

produce a parallel vector, i.e.
$$TX = \tau X \tag{3.84}$$
for some scalar τ. Such vectors are called **eigenvectors** of **T**, and the corresponding scalars τ **eigenvalues** of **T**. Since **T** is a linear operator, any scalar multiple of an eigenvectors is still an eigenvector.

The contravariant component form of (3.84) may be written
$$T^i_{\cdot j} X^j = \tau X^i,$$
and this equation leads to the familiar (cubic) characteristic equation
$$\det(T - \tau I) = 0 \tag{3.85}$$
for the three eigenvalues, where T is the matrix array of mixed components $T^i_{\cdot j}$. Let us suppose that the three eigenvalues τ_1, τ_2, τ_3 are all real (which will not necessarily be the case, unless **T** is symmetric) and distinct. The corresponding real eigenvectors will be $\mathbf{X}_1, \mathbf{X}_2, \mathbf{X}_3$, say, and we shall assume that these vectors form a **basis** for our three-dimensional space.

The tensor transpose $\mathbf{T}^T$ has eigenvectors $\mathbf{Y}^1, \mathbf{Y}^2, \mathbf{Y}^3$, defined by the equations
$$\mathbf{T}^T \mathbf{Y} = \tau \mathbf{Y}, \tag{3.86}$$
which has the contravariant form (according to (3.68))
$$T^{\cdot i}_j Y^j = \tau Y^i,$$
i.e.
$$g_{jk} g^{il} T^k_{\cdot l} Y^j - \tau Y^i \quad \text{(raising and lowering the indices in } T^{\cdot i}_j\text{)}$$
$$= g_{jk} g^{il} T^k_{\cdot l} Y^j - \tau \delta^i_j Y^j$$
$$= g_{jk} (T^k_{\cdot l} - \tau \delta^k_l) g^{il} Y^j \quad \text{using (1.74)}$$
$$= 0.$$

This equation has non-trivial solutions $\mathbf{Y}$ if the determinant of the coefficients $g_{jk}(T^k_{\cdot l} - \tau \delta^k_l) g^{il}$ is zero. This determinant is equal to the product of the determinants
$$\det(g_{jk}) \det(T^k_{\cdot l} - \tau \delta^k_l) \det(g^{il})$$
of the matrices involved. So the characteristic equation again reduces to (3.85), and the eigenvalues of $\mathbf{T}^T$ must be identical with those of **T**.

Now we have
$$\mathbf{TX}_r = \tau_r \mathbf{X}_r \quad \text{(n.s.)}$$
and
$$\mathbf{T}^T \mathbf{Y}^s = \tau_s \mathbf{Y}^s \quad \text{(n.s.)}.$$

(Note the use of the symbol (n.s.), indicating that there is no summation over the repeated index, i.e. the summation convention does not apply to this equation.)

It follows that
$$\mathbf{Y}^s \cdot \mathbf{TX}_r = \tau_r \mathbf{Y}^s \cdot \mathbf{X}_r \quad \text{(n.s.)}$$
and, by (3.69),
$$\mathbf{Y}^s \cdot \mathbf{TX}_r = \mathbf{X}_r \cdot \mathbf{T}^T \mathbf{Y}^s = \tau_s \mathbf{X}_r \cdot \mathbf{Y}^s \quad \text{(n.s.)}$$
$$= \tau_s \mathbf{Y}^s \cdot \mathbf{X}_r \quad \text{(n.s.)}.$$

If the eigenvalues are distinct, i.e. $\tau_r \neq \tau_s$ for $r \neq s$, these equations are compatible only if

$$\mathbf{Y}^s \cdot \mathbf{X}_r = 0 \qquad \text{if } r \neq s. \tag{3.87}$$

Thus $\mathbf{Y}^1$ is orthogonal to $\mathbf{X}_2$ and $\mathbf{X}_3$, etc. If the lengths of the eigenvectors $\mathbf{X}_1, \mathbf{X}_2, \mathbf{X}_3$ are fixed, we may now adjust the lengths of the eigenvectors $\mathbf{Y}^1, \mathbf{Y}^2, \mathbf{Y}^3$ (by multiplying each by some appropriate scalar) so that

$$\mathbf{Y}^s \cdot \mathbf{X}_r = \delta_r^s. \tag{3.88}$$

Then if $\{\mathbf{X}_1, \mathbf{X}_2, \mathbf{X}_3\}$ is regarded as a basis, the set $\{\mathbf{Y}^1, \mathbf{Y}^2, \mathbf{Y}^3\}$ behaves like a reciprocal basis.

Since we also have now

$$\mathbf{Y}^s \cdot \mathbf{T} \mathbf{X}_r = \tau_r \delta_r^s \quad \text{(n.s.)},$$

it follows from (3.67) that the mixed components $T^s_{\cdot r}$ of the tensor $\mathbf{T}$ with respect to the basis of eigenvectors have the diagonalized matrix array

$$\begin{bmatrix} \tau_1 & 0 & 0 \\ 0 & \tau_2 & 0 \\ 0 & 0 & \tau_3 \end{bmatrix}. \tag{3.89}$$

The matrix array of the associated matrix components $T_s^{\cdot r}$, however, are not, in general, diagonalized with respect to this basis. We thus have the linear operator form for $\mathbf{T}$, by (3.68),

$$\mathbf{T} = \sum_{r=1}^{3} \tau_r \mathbf{X}_r \otimes \mathbf{Y}^r. \tag{3.90}$$

When $\mathbf{T}$ is a symmetric second-order tensor, we have $\mathbf{T}^T = \mathbf{T}$. Hence the eigenvectors of $\mathbf{T}$ and $\mathbf{T}^T$ are the same. As long as the eigenvalues are distinct, equation (3.88) then shows that the eigenvectors are mutually orthogonal, and we can adjust their lengths so that they form an orthonormal set, i.e.

$$\mathbf{X}_r \cdot \mathbf{X}_s = \delta_{rs}.$$

The representation (3.90) reduces to

$$\mathbf{T} = \sum_{r=1}^{3} \tau_r \mathbf{X}_r \otimes \mathbf{X}_r, \tag{3.90}$$

which is called the **spectral representation** of $\mathbf{T}$. We have here a rectangular cartesian reference system (of eigenvectors of $\mathbf{T}$), and with respect to this system the contravariant, covariant and mixed components of $\mathbf{T}$ are indistinguishable, all reducing to the diagonalized form (3.89).

4

Covariant differentiation

4.1 INTRODUCTION

The transformation properties of tensors considered in Chapter 3 apply to the components of tensor fields, e.g. the stress tensor in a continuum, evaluated at a particular point in space. Partial derivatives $\partial \bar{x}_i/\partial x_j$ occurring in the transformation rules are also evaluated at the point under consideration. The fact that tensor field components can be evaluated at different point of space, however, leads us to consider the differentiation of tensors with respect to spatial co-ordinates.

We have already considered the gradient grad Φ of a scalar field Φ and concluded that grad Φ is itself a vector field. With respect to a set of cartesian co-ordinates $Oy_1 y_2 y_3$, this means that, if Φ is a function of y_1, y_2 and y_3, the partial derivatives $\partial \Phi/\partial y_i$, $i = 1, 2, 3$, form a CT1. Similarly, the set of second derivatives $\partial^2 \Phi/\partial y_i \partial y_j$ forms a CT2. We have also seen that for a vector field **u** with cartesian components u_i, the set of derivatives $\partial u_i/\partial y_j$ forms a CT2, while the contracted form $\partial u_i/\partial y_i$ is the scalar field div **u**. The set of components $e_{ijk} \partial u_k/\partial y_j$ forms the vector field curl **u**.

With respect to a set of general curvilinear co-ordinates (x^1, x^2, x^3), the situation is more complicated. For example, for scalar fields Φ the set of partial derivatives $\partial \Phi/\partial x^i$ constitutes a covariant vector field (according to section 3.3.2), i.e. the covariant components of grad Φ. However, problem 3.3.2 shows that the set of second derivatives $\partial^2 \Phi/\partial x^i \partial x^j$ transforms according to the rule

$$\frac{\partial^2 \Phi}{\partial \bar{x}^i \partial \bar{x}^j} = \frac{\partial x^k}{\partial \bar{x}^i} \frac{\partial x^l}{\partial \bar{x}^j} \frac{\partial^2 \Phi}{\partial x^k \partial x^l} + \frac{\partial^2 x^l}{\partial \bar{x}^i \partial \bar{x}^j} \frac{\partial \Phi}{\partial x^l},$$

and the presence of the final term indicates that these second derivatives do not constitute a covariant second-order tensor.

This complication arises from the fact that the base vectors in a curvilinear co-ordinate system may vary in orientation and magnitude from one point of space to another. A vector field **u** may be expressed in terms of the local base vectors $\mathbf{g}_1, \mathbf{g}_2, \mathbf{g}_3$, as

$$\mathbf{u} = u^i \mathbf{g}_i,$$

with contravariant components u^i. It follows that in general **u** would vary in space owing to the variation in the base vectors, even if the components u^i did not vary.

In this chapter we define derivatives of tensor fields in such a way that they constitute tensors; these derivatives are called **covariant derivatives**. For scalar fields Φ we have seen that the set of partial derivatives $\partial \Phi / \partial x^i$ forms a vector field without any need for modification. We may call this vector field the **covariant derivative** of Φ and write it as $\Phi_{,i}$.

4.2 DIFFERENTIATION OF BASE VECTORS AND CHRISTOFFEL SYMBOLS

Since the base vectors are given in terms of the position vector **r** by

$$\mathbf{g}_i = \frac{\partial \mathbf{r}}{\partial x^i},$$

further partial differentiation with respect to the co-ordinate x^j gives

$$\frac{\partial \mathbf{g}_i}{\partial x^j} = \frac{\partial^2 \mathbf{r}}{\partial x^j \partial x^i}.$$

The resulting second derivative is independent of the order in which the differentiations are carried out. Hence

$$\frac{\partial \mathbf{g}_i}{\partial x^j} = \frac{\partial^2 \mathbf{r}}{\partial x^j \partial x^i} = \frac{\partial^2 \mathbf{r}}{\partial x^i \partial x^j} = \frac{\partial \mathbf{g}_j}{\partial x^i}. \tag{4.1}$$

These derivatives are *vectors* and form a vector field. At any point P with co-ordinates x^i they are equal to

$$\lim_{\delta x^j \to 0} \left(\frac{\delta \mathbf{g}_i}{\delta x^j} \right),$$

where $\delta \mathbf{g}_i$ is the increment in the vector $\mathbf{g}_i$ between the point P and the neighbouring point P′ situated on the x^j-co-ordinate curve through P, so that the co-ordinates of P′ are identical with those of P, except for the x^j co-ordinate, which differs by the increment δx^j. The vector $\partial \mathbf{g}_i / \partial x^j$ is associated with the point P and may be expressed as a linear combination of the base vectors $\mathbf{g}_i$ at P. Thus

$$\frac{\partial \mathbf{g}_i}{\partial x^j} = \Gamma^1_{ij} \mathbf{g}_1 + \Gamma^2_{ij} \mathbf{g}_2 + \Gamma^3_{ij} \mathbf{g}_3 = \Gamma^k_{ij} \mathbf{g}_k, \tag{4.2}$$

where the coefficients Γ^k_{ij} (evaluated at P) are called **Christoffel symbols of the second kind**. Taking scalar products of both sides of (4.2) with the contravariant base vector $\mathbf{g}^l$ gives

$$\Gamma^k_{ij} \mathbf{g}_k \cdot \mathbf{g}^l = \Gamma^k_{ij} \delta^l_k = \frac{\partial \mathbf{g}_i}{\partial x^j} \cdot \mathbf{g}^l.$$

Base vectors and Christoffel symbols

Hence, using the substitution operator property and writing k for l,

$$\Gamma^k_{ij} = \frac{\partial \mathbf{g}_i}{\partial x^j} \cdot \mathbf{g}^k. \tag{4.3}$$

Now from equation (1.72) we know that the components of $\mathbf{g}^k$ with respect to a background cartesian system y_i are $\partial x^k/\partial y_l$, $l = 1, 2, 3$, and from (4.1) the cartesian components of $\partial \mathbf{g}_i/\partial x^j$ are $\partial^2 y_l/\partial x^i \partial x^j$, $l = 1, 2, 3$. Thus, evaluating the scalar product (4.3) in terms of cartesian components gives

$$\Gamma^k_{ij} = \frac{\partial^2 y_l}{\partial x^i \partial x^j} \frac{\partial x^k}{\partial y_l}. \tag{4.4}$$

Since (4.1) must hold, we always have

$$\Gamma^k_{ij} = \Gamma^k_{ji}. \tag{4.5}$$

As an alternative to (4.2) the derivatives of $\mathbf{g}_i$ may be expressed as linear combinations of the *contravariant* base vectors at P, giving

$$\frac{\partial \mathbf{g}_i}{\partial x^j} = \Gamma_{ijk}\mathbf{g}^k, \tag{4.6}$$

where the new coefficients Γ_{ijk} are called **Christoffel symbols of the first kind**. Formulas for Γ_{ijk} corresponding to (4.3) and (4.4) are

$$\Gamma_{ijk} = \frac{\partial \mathbf{g}_i}{\partial x^j} \cdot \mathbf{g}_k = \frac{\partial^2 y_l}{\partial x^i \partial x^j} \frac{\partial y_l}{\partial x^k} \tag{4.7}$$

and are derived similarly. Moreover,

$$\Gamma_{ijk} = \Gamma_{jik}. \tag{4.8}$$

The relationship between the Christoffel symbols may be deduced using (1.75) since, by (4.3),

$$\Gamma^k_{ij} = \frac{\partial \mathbf{g}_i}{\partial x^j} \cdot \mathbf{g}^k = \frac{\partial \mathbf{g}_i}{\partial x^j} \cdot g^{kl}\mathbf{g}_l = g^{kl}\left(\frac{\partial \mathbf{g}_i}{\partial x^j} \cdot \mathbf{g}_l\right) = g^{kl}\Gamma_{ijl}. \tag{4.9}$$

Similarly,

$$\Gamma_{ijk} = g_{kl}\Gamma^l_{ij}. \tag{4.10}$$

Formulas for the derivatives of the contravariant base vectors follow from differentiating equation (1.59), giving

$$\frac{\partial \mathbf{g}_i}{\partial x^k} \cdot \mathbf{g}^j + \mathbf{g}_i \cdot \frac{\partial \mathbf{g}^j}{\partial x^k} = 0,$$

and hence, by (4.3),

$$\Gamma^j_{ik} = -\mathbf{g}_i \cdot \frac{\partial \mathbf{g}^j}{\partial x^k}.$$

Thus

$$\frac{\partial \mathbf{g}^j}{\partial x^k} = -\Gamma^j_{ik}\mathbf{g}^i, \qquad (4.11)$$

after a little manipulation.

It is important to realize that the Christoffel symbols do not constitute third-order tensors. This fact becomes apparent once it is appreciated that in rectangular (or non-rectangular) cartesian reference systems the base vectors do not vary with position and thus all components of the Christoffel symbols are zero. If the Christoffel symbols transformed like tensors, it would follow that their components would remain zero with respect to any co-ordinate system, which is not the case.

4.3 EXAMPLES OF CHRISTOFFEL SYMBOLS

4.3.1 Cylindrical polars

With $x^1 = \rho$, $x^2 = \phi$, $x^3 = z$, the covariant base vectors are given by (1.53) in terms of a fixed orthonormal set of vectors. Term-by-term differentiation gives the nine derivatives

$$\left.\begin{array}{l} \dfrac{\partial \mathbf{g}_1}{\partial x^1} = \mathbf{0}, \quad \dfrac{\partial \mathbf{g}_1}{\partial x^2} = -\sin\phi\,\mathbf{i}_1 + \cos\phi\,\mathbf{i}_2 = \dfrac{1}{\rho}\mathbf{g}_2, \quad \dfrac{\partial \mathbf{g}_1}{\partial x^3} = \mathbf{0}, \\[1em] \dfrac{\partial \mathbf{g}_2}{\partial x^1} = -\sin\phi\,\mathbf{i}_1 + \cos\phi\,\mathbf{i}_2 = \dfrac{1}{\rho}\mathbf{g}_2, \\[1em] \dfrac{\partial \mathbf{g}_2}{\partial x^2} = \rho(-\cos\phi\,\mathbf{i}_1 - \sin\phi\,\mathbf{i}_2) = -\rho\mathbf{g}_1, \quad \dfrac{\partial \mathbf{g}_2}{\partial x^3} = \mathbf{0}, \\[1em] \dfrac{\partial \mathbf{g}_3}{\partial x^1} = \mathbf{0}, \quad \dfrac{\partial \mathbf{g}_3}{\partial x^2} = \mathbf{0}, \quad \dfrac{\partial \mathbf{g}_3}{\partial x^3} = \mathbf{0}. \end{array}\right\} \qquad (4.12)$$

Note the symmetrical array, due to (4.1). The coefficients in the array are the Christoffel symbols of the second kind, and we see from (4.2) that all the components of Γ^k_{ij} are zero except for

$$\Gamma^2_{12} = \Gamma^2_{21} = \frac{1}{\rho} \quad \text{and} \quad \Gamma^1_{22} = -\rho. \qquad (4.13)$$

Since the contravariant base vectors satisfy (1.62), we also have

$$\frac{\partial \mathbf{g}^1}{\partial x^1} = \mathbf{0}, \quad \frac{\partial \mathbf{g}^1}{\partial x^2} = \rho\mathbf{g}^2, \quad \frac{\partial \mathbf{g}^1}{\partial x^3} = \mathbf{0},$$

$$\frac{\partial \mathbf{g}^2}{\partial x^1} = \rho\mathbf{g}^2, \quad \frac{\partial \mathbf{g}^2}{\partial x^2} = -\rho\mathbf{g}^1, \quad \frac{\partial \mathbf{g}^2}{\partial x^3} = \mathbf{0}, \text{ etc.,}$$

and hence all components of Γ_{ijk}, the Christoffel symbol of the first kind, vanish, except for

$$\Gamma_{122} = \Gamma_{212} = \rho \quad \text{and} \quad \Gamma_{221} = -\rho. \qquad (4.14)$$

Sec. 4.3]   **Examples of Christoffel symbols**   115

4.3.2 Spherical polars

Now, with $x^1 = r$, $x^2 = \theta$, $x^3 = \phi$, we may differentiate the components of (1.55) term by term to obtain

$$\frac{\partial \mathbf{g}_1}{\partial x^1} = \mathbf{0},$$

$$\frac{\partial \mathbf{g}_1}{\partial x^2} = [\cos\theta\cos\phi \quad \cos\theta\sin\phi \quad -\sin\theta] = \frac{1}{r}\mathbf{g}_2,$$

$$\frac{\partial \mathbf{g}_1}{\partial x^3} = [-\sin\theta\sin\phi \quad \sin\theta\cos\phi \quad 0] = \frac{1}{r}\mathbf{g}_3,$$

$$\frac{\partial \mathbf{g}_2}{\partial x^1} = [\cos\theta\cos\phi \quad \cos\theta\sin\phi \quad -\sin\theta] = \frac{1}{r}\mathbf{g}_2,$$

$$\frac{\partial \mathbf{g}_2}{\partial x^2} = [-r\sin\theta\cos\phi \quad -r\sin\theta\sin\phi \quad -r\cos\theta) = -r\mathbf{g}_1,$$

$$\frac{\partial \mathbf{g}_2}{\partial x^3} = [-r\cos\theta\sin\phi \quad r\cos\theta\cos\phi \quad 0] = \frac{\cos\theta}{\sin\theta}\mathbf{g}_3,$$

$$\frac{\partial \mathbf{g}_3}{\partial x^1} = [-\sin\theta\sin\phi \quad \sin\theta\cos\phi \quad 0] = \frac{1}{r}\mathbf{g}_3,$$

$$\frac{\partial \mathbf{g}_3}{\partial x^2} = [-r\cos\theta\sin\phi \quad r\cos\theta\cos\phi \quad 0] = \frac{\cos\theta}{\sin\theta}\mathbf{g}_3,$$

and $\dfrac{\partial \mathbf{g}_3}{\partial x^3} = [-r\sin\theta\cos\phi \quad -r\sin\theta\sin\phi \quad 0].$

The last derivative requires a little manipulation to express it in terms of the base vectors. We note that

$$r\sin\theta\,\mathbf{g}_1 + \cos\theta\,\mathbf{g}_2 = [r(\sin^2\theta + \cos^2\theta)\sin\phi \quad r(\sin^2\theta + \cos^2\theta)\cos\phi \quad 0]$$
$$= [r\sin\phi \quad r\cos\phi \quad 0],$$

and hence

$$\frac{\partial \mathbf{g}_3}{\partial x^3} = -\sin\theta\,(r\sin\theta\,\mathbf{g}_1 + \cos\theta\,\mathbf{g}_2).$$

It follows that the only non-vanishing components of Γ^k_{ij} for spherical polar co-ordinates are

$$\left.\begin{array}{l}\Gamma^2_{12} = \Gamma^2_{21} = \dfrac{1}{r}, \quad \Gamma^3_{13} = \Gamma^3_{31} = \dfrac{1}{r}, \\[6pt] \Gamma^1_{22} = -r, \quad \Gamma^3_{23} = \Gamma^3_{32} = \cot\theta, \quad \Gamma^1_{33} = -r\sin^2\theta, \quad \Gamma^2_{33} = -\sin\theta\cos\theta.\end{array}\right\} \quad (4.15)$$

Making use of (1.63), we can also show that the only non-vanishing components of Γ_{ijk} are

$$\left.\begin{array}{l}\Gamma_{122}=\Gamma_{212}=r, \quad \Gamma_{133}=\Gamma_{313}=r\sin^2\theta,\\ \Gamma_{221}=-r, \quad \Gamma_{233}=\Gamma_{323}=r^2\sin\theta\cos\theta,\\ \Gamma_{331}=-r\sin^2\theta, \quad \Gamma_{332}=-r^2\sin\theta\cos\theta.\end{array}\right\} \quad (4.16)$$

4.4 OTHER PROPERTIES OF CHRISTOFFEL SYMBOLS

An alternative formula for Γ_{ijk} may be obtained by differentiating the first of (1.65) to give

$$\frac{\partial g_{ij}}{\partial x^k} = \mathbf{g}_i \cdot \frac{\partial \mathbf{g}_j}{\partial x^k} + \frac{\partial \mathbf{g}_i}{\partial x^k} \cdot \mathbf{g}_j = \Gamma_{jki} + \Gamma_{ikj}$$

by (4.7). Since i, j, k are free indices, we also have

$$\frac{\partial g_{jk}}{\partial x^i} = \Gamma_{kij} + \Gamma_{jik}$$

and

$$\frac{\partial g_{ik}}{\partial x^j} = \Gamma_{kji} + \Gamma_{ijk}.$$

Adding the last two equations, and using (4.8),

$$2\Gamma_{ijk} + (\Gamma_{ikj} + \Gamma_{jki}) = \frac{\partial g_{jk}}{\partial x^i} + \frac{\partial g_{ik}}{\partial x^j}.$$

Hence we have the formula

$$\Gamma_{ijk} = \frac{1}{2}\left(\frac{\partial g_{jk}}{\partial x^i} + \frac{\partial g_{ik}}{\partial x^j} - \frac{\partial g_{ij}}{\partial x^k}\right). \quad (4.17)$$

giving a direct connection between the metric tensor g_{ij} and the Christoffel symbols. If Γ_{ijk} is calculated from this formula, the Christoffel symbols of the second kind may then be calculated using (4.9):

$$\Gamma^k_{ij} = \tfrac{1}{2}g^{kl}\left(\frac{\partial g_{jl}}{\partial x^i} + \frac{\partial g_{il}}{\partial x^j} - \frac{\partial g_{ij}}{\partial x^l}\right). \quad (4.18)$$

Another useful formula emerges if we contract equation (4.18) to get

$$\Gamma^i_{ij} = \tfrac{1}{2}g^{il}\frac{\partial g_{jl}}{\partial x^i} + \tfrac{1}{2}g^{il}\frac{\partial g_{il}}{\partial x^j} - \tfrac{1}{2}g^{il}\frac{\partial g_{ij}}{\partial x^l}$$

$$= \tfrac{1}{2}g^{il}\frac{\partial g_{jl}}{\partial x^i} + \tfrac{1}{2}g^{il}\frac{\partial g_{il}}{\partial x^j} - \tfrac{1}{2}g^{li}\frac{\partial g_{lj}}{\partial x^i}$$

writing i for l and l for i in the last term. The first and last terms now cancel because of the symmetry of g_{ij} and g^{ij}, giving

$$\Gamma^i_{ij} = \tfrac{1}{2}g^{il}\frac{\partial g_{il}}{\partial x^j}. \quad (4.19)$$

It is possible to express the RHS in terms of the determinant g of the matrix G of

Sec. 4.4] Other properties of Christoffel symbols

the metric tensor. If we formally regard g as a function of the nine quantities g_{ij}, then we can consider the partial derivatives $\partial g/\partial g_{ij}$. Now equation (2.16) shows that, for a general matrix A,

$$\frac{\partial}{\partial a_{11}}(\det A) = A_{11}, \quad \frac{\partial}{\partial a_{12}}(\det A) = A_{12}, \text{ etc.,}$$

since the cofactors A_{11} and A_{12} do not contain terms in a_{11} and a_{12}, respectively. In general we have

$$\frac{\partial}{\partial a_{ij}}(\det A) = A_{ij}. \tag{4.20}$$

The matrix of cofactors adj A is given by $(\det A)(A^{-1})^{\mathrm{T}}$, according to (2.27), and we may deduce from (1.74) that the inverse of G is the matrix whose elements are g^{ij}. Thus we have

$$\frac{\partial g}{\partial g_{ij}} = gg^{ij}, \tag{4.21}$$

using the symmetry of G and G^{-1}. The chain rule for partial differentiation now gives

$$\frac{\partial g}{\partial x^j} = \frac{\partial g}{\partial g_{il}}\frac{\partial g_{il}}{\partial x^j} = gg^{il}\frac{\partial g_{il}}{\partial x^j}, \tag{4.22}$$

and hence, by comparison with (4.19),

$$\Gamma^i_{ij} = \frac{1}{2}\frac{1}{g}\frac{\partial g}{\partial x^j} = \frac{1}{\sqrt{g}}\frac{\partial}{\partial x^j}(\sqrt{g}) = \frac{\partial}{\partial x^j}(\log \sqrt{g}). \tag{4.23}$$

The word 'formally' was used above because G is always a symmetric matrix and it might be objected that g should be regarded as a function of six quantities only, e.g. $g_{11}, g_{22}, g_{33}, g_{12}, g_{23}$ and g_{31}. If we insist on doing this, it may be seen that, while we retain

$$\frac{\partial g}{\partial g_{11}} = gg^{11},$$

as in (4.21), we also have

$$\frac{\partial g}{\partial g_{12}} = 2gg^{12}, \text{ etc..}$$

However, the chain rule now involves summation over six terms only, i.e.

$$\frac{\partial g}{\partial x^j} = \frac{\partial g}{\partial g_{11}}\frac{\partial g_{11}}{\partial x^j} + \frac{\partial g}{\partial g_{22}}\frac{\partial g_{22}}{\partial x^j} + \frac{\partial g}{\partial g_{33}}\frac{\partial g_{33}}{\partial x^j} + \frac{\partial g}{\partial g_{12}}\frac{\partial g_{12}}{\partial x^j} + \frac{\partial g}{\partial g_{23}}\frac{\partial g_{23}}{\partial x^j} + \frac{\partial g}{\partial g_{31}}\frac{\partial g_{31}}{\partial x^j}$$

$$= gg^{11}\frac{\partial g_{11}}{\partial x^j} + gg^{22}\frac{\partial g_{22}}{\partial x^j} + gg^{33}\frac{\partial g_{33}}{\partial x^j} + 2gg^{12}\frac{\partial g_{12}}{\partial x^j} + 2gg^{23}\frac{\partial g_{23}}{\partial x^j} + 2gg^{31}\frac{\partial g_{31}}{\partial x^j}$$

and this agrees precisely with (4.22).

We have already indicated that Γ^k_{ij} and Γ_{ijk} do not constitute third-order tensors. The transformation rules that they obey may be determined as follows. With respect to a set of curvilinear co-ordinates $\bar{x}^i$ we have, from (4.3),

$$\bar{\Gamma}^k_{ij} = \frac{\partial \bar{\mathbf{g}}_i}{\partial \bar{x}^j} \cdot \bar{\mathbf{g}}^k = \frac{\partial}{\partial \bar{x}^j}\left(\frac{\partial x^m}{\partial \bar{x}^i}\mathbf{g}_m\right)\cdot\frac{\partial \bar{x}^k}{\partial x^n}\mathbf{g}^n,$$

using equations (3.42) and (3.46).
Hence

$$\bar{\Gamma}^k_{ij} = \frac{\partial^2 x^m}{\partial \bar{x}^i \partial \bar{x}^j}\frac{\partial \bar{x}^k}{\partial x^n}\mathbf{g}_m\cdot\mathbf{g}^n + \frac{\partial x^m}{\partial \bar{x}^i}\frac{\partial \bar{x}^k}{\partial x^n}\frac{\partial \mathbf{g}_m}{\partial \bar{x}^j}\cdot\mathbf{g}^n$$

$$= \frac{\partial^2 x^m}{\partial \bar{x}^i \partial \bar{x}^j}\frac{\partial \bar{x}^k}{\partial x^n}\delta^n_m + \frac{\partial x^m}{\partial \bar{x}^i}\frac{\partial \bar{x}^k}{\partial x^n}\frac{\partial x^l}{\partial \bar{x}^j}\frac{\partial \mathbf{g}_m}{\partial x^l}\cdot\mathbf{g}^n$$

$$= \frac{\partial^2 x^m}{\partial \bar{x}^i \partial \bar{x}^j}\frac{\partial \bar{x}^k}{\partial x^m} + \frac{\partial \bar{x}^k}{\partial x^n}\frac{\partial x^m}{\partial \bar{x}^i}\frac{\partial x^l}{\partial \bar{x}^j}\Gamma^n_{ml}. \tag{4.24}$$

The presence of the first term on the RHS demonstrates that Γ^k_{ij} does not transform like a third-order tensor.

Using (4.10) we may show similarly that the transformation law for Γ_{ijk} is

$$\bar{\Gamma}_{ijk} = g_{lm}\frac{\partial^2 x^m}{\partial \bar{x}^i \partial \bar{x}^j}\frac{\partial x^l}{\partial \bar{x}^k} + \frac{\partial x^l}{\partial \bar{x}^i}\frac{\partial x^m}{\partial \bar{x}^j}\frac{\partial x^n}{\partial \bar{x}^k}\Gamma_{lmn}. \tag{4.25}$$

Problem

4.4.1 Verify (4.23) for the cases of circular cylindrical and spherical polar co-ordinates.

4.5 COVARIANT DERIVATIVES OF VECTOR FIELDS

If $\mathbf{u}$ is a vector field, then so is each partial derivative $\partial \mathbf{u}/\partial x^i$, $i = 1, 2, 3$, with respect to each of a set of curvilinear co-ordinates x^i. These partial derivatives may be expressed in terms of local base vectors, and the coefficients form a set of nine quantities which we shall show constitute a second-order tensor. There are two ways of proceeding.

First, we express $\mathbf{u}$ locally in terms of covariant base vectors as $\mathbf{u} = u^i \mathbf{g}_i$. Partial differentiation gives

$$\frac{\partial \mathbf{u}}{\partial x^j} = u^i \frac{\partial \mathbf{g}_i}{\partial x^j} + \frac{\partial u^i}{\partial x^j}\mathbf{g}_i$$

$$= u^i \Gamma^k_{ij}\mathbf{g}_k + \frac{\partial u^i}{\partial x^j}\mathbf{g}_i = \left(\frac{\partial u^i}{\partial x^j} + \Gamma^i_{kj}u^k\right)\mathbf{g}_i,$$

where we have used (4.2). The coefficient here is called the **covariant derivative of the contravariant vector** u^i and will be denoted by $u^i{}_{,j}$. Thus

$$u^i{}_{,j} = \frac{\partial u^i}{\partial x^j} + \Gamma^i_{kj}u^k \tag{4.26}$$

and

$$\frac{\partial \mathbf{u}}{\partial x^j} = u^i{}_{,j}\mathbf{g}_i. \tag{4.27}$$

Covariant derivatives of vector fields

Secondly, in terms of covariant components of **u**, we have $u = u_i g^i$, and

$$\frac{\partial \mathbf{u}}{\partial x^j} = u_i \frac{\partial \mathbf{g}^i}{\partial x^j} + \frac{\partial u_i}{\partial x^j} \mathbf{g}^i = -u_i \Gamma^i_{kj} \mathbf{g}^k + \frac{\partial u_i}{\partial x^j} \mathbf{g}^i \qquad \text{by (4.11)},$$

$$= \left(\frac{\partial u_i}{\partial x^j} - \Gamma^k_{ij} u_k \right) \mathbf{g}^i = u_{i,j} \mathbf{g}^i,$$

where the **covariant derivative of the covariant vector** u_i is given by

$$u_{i,j} = \frac{\partial u_i}{\partial x^j} - \Gamma^k_{ij} u_k. \tag{4.28}$$

It is possible to show that $u^i{}_{,j}$ and $u_{i,j}$ transform as second-order tensors by transforming all the terms in the definitions (4.26) and (4.28), making use of (4.24)—see problem 4.5.4. Instead, transforming (4.27) directly, using the chain rule and (3.41), gives

$$\frac{\partial \mathbf{u}}{\partial \bar{x}^j} = \bar{u}^i{}_{,j} \bar{\mathbf{g}}_i = \frac{\partial \mathbf{u}}{\partial x^k} \frac{\partial x^k}{\partial \bar{x}^j} = (u^l{}_{,k} \mathbf{g}_l) \frac{\partial x^k}{\partial \bar{x}^j}$$

$$= u^l{}_{,k} \left(\frac{\partial \bar{x}^i}{\partial x^l} \bar{\mathbf{g}}_i \right) \frac{\partial x^k}{\partial \bar{x}^j}.$$

Identifying coefficients of $\bar{\mathbf{g}}_i$,

$$\bar{u}^i{}_{,j} = \frac{\partial \bar{x}^i}{\partial x^l} \frac{\partial x^k}{\partial \bar{x}^j} u^l{}_{,k},$$

showing that $u^i{}_{,j}$ transforms according to the law (3.55) for a mixed second-order tensor.

Similarly we can show that $u_{i,j}$ transforms according to the law (3.54) for a covariant second-order tensor. This is more simply demonstrated, however, by noting that

$$\frac{\partial \mathbf{u}}{\partial x^j} = u_{i,j} \mathbf{g}^i = u^i{}_{,j} \mathbf{g}_i = u^i{}_{,j} g_{ik} \mathbf{g}^k = u^k{}_{,j} g_{ki} \mathbf{g}^i,$$

and comparison of coefficients leads to

$$u_{i,j} = g_{ik} u^k{}_{,j}. \tag{4.29}$$

The tensor character of $u_{i,j}$ follows from the fact that g_{ij} and $u^i{}_{,j}$ are both tensors. Moreover, (4.29) demonstrates that $u_{i,j}$ and $u^i{}_{,j}$ are *associated* components of the same tensor. Note that the associated contravariant components of this tensor are $g^{jk} u^i{}_{,k}$, which must *not* be written as $u^{i,j}$.

The product rule for differentiating products remains valid for covariant differentiation, provided that properly formed tensor expressions are involved. For example, if **u** and **v** are vector fields, the scalar product **u**·**v** may be written $u^i v_i$, and since this is a scalar field we may write

$$(u^i v_i)_{,j} = \frac{\partial}{\partial x^j} (u^i v_i)$$

$$= u^i \frac{\partial v_i}{\partial x^j} + \frac{\partial u^i}{\partial x^j} v_i \qquad \text{by the usual product rule}$$

$$= u^i(v_{i,j} + \Gamma^k_{ij}v_k) + (u^i_{,j} - \Gamma^i_{kj}u^k)v_i$$
$$= u^i v_{i,j} + u^i_{,j} v_i + \Gamma^k_{ij} u^i v_k - \Gamma^i_{kj} u^k v_i,$$

using (4.26) and (4.28). The last two terms cancel, leaving us with

$$(u^i v_i)_{,j} = u^i v_{i,j} + u^i_{,j} v_i. \qquad (4.30)$$

Equation (4.30) may be regarded as the generalization to arbitrary co-ordinate systems of the product rule for differentiation of scalar products in cartesian co-ordinates, since covariant derivatives reduce to ordinary partial derivatives in cartesian systems.

Problems

4.5.1 Verify that, for vectors **u**, **v**,

$$(u^i + v^i)_{,j} = u^i_{,j} + v^i_{,j}.$$

4.5.2 By putting $\mathbf{u} = \text{grad } \Phi$ in (4.28), show that the covariant second derivative $(\Phi_{,i})_{,j}$, of a scalar Φ, which we write as $\Phi_{,ij}$, is given by

$$\Phi_{,ij} = \frac{\partial^2 \Phi}{\partial x^i \partial x^j} - \Gamma^k_{ij} \frac{\partial \Phi}{\partial x^k},$$

and show explicitly that this transforms as a covariant second-order tensor.

4.5.3 Transform the identity $\partial \mathbf{u}/\partial x^j = u_{i,j}\mathbf{g}^i$ to prove that $u_{i,j}$ is a covariant second-order tensor.

4.5.4 Show that the partial derivatives $\partial u^i/\partial x^j$, where u^i is a contravariant vector field, transform according to the law

$$\frac{\partial \bar{u}^i}{\partial \bar{x}^j} = \frac{\partial \bar{x}^i}{\partial x^k} \frac{\partial x^l}{\partial \bar{x}^j} \frac{\partial u^k}{\partial x^l} + \frac{\partial^2 \bar{x}^i}{\partial x^k \partial x^l} \frac{\partial x^l}{\partial \bar{x}^j} u^k.$$

Deduce that

$$\frac{\partial \bar{u}^i}{\partial \bar{x}^j} - \bar{\Gamma}^i_{kj} \bar{u}^k = \frac{\partial \bar{x}^i}{\partial x^k} \frac{\partial x^l}{\partial \bar{x}^j} \left(\frac{\partial u^k}{\partial x^l} - \Gamma^k_{ml} u^m \right),$$

thus demonstrating that (4.26) constitutes a mixed second-order tensor. (Hint: it may be necessary to use the identity

$$\frac{\partial^2 \bar{x}^i}{\partial x^k \partial x^l} \frac{\partial x^k}{\partial \bar{x}^j} + \frac{\partial^2 x^k}{\partial \bar{x}^j \partial \bar{x}^m} \frac{\partial \bar{x}^i}{\partial x^k} \frac{\partial \bar{x}^m}{\partial x^l} = 0,$$

which may be obtained by differentiating the identity

$$\frac{\partial \bar{x}^i}{\partial x^k} \frac{\partial x^k}{\partial \bar{x}^j} = \delta^i_j$$

with respect to x^l.)

4.6 DIV AND CURL

Since $u^i_{,j}$, for vector fields **u**, is a mixed second-order tensor, contraction yields the scalar $u^i_{,i}$. In a rectangular cartesian system the Christoffel symbols vanish, there is

Div and curl

no need to distinguish between covariant and contravariant components, and this scalar consequently becomes $\partial u_i/\partial y_i$, which is just div **u**. Since the value of the scalar at a point is independent of the co-ordinate system, we have

$$\text{div } \mathbf{u} = u^i_{,i}. \tag{4.31}$$

It is important to note that the contracted form $u_{i,i}$ does *not* represent a scalar in general co-ordinate systems and thus cannot be equal to div **u**.

Pursuing this point, suppose $\mathbf{u} = \text{grad } \Phi$, for some scalar field Φ. The covariant components of **u** are just the partial derivatives $\partial \Phi/\partial x^i$, which we have called the covariant derivative of Φ, and written $\Phi_{,i}$. Now div $\mathbf{u} = $ div grad $\Phi = \nabla^2 \Phi$, where ∇^2 is the laplacian operator, but it is *not* permissible to write

$$\nabla^2 \Phi = (\Phi_{,i})_{,i}$$

in general co-ordinate systems, since summation must be carried out on indices at different levels. Instead, it is necessary to obtain the associated contravariant components of grad Φ:

$$(\text{grad } \Phi)^i = g^{ij} \Phi_{,j},$$

and use these in (4.31), giving

$$\nabla^2 \Phi = (g^{ij} \Phi_{,j})_{,i}. \tag{4.32}$$

Using (4.26), this may be written more fully as

$$\nabla^2 \Phi = \frac{\partial}{\partial x^i}\left(g^{ij} \frac{\partial \Phi}{\partial x^j}\right) + \Gamma^i_{ki} g^{kj} \frac{\partial \Phi}{\partial x^j}. \tag{4.33}$$

Moreover, (4.31) may be written as

$$\text{div } \mathbf{u} = \frac{\partial u^i}{\partial x^i} + \Gamma^i_{ki} u^k = \frac{\partial u^i}{\partial x^i} + \Gamma^i_{ik} u^k$$

$$= \frac{\partial u^i}{\partial x^i} + \frac{\partial}{\partial x^k}(\log \sqrt{g}) u^k = \frac{1}{\sqrt{g}} \frac{\partial}{\partial x^i}(\sqrt{g} u^i), \tag{4.34}$$

using (4.5) and (4.23).

To form the vector curl **u**, we look for a valid tensor expression which reduces to (1.48) for rectangular cartesian co-ordinates. Use of the alternating tensor and covariant derivative yields contravariant components

$$(\text{curl } \mathbf{u})^i = \varepsilon^{ijk} u_{k,j} \tag{4.35}$$

or

$$\text{curl } \mathbf{u} = \varepsilon^{ijk} u_{k,j} \mathbf{g}_i. \tag{4.36}$$

The covariant components are given by the equivalent forms

$$(\text{curl } \mathbf{u})_i = g_{il} \varepsilon^{ljk} u_{k,j} = g^{jl} \varepsilon_{ijk} u^k_{,l}. \tag{4.37}$$

From (4.35),

$$(\text{curl } \mathbf{u})^i = \frac{1}{\sqrt{g}} e^{ijk} \left(\frac{\partial u_k}{\partial x^j} - \Gamma^l_{kj} u_l\right) = \frac{1}{\sqrt{g}} e^{ijk} \frac{\partial u_k}{\partial x^j},$$

since $e^{ijk} \Gamma^l_{kj} = 0$ on account of the skew symmetry of e^{ijk} and the symmetry of Γ^l_{kj}

in j and k. Hence the contravariant components of curl **u** are given by

$$\frac{1}{\sqrt{g}}\left(\frac{\partial u_3}{\partial x^2}-\frac{\partial u_2}{\partial x^3},\ \frac{\partial u_1}{\partial x^3}-\frac{\partial u_3}{\partial x^1},\ \frac{\partial u_2}{\partial x^1}-\frac{\partial u_1}{\partial x^2}\right). \tag{4.38}$$

Problems

4.6.1 Show that, for a scalar Φ,

$$\nabla^2\Phi = \frac{1}{\sqrt{g}}\frac{\partial}{\partial x^i}\left(\sqrt{g}\,g^{ij}\frac{\partial\Phi}{\partial x^j}\right).$$

Hence show that in cylindrical polar co-ordinates and spherical polar co-ordinates, respectively,

$$\nabla^2\Phi = \frac{1}{\rho}\frac{\partial}{\partial\rho}\left(\rho\frac{\partial\Phi}{\partial\rho}\right)+\frac{1}{\rho^2}\frac{\partial^2\Phi}{\partial\theta^2}+\frac{\partial^2\Phi}{\partial z^2}$$

and

$$\nabla^2\Phi = \frac{1}{r^2}\frac{\partial}{\partial r}\left(r^2\frac{\partial\Phi}{\partial r}\right)+\frac{1}{r^2\sin\theta}\frac{\partial}{\partial\theta}\left(\sin\theta\frac{\partial\Phi}{\partial\theta}\right)+\frac{1}{r^2\sin\theta}\frac{\partial^2\Phi}{\partial\phi^2}.$$

4.7 COVARIANT DERIVATIVES OF HIGHER-ORDER TENSORS

Suppose **T** is a second-order tensor field, expressible locally in terms of the dyadic products of base vectors (3.63). We would like to be able to write

$$\frac{\partial \mathbf{T}}{\partial x^k} = T^{ij}{}_{,k}\mathbf{g}_i\otimes\mathbf{g}_j = T_{ij,k}\mathbf{g}^i\otimes\mathbf{g}^j = T^i{}_{.j,k}\mathbf{g}_i\otimes\mathbf{g}^j = T_i{}^{.j}{}_{,k}\mathbf{g}^i\otimes\mathbf{g}_j, \tag{4.39}$$

in terms of the four possible covariant derivatives $T^{ij}{}_{,k}$, $T_{ij,k}$, $T^i{}_{.j,k}$ and $T_i{}^{.j}{}_{,k}$.

Let us begin with $\mathbf{T}=T^{ij}\mathbf{g}_i\otimes\mathbf{g}_j$ and take partial derivatives of both sides with respect to x^k, assuming that it is permissible to apply the usual product rule to the RHS. We obtain

$$\frac{\partial \mathbf{T}}{\partial x^k} = \frac{\partial T^{ij}}{\partial x^k}\mathbf{g}_i\otimes\mathbf{g}_j + T^{ij}\frac{\partial \mathbf{g}_i}{\partial x^k}\otimes\mathbf{g}_j + T^{ij}\mathbf{g}_i\otimes\frac{\partial \mathbf{g}_j}{\partial x^k}$$

$$= \frac{\partial T^{ij}}{\partial x^k}\mathbf{g}_i\otimes\mathbf{g}_j + T^{ij}\Gamma^l_{ik}\mathbf{g}_l\otimes\mathbf{g}_j + T^{ij}\mathbf{g}_i\otimes\Gamma^l_{jk}\mathbf{g}_l$$

$$= \frac{\partial T^{ij}}{\partial x^k}\mathbf{g}_i\otimes\mathbf{g}_j + \Gamma^i_{lk}T^{lj}\mathbf{g}_i\otimes\mathbf{g}_j + \Gamma^j_{lk}T^{il}\mathbf{g}_i\otimes\mathbf{g}_j$$

using (4.2) and reorganizing the dummy indices. Comparison with (4.39) gives

$$T^{ij}{}_{,k} = \frac{\partial T^{ij}}{\partial x^k} + \Gamma^i_{lk}T^{lj} + \Gamma^j_{lk}T^{il}. \tag{4.40}$$

It may be checked (see problem 4.7.1) that the RHS of (4.40) transforms as a genuine third-order tensor.

Derivatives of higher-order tensors

As an alternative route to (4.40), we could consider the scalar quantity $T^{ij}u_i v_j$, where **u** and **v** are arbitrary vector fields. Since the covariant derivative of a scalar is identical with the set of partial derivatives, we have, assuming the usual product rules,

$$(T^{ij}u_i v_j)_{,k} = T^{ij}{}_{,k}u_i v_j + T^{ij}u_{i,k}v_j + T^{ij}u_i v_{j,k}$$

$$= T^{ij}{}_{,k}u_i v_j + T^{ij}\left(\frac{\partial u_i}{\partial x^k} - \Gamma^l_{ik}u_l\right)v_j + T^{ij}u_i\left(\frac{\partial v_j}{\partial x^k} - \Gamma^l_{jk}v_l\right)$$

$$= \frac{\partial}{\partial x^k}(T^{ij}u_i v_j) = \frac{\partial T^{ij}}{\partial x^k}u_i v_j + T^{ij}\frac{\partial u_i}{\partial x^k}v_j + T^{ij}u_i\frac{\partial v_j}{\partial x^k}.$$

Cancellation of terms and reorganization of dummy indices yield

$$\left(T^{ij}{}_{,k} - T^{lj}\Gamma^i_{lk} - T^{il}\Gamma^j_{lk} - \frac{\partial T^{ij}}{\partial x^k}\right)u_i v_j = 0.$$

Since **u** and **v** are arbitrary, (4.40) follows

In a similar manner we can show that

$$T_{ij,k} = \frac{\partial T_{ij}}{\partial x^k} - \Gamma^l_{ik}T_{lj} - \Gamma^l_{jk}T_{il}, \tag{4.41}$$

$$T^i{}_{.j,k} = \frac{\partial T^i{}_{.j}}{\partial x^k} + \Gamma^i_{lk}T^l{}_{.j} - \Gamma^l_{jk}T^i{}_{.l}, \tag{4.42}$$

and

$$T_i{}^j{}_{,k} = \frac{\partial T_i{}^j}{\partial x^k} - \Gamma^l_{ik}T_l{}^j + \Gamma^j_{lk}T_i{}^l. \tag{4.43}$$

Equations (4.40)–(4.43) are not difficult to memorize if one realizes that the dummy index l replaces each index of the tensor **T** in turn and that this fixes its position in the corresponding Christoffel symbol. The positive and negative signs depend on whether l replaces a contravariant or covariant index, respectively.

Rewriting (4.43) as

$$T^{.i}_{j,k} = \frac{\partial T^{.i}_j}{\partial x^i} + \Gamma^i_{lk}T^{.l}_j - \Gamma^l_{jk}T^{.i}_l$$

shows that it is essentially the same as (4.42), so that the covariant derivatives of $T^i{}_{.j}$ and $T^{.i}_j$ are given by the same formula. This observation allows us to write the generalization of these formulas to the case of the covariant derivative of a tensor of order $M+N$ with M contravariant and N covariant components as

$$U^{ij...k}_{pq...r,t} = \frac{\partial U^{ij...k}_{pq...r}}{\partial x^t} + \Gamma^i_{st}U^{sj...k}_{pq...r} + \Gamma^j_{st}U^{is...k}_{pq...r} + \cdots \tag{4.44}$$

$$+ \Gamma^k_{st}U^{ij...s}_{pq...r} - \Gamma^s_{pt}U^{ij...k}_{sq...r} - \Gamma^s_{qt}U^{ij...k}_{ps...r} - \cdots - \Gamma^s_{rt}U^{ij...k}_{pq...s}$$

where the precise positions (before or after any relevant dots) of the M indices $i, j, \ldots, k$ and the N indices $p, q, \ldots, r$ do not affect the formula. The covariant derivative (4.44) is a tensor of order $M+N+1$ with M contravariant and $N+1$ covariant components.

Problems

4.7.1 Show that, for second-order contravariant tensors T^{ij}, $T^{ij}{}_{,k}$ satisfies the transformation rule

$$\left(\frac{\partial \bar{T}^{ij}}{\partial \bar{x}^k} + \bar{\Gamma}^i_{lk}\bar{T}^{lj} + \bar{\Gamma}^j_{lk}\bar{T}^{il}\right) = \frac{\partial \bar{x}^i}{\partial x^p}\frac{\partial \bar{x}^j}{\partial x^q}\frac{\partial x^r}{\partial \bar{x}^k}\left(\frac{\partial T^{pq}}{\partial x^r} + \Gamma^p_{sr}T^{sq} + \Gamma^q_{sr}T^{ps}\right).$$

4.7.2 Prove the identities (4.41)–(4.43).

4.7.3 Verify that, if T_{ij} and u^i are a second-order tensor and vector, respectively, then

$$(T_{ij}u^j)_{,k} = T_{ij,k}u^j + T_{ij}u^j{}_{,k}.$$

4.8 SPECIAL CASES

4.8.1 The metric tensor

If we consider the covariant derivative $g_{ij,k}$ of the metric tensor g_{ij} which must be a third-order tensor covariant in all its indices, we see that in rectangular cartesian co-ordinate systems g_{ij} reduces to the Kronecker delta δ_{ij}, and thus the covariant derivative $g_{ij,k}$ reduces to the partial derivative

$$\frac{\partial}{\partial y_k}(\delta_{ij}),$$

which must be zero, since the δ_{ij} are constants. A tensor whose components are all zero in one system of co-ordinates must have all its components zero with respect to any other system, according to the tensor transformation rules. Hence

$$g_{ij,k} = 0. \tag{4.45}$$

To verify this result by direct calculation, consider

$$\frac{\partial}{\partial x^k}(g_{ij}) = \frac{\partial}{\partial x^k}(\mathbf{g}_i \cdot \mathbf{g}_j) = \mathbf{g}_i \cdot \frac{\partial \mathbf{g}_j}{\partial x^k} + \frac{\partial \mathbf{g}_i}{\partial x^k} \cdot \mathbf{g}_j$$

$$= \mathbf{g}_i \cdot \Gamma_{jkl}\mathbf{g}^l + \Gamma_{ikl}\mathbf{g}^l \cdot \mathbf{g}_j = \Gamma_{jkl}\delta^l_i + \Gamma_{ikl}\delta^l_j$$

$$= \Gamma_{jki} + \Gamma_{ikj},$$

using (4.6) and (1.59). Hence, by (4.41),

$$g_{ij,k} = \frac{\partial g_{ij}}{\partial x^k} - \Gamma^l_{ik}g_{lj} - \Gamma^l_{jk}g_{il}$$

$$= \Gamma_{jki} + \Gamma_{ikj} - \Gamma_{ikj} - \Gamma_{jki} = 0,$$

using (4.10).

This result is directly connected to equation (4.29), which could now be deduced, using the product rule, by the following:

$$u_{i,j} = (g_{ik}u^k)_{,j} = g_{ik,j}u^k + g_{ik}u^k{}_{,j} = 0 + g_{ik}u^k{}_{,j}.$$

We also have, again from considering its value in cartesian systems,

$$g^{ij}{}_{,k} = 0, \qquad (4.46)$$

and this result may be verified by evaluating (4.40) for g^{ij}.

4.8.2 The alternating tensor

Since the alternating tensor ε_{ijk} and ε^{ijk} in covariant and contravariant form both reduce to the alternating symbol e_{ijk} in rectangular cartesian systems, it follows that

$$\varepsilon_{ijk,l} = 0 \qquad (4.47)$$

and

$$\varepsilon^{ijk}{}_{,l} = 0. \qquad (4.48)$$

We can verify (4.47), for example, by direct calculation on a typical non-zero component such as $\varepsilon_{123} = \sqrt{g}$, by (3.75). We have, by (4.44)

$$\varepsilon_{123,l} = \frac{\partial \varepsilon_{123}}{\partial x^l} - \Gamma^s_{1l}\varepsilon_{s23} - \Gamma^s_{2l}\varepsilon_{1s3} - \Gamma^s_{3l}\varepsilon_{12s}$$

$$= \frac{\partial(\sqrt{g})}{\partial x^l} - \Gamma^1_{1l}\sqrt{g} - \Gamma^2_{2l}\sqrt{g} - \Gamma^3_{3l}\sqrt{g}$$

$$= \frac{\partial}{\partial x^l}(\sqrt{g}) - \Gamma^i_{il}\sqrt{g} = 0 \qquad \text{by (4.23)}.$$

Problems

4.8.1 Using (4.40), (1.65) and (4.11), establish (4.46).

4.8.2 Verify that $\varepsilon^{123}{}_{,l} = 0$ by using (4.44) together with (3.75) and (4.23).

4.8.3 Verify the identity div curl $\mathbf{u} = 0$ for vector fields $\mathbf{u}$ using generalized co-ordinates.

4.8.4 Using (4.42), verify that $\delta^i_{j,k} = 0$.

4.9 SECOND DERIVATIVES

For a scalar field Φ the usual assumptions about continuity and differentiability guarantee that

$$\frac{\partial^2 \Phi}{\partial x^i \partial x^j} = \frac{\partial^2 \Phi}{\partial x^j \partial x^i}.$$

From problem 4.5.2 and from equation (4.5) it follows that the second covariant derivatives satisfy

$$\Phi_{,ij} = \Phi_{,ji}. \qquad (4.49)$$

However, we could argue that $\Phi_{,ij}$ is a covariant second-order tensor which reduces to

$\partial^2 \Phi / \partial y_i \partial y_j$ in cartesian co-ordinates. Since this is a symmetric CT2 and the property of symmetry is preserved under general tensor transformations, (4.49) must hold.

We can use the same argument with a vector field **u** to show that

$$u^i{}_{,jk} = u^i{}_{,kj} \qquad (4.50)$$

and
$$u_{i,jk} = u_{i,kj}. \qquad (4.51)$$

Writing equation (4.50) in more detail using (4.26) and (4.42), we obtain

$$(u^i{}_{,j})_{,k} = \left(\frac{\partial u^i}{\partial x^j} + \Gamma^i_{lj} u^l\right)_{,k}$$

$$= \frac{\partial}{\partial x^k}\left(\frac{\partial u^i}{\partial x^j} + \Gamma^i_{lj} u^l\right) + \Gamma^i_{lk}\left(\frac{\partial u^l}{\partial x^j} + \Gamma^l_{mj} u^m\right) - \Gamma^l_{jk}\left(\frac{\partial u^i}{\partial x^l} + \Gamma^i_{ml} u^m\right)$$

$$= \frac{\partial^2 u^i}{\partial x^k \partial x^j} + \frac{\partial u^l}{\partial x^k}\Gamma^i_{lj} + \frac{\partial u^l}{\partial x^j}\Gamma^i_{lk} - \frac{\partial u^i}{\partial x^l}\Gamma^l_{jk} + u^l\frac{\partial \Gamma^i_{lj}}{\partial x^k}$$
$$+ u^m \Gamma^i_{lk}\Gamma^l_{mj} - u^m \Gamma^l_{jk}\Gamma^i_{ml}$$

$$= (u^i{}_{,k})_{,j}$$

$$= \frac{\partial^2 u^i}{\partial x^j \partial x^k} + \frac{\partial u^l}{\partial x^j}\Gamma^i_{lk} + \frac{\partial u^l}{\partial x^k}\Gamma^i_{lj} - \frac{\partial u^i}{\partial x^l}\Gamma^l_{kj} + u^l\frac{\partial \Gamma^i_{lk}}{\partial x^j}$$
$$+ u^m \Gamma^i_{lj}\Gamma^l_{mk} - u^m \Gamma^l_{jk}\Gamma^i_{ml}. \qquad (4.52)$$

Cancelling terms gives

$$u^l \frac{\partial \Gamma^i_{lj}}{\partial x^k} + u^m \Gamma^i_{lk}\Gamma^l_{mj} = u^l \frac{\partial \Gamma^i_{lk}}{\partial x^j} + u^m \Gamma^i_{lj}\Gamma^l_{mk},$$

i.e.

$$u^l\left(\frac{\partial \Gamma^i_{lj}}{\partial x^k} - \frac{\partial \Gamma^i_{lk}}{\partial x^j} + \Gamma^i_{mk}\Gamma^m_{lj} - \Gamma^i_{mj}\Gamma^m_{lk}\right) = 0.$$

Since **u** is arbitrary, we have the identity

$$\frac{\partial \Gamma^i_{lj}}{\partial x^k} - \frac{\partial \Gamma^i_{lk}}{\partial x^j} + \Gamma^i_{mk}\Gamma^m_{lj} - \Gamma^i_{mj}\Gamma^m_{lk} = 0. \qquad (4.53)$$

It turns out that this identity is a feature of euclidean, or 'flat', spaces, in which it is possible to have a rectangular cartesian reference system. One can develop tensor analysis to deal with non-euclidean, or 'curved', spaces. For example, our space might be the two-dimensional surface of a sphere, on which it is not possible to lay down a reference system containing two mutually perpendicular unit vectors $\mathbf{i}_1$ and $\mathbf{i}_2$ at every point such that the set of vectors $\{\mathbf{i}_1\}$ are parallel to each other. The above arguments, involving the reduction of generalized tensors to cartesian form, then do not apply, and (4.53), and, indeed, (4.50) and (4.51), are no longer valid. The LHS of (4.53) turns out to be a fourth-order tensor, the mixed Riemann–Christoffel tensor, which is not identically zero for non-euclidean spaces.

5

Orthogonal curvilinear co-ordinates and physical components

5.1 GENERAL THEORY

5.1.1 General orthogonal co-ordinate systems

Cylindrical polar co-ordinates and spherical polar co-ordinates are common examples of orthogonal curvilinear co-ordinate systems, in which the co-ordinate surfaces and co-ordinate curves passing through any given point intersect orthogonally. The covariant base vectors $\mathbf{g}_1, \mathbf{g}_2, \mathbf{g}_3$ at any point are tangential to the co-ordinate curves passing through that point and are thus mutually orthogonal. Hence for such systems

$$g_{ij} = \mathbf{g}_i \cdot \mathbf{g}_j = 0 \quad \text{for} \quad i \neq j, \tag{5.1}$$

and the metric (1.68) is given by

$$\begin{aligned} ds^2 &= g_{11}(dx^1)^2 + g_{22}(dx^2)^2 + g_{33}(dx^3)^2 \\ &= h_1^{\,2}(dx^1)^2 + h_2^{\,2}(dx^2)^2 + h_3^{\,2}(dx^3)^2, \end{aligned} \tag{5.2}$$

where the positive quantities h_1, h_2, h_3, called **scale factors**, are given by

$$h_i = \sqrt{g_{ii}} \quad \text{(n.s.)}. \tag{5.3}$$

It will be observed that the summation convention is of limited use in this section and that we consequently need to employ the 'no summation' abbreviation to avoid confusion. Clearly, from (5.2), $h_1 \, dx^1, h_2 \, dx^2$ and $h_3 \, dx^3$ are measures of *length* along the co-ordinate curves, and the scale factors are equal to the *magnitudes* of the covariant base vectors.

The magnitudes of the contravariant base vectors must be equal to the reciprocals of the scale factors if (1.59) is to hold, and

$$h_i^{\,-1} = \sqrt{g^{ii}} \quad \text{(n.s.)}. \tag{5.4}$$

Thus we have

$$g_{ij} = \begin{bmatrix} h_1^2 & 0 & 0 \\ 0 & h_2^2 & 0 \\ 0 & 0 & h_3^2 \end{bmatrix} \quad \text{and} \quad g^{ij} = \begin{bmatrix} h_1^{-2} & 0 & 0 \\ 0 & h_2^{-2} & 0 \\ 0 & 0 & h_3^{-2} \end{bmatrix}. \quad (5.5)$$

Equation (4.17) can be used to obtain expressions for the Christoffel symbols. If i, j, k take the same value,

$$\Gamma_{iii} = \frac{1}{2}\left(\frac{\partial g_{ii}}{\partial x^i} + \frac{\partial g_{ii}}{\partial x^i} - \frac{\partial g_{ii}}{\partial x^i}\right) \quad \text{(n.s.)}$$

$$= \frac{1}{2}\frac{\partial g_{ii}}{\partial x^i} = \frac{1}{2}\frac{\partial}{\partial x^i}(h_i^2) = h_i \frac{\partial h_i}{\partial x^i} \quad \text{(n.s.).} \quad (5.6)$$

If i and j take the same value, but k is different, we have, by (5.1),

$$\Gamma_{iik} = \frac{1}{2}\left(\frac{\partial g_{ik}}{\partial x^i} + \frac{\partial g_{ik}}{\partial x^i} - \frac{\partial g_{ii}}{\partial x^k}\right) \quad \text{(n.s.)}$$

$$= -\frac{1}{2}\frac{\partial g_{ii}}{\partial x^k} = -\frac{1}{2}\frac{\partial}{\partial x^k}(h_i^2) = -h_i \frac{\partial h_i}{\partial x^k} \quad \text{(n.s.),} \quad (5.7)$$

while, similarly, if j and k take the same value, but i is different,

$$\Gamma_{jij} = \Gamma_{ijj} = \frac{1}{2}\frac{\partial g_{jj}}{\partial x^i} = h_j \frac{\partial h_j}{\partial x^i} \quad \text{(n.s.).} \quad (5.8)$$

If i, j and k are all different, it follows from (5.1) that

$$\Gamma_{ijk} = 0. \quad (5.9)$$

Using (4.18), together with (5.5), we can now obtain

$$\begin{aligned} \Gamma_{ij}^k &= 0 \quad \text{if } i, j, k \text{ are all different,} \\ \Gamma_{ik}^k &= \Gamma_{ki}^k = \frac{1}{h_k}\frac{\partial h_k}{\partial x^i} \quad \text{(n.s.),} \\ \Gamma_{kk}^k &= \frac{1}{h_k}\frac{\partial h_k}{\partial x^k} \quad \text{(n.s.),} \\ \text{and} \quad \Gamma_{ii}^k &= -\frac{h_i}{h_k^2}\frac{\partial h_i}{\partial x^k} \quad \text{(n.s.),} \quad \text{if } i \text{ and } k \text{ are different.} \end{aligned} \quad (5.10)$$

We also note that

$$g = \det G = g_{11}g_{22}g_{33} = h_1^2 h_2^2 h_3^2,$$

and

$$\sqrt{g} = h_1 h_2 h_3. \quad (5.11)$$

It is clear that, since $h_1 \, dx^1$, etc., are infinitesimal lengths along the co-ordinate curves, $(h_1 h_2 h_3 \, dx^1 \, dx^2 \, dx^3)$ represents the *volume* of an infinitesimal rectangular parallelepiped whose sides are parallel to the co-ordinate curves.

5.1.2 Physical components

At any point P a set of mutually orthogonal unit vectors, denoted $\mathbf{g}_{(1)}$, $\mathbf{g}_{(2)}$, $\mathbf{g}_{(3)}$, is obtained by dividing each of the covariant base vectors by its magnitude. Thus, by definition,

$$\mathbf{g}_{(i)} = \frac{1}{h_i}\mathbf{g}_i \quad \text{(n.s.)}. \tag{5.12}$$

A vector field **u** having contravariant components u^i can now be expanded in terms of these vectors, giving

$$\mathbf{u} = u^i \mathbf{g}_i = \sum_{i=0}^{3} h_i u^i \mathbf{g}_{(i)}.$$

The components $h_i u^i$ (n.s.) have the same physical dimensions as the vector **u** itself, since $\mathbf{g}_{(i)}$ is a unit vector, and we shall call these the **physical components** $u_{(i)}$ of **u**. Thus

$$u_{(i)} = h_i u^i \quad \text{(n.s.)}. \tag{5.13}$$

We may deduce that

$$u_{(i)} = h_i g^{ij} u_j \quad \text{(n.s. on } i\text{)}$$
$$= h_i g^{ii} u_i \quad \text{(n.s)}$$
$$= h_i h_i^{-2} u_i = \frac{1}{h_i} u_i \quad \text{(n.s.)}. \tag{5.14}$$

For particular co-ordinate systems it is convenient to replace the (numerical) suffix of physical components by the corresponding co-ordinate itself. For example, in the case of spherical polar co-ordinates, with $x^1 = r$, $x^2 = \theta$, $x^3 = \phi$, we can put $u_{(1)} = u_r$, $u_{(2)} = u_\theta$ and $u_{(3)} = u_\phi$. Since, by (1.56), we have $h_1 = 1$, $h_2 = r$, $h_3 = r\sin\theta$, it follows that

$$u_r = u^1 = u_1, \quad u_\theta = ru^2 = \frac{1}{r}u_2, \quad u_\phi = r\sin\theta\, u^3 = \frac{1}{r\sin\theta}u_3.$$

For second-order tensors **T** we have, in general,

$$\mathbf{T} = T^{ij}\mathbf{g}_i \otimes \mathbf{g}_j = \sum_i \sum_j T^{ij}(h_i \mathbf{g}_{(i)} \otimes h_j \mathbf{g}_{(j)})$$
$$= \sum_i \sum_j T^{ij} h_i h_j \mathbf{g}_{(i)} \otimes \mathbf{g}_{(j)}$$

in terms of dyadic products of the unit vectors $\mathbf{g}_{(i)}$. Hence we may define the physical components of **T** to be $T_{(ij)}$, where

$$T_{(ij)} = T^{ij} h_i h_j \quad \text{(n.s.)}. \tag{5.15}$$

By considering the alternative expansions of **T** in (3.63), we obtain, similarly,

$$T_{(ij)} = \frac{1}{h_i h_j} T_{ij} = \frac{h_i}{h_j} T^i_{\cdot j} = \frac{h_j}{h_i} T_i^{\cdot j} \quad \text{(n.s.)} \tag{5.16}$$

For third-order tensors

$$\mathbf{U} = U^{ijk}\mathbf{g}_i \otimes \mathbf{g}_j \otimes \mathbf{g}_k,$$

the physical components are given by

$$U_{(ijk)} = h_i h_j h_k U^{ijk} = \frac{h_j h_k}{h_i} U_i{}^{jk} = \frac{h_i h_k}{h_j} U^{i\cdot k}_{\cdot j} = \frac{h_j}{h_i h_k} U_{i\cdot k}^{\cdot j}$$

$$= \frac{h_i}{h_j h_k} U^i{}_{\cdot jk} = \frac{h_k}{h_i h_j} U_{ij\cdot}{}^k = \frac{h_i h_j}{h_k} U^{ij\cdot}{}_k = \frac{1}{h_i h_j h_k} U_{ijk} \quad \text{(n.s.)}, \qquad (5.17)$$

and it is easy to see how to generalize this result to higher-order tensors.

5.1.3 Grad, div and curl

It is convenient to present here general expressions for grad, div and curl in terms of physical components of vectors with respect to orthogonal curvilinear co-ordinates.

Firstly, for scalar fields Φ, grad Φ is a vector with covariant components $\partial \Phi / \partial x^i$. Hence by (5.14) the physical components of grad Φ are given by

$$(\text{grad } \Phi)_{(i)} = \frac{1}{h_i} \frac{\partial \Phi}{\partial x^i}. \qquad (5.18)$$

Secondly, for vector fields $\mathbf{u}$, div $\mathbf{u}$ is equal to the covariant derivative $u^i{}_{;i}$, which by (4.33) is equal to

$$\frac{1}{\sqrt{g}} \frac{\partial}{\partial x^i}(\sqrt{g}\, u^i),$$

and this, using (5.13) and (5.11), can be expressed as

$$\text{div } \mathbf{u} = \sum_{i=1}^{3} \frac{1}{h_1 h_2 h_3} \frac{\partial}{\partial x^i}\left(\frac{h_1 h_2 h_3}{h_i} u_{(i)}\right). \qquad (5.19)$$

It follows from (5.18) and (5.19) that the Laplacian operator in orthogonal curvilinear co-ordinates is given by

$$\nabla^2 \Phi = \sum_{i=1}^{3} \frac{1}{h_1 h_2 h_3} \frac{\partial}{\partial x_i}\left(\frac{h_1 h_2 h_3}{h_i^2} \frac{\partial \Phi}{\partial x_i}\right)$$

$$= \frac{1}{h_1 h_2 h_3}\left[\frac{\partial}{\partial x^1}\left(\frac{h_2 h_3}{h_1} \frac{\partial \Phi}{\partial x^1}\right) + \frac{\partial}{\partial x^2}\left(\frac{h_3 h_1}{h_2} \frac{\partial \Phi}{\partial x^2}\right) + \frac{\partial}{\partial x^3}\left(\frac{h_1 h_2}{h_3} \frac{\partial \Phi}{\partial x^3}\right)\right]. \qquad (5.20)$$

Thirdly, the contravariant components of curl $\mathbf{u}$ are given by (4.38). Hence the physical components of curl $\mathbf{u}$ are

$$(\text{curl } \mathbf{u})_{(1)} = h_1\left[\frac{1}{h_1 h_2 h_3}\left(\frac{\partial}{\partial x^2}(h_3 u_{(3)}) - \frac{\partial}{\partial x^3}(h_2 u_{(2)})\right)\right]$$

$$= \frac{1}{h_2 h_3}\left(\frac{\partial}{\partial x^2}(h_3 u_{(3)}) - \frac{\partial}{\partial x^3}(h_2 u_{(2)})\right), \qquad (5.21)$$

together with two similar components, obtained by cyclic permutation of the indices.

5.2 EXAMPLES OF ORTHOGONAL CURVILINEAR CO-ORDINATES

5.2.1 Circular cylindrical co-ordinates

Taking, as usual, $x^1 = \rho, x^2 = \phi, x^3 = z$, we obtain covariant base vectors as in (1.53). The unit vectors $\mathbf{i}_\rho, \mathbf{i}_\phi$ shown in Fig. 1.21 are identical with $\mathbf{g}_{(1)}$ and $\mathbf{g}_{(2)}$ as defined in (5.12), and the scale factors are $h_1 = 1, h_2 = \rho, h_3 = 1$. The reader should verify that use of equations (5.6)–(5.10) produces the same set of Christoffel symbols given in (4.13).

Equations (5.18) show that the physical components of grad Φ are given by

$$\text{grad } \Phi = \left[\frac{\partial \Phi}{\partial \rho} \quad \frac{1}{\rho} \frac{\partial \Phi}{\partial \phi} \quad \frac{\partial \Phi}{\partial z} \right]. \tag{5.22}$$

Writing the physical components of the vector field $\mathbf{u}$ as u_ρ, u_ϕ and u_z, we obtain, from (5.19),

$$\text{div } \mathbf{u} = \frac{1}{\rho} \frac{\partial}{\partial \rho} (\rho u_\rho) + \frac{1}{\rho} \frac{\partial u_\phi}{\partial \phi} + \frac{\partial u_z}{\partial z} \tag{5.23}$$

and, from (5.21),

$$\text{curl } \mathbf{u} = \left[\frac{1}{\rho} \left(\frac{\partial u_z}{\partial \phi} - \frac{\partial}{\partial z}(\rho u_\phi) \right) \quad \frac{\partial u_\rho}{\partial z} - \frac{\partial u_z}{\partial \rho} \quad \frac{1}{\rho} \left(\frac{\partial}{\partial \rho}(\rho u_\phi) - \frac{\partial u_\rho}{\partial \phi} \right) \right]$$

$$= \left[\frac{1}{\rho} \frac{\partial u_z}{\partial \phi} - \frac{\partial u_\phi}{\partial z} \quad \frac{\partial u_\rho}{\partial z} - \frac{\partial u_z}{\partial \rho} \quad \frac{1}{\rho} \frac{\partial}{\partial \rho}(\rho u_\phi) - \frac{1}{\rho} \frac{\partial u_\rho}{\partial \phi} \right]. \tag{5.24}$$

5.2.2 Spherical polar co-ordinates

With $x^1 = r, x^2 = \theta, x^3 = \phi$, covariant base vectors are given by (1.55), and the unit vectors $\mathbf{i}_r, \mathbf{i}_\theta$ and $\mathbf{i}_\phi$ in (1.56) are equivalent to $\mathbf{g}_{(1)}, \mathbf{g}_{(2)}$ and $\mathbf{g}_{(3)}$. Hence the scale factors are $h_1 = 1, h_2 = r, h_3 = r \sin \theta$, and it should be verified that use of (5.6)–(5.10) gives the Christoffel symbols shown in (4.15) and (4.16).

From (5.18) the physical components of grad Φ are

$$\text{grad } \Phi = \left[\frac{\partial \Phi}{\partial r} \quad \frac{1}{r} \frac{\partial \Phi}{\partial \theta} \quad \frac{1}{r \sin \theta} \frac{\partial \Phi}{\partial \phi} \right], \tag{5.25}$$

and writing the physical components of a vector $\mathbf{u}$ as u_r, u_θ, u_ϕ, we obtain from (5.19)

$$\text{div } \mathbf{u} = \frac{1}{r^2 \sin \theta} \left(\frac{\partial}{\partial r}(r^2 \sin \theta \, u_r) + \frac{\partial}{\partial \theta}(r \sin \theta \, u_\theta) + \frac{\partial}{\partial \phi}(r u_\phi) \right)$$

$$= \frac{1}{r^2} \frac{\partial}{\partial r}(r^2 u_r) + \frac{1}{r \sin \theta} \frac{\partial}{\partial \theta}(\sin \theta \, u_\theta) + \frac{1}{r \sin \theta} \frac{\partial u_\phi}{\partial \phi}. \tag{5.26}$$

The physical components of curl $\mathbf{u}$, from (5.21), are

$$\text{curl } \mathbf{u} = \left[\frac{1}{r^2 \sin \theta} \left(\frac{\partial}{\partial \theta}(r \sin \theta \, u_\phi) - \frac{\partial}{\partial \phi}(r u_\theta) \right) \quad \frac{1}{r \sin \theta} \left(\frac{\partial u_r}{\partial \phi} - \frac{\partial}{\partial r}(r \sin \theta \, u_\phi) \right) \right.$$

$$\left. \frac{1}{r} \left(\frac{\partial}{\partial r}(r u_\theta) - \frac{\partial u_r}{\partial \theta} \right) \right]$$

$$= \left[\frac{1}{r \sin \theta}\left(\frac{\partial}{\partial \theta}(u_\phi \sin \theta) - \frac{\partial u_\theta}{\partial \phi}\right) \quad \frac{1}{r \sin \theta}\frac{\partial u_r}{\partial \phi} - \frac{1}{r}\frac{\partial}{\partial r}(r u_\phi) \quad \frac{1}{r}\frac{\partial}{\partial r}(r u_\theta) - \frac{1}{r}\frac{\partial u_r}{\partial \theta}\right]. \quad (5.27)$$

We apply our results next to a number of standard orthogonal curvilinear co-ordinate systems. The examples in sections 5.2.3–5.2.5 are essentially transformations in the plane, involving two-dimensional transformation equations between rectangular cartesian co-ordinates (y_1, y_2) and two new co-ordinates in the plane. The third cartesian co-ordinate y_3 is unaffected by the transformation, the final transformation equation being $y_3 = z$ in each case (as in (1.52) for cylindrical polars).

5.2.3 Parabolic cylindrical co-ordinates

The new co-ordinates in the y_1–y_2 plane are ξ, η, where

and
$$\left.\begin{array}{l} y_1 = \xi\eta \\ y_2 = \tfrac{1}{2}(\eta^2 - \xi^2) \end{array}\right\}. \quad (5.28)$$

We could add the equation

$$y_3 = z$$

for completeness. There are co-ordinate curves in the Oy_1y_2 plane given by $\eta =$ constant which are parabolas given by

$$y_2 = \frac{1}{2}\left(\eta^2 - \frac{y_1^2}{\eta^2}\right),$$

and co-ordinates curves $\xi =$ constant which are also parabolas given by

$$y_2 = \frac{1}{2}\left(\frac{y_1^2}{\xi^2} - \xi^2\right).$$

We illustrate some of these in Fig. 5.1 and indicate the shape of the co-ordinate surfaces in Fig. 5.2.

If we take $x^1 = \xi, x^2 = \eta, x^3 = z$, the covariant base vectors are, in cartesian components,

and
$$\left.\begin{array}{l} \mathbf{g}_1 = [\eta \quad -\xi \quad 0], \\ \mathbf{g}_2 = [\xi \quad \eta \quad 0], \\ \mathbf{g}_3 = [0 \quad 0 \quad 1]. \end{array}\right\} \quad (5.29)$$

It is clear that (5.1) is satisfied. Moreover,

$$h_1 = h_2 = \sqrt{\xi^2 + \eta^2}, \quad h_3 = 1. \quad (5.30)$$

We see that, as for cylindrical polars, there is a singularity at the origin, where $\mathbf{g}_1 = \mathbf{g}_2 = \mathbf{0}$, and $h_1, h_2 = 0$. A one-to-one correspondence between points (y_1, y_2) in the Oy_1y_2 plane and the co-ordinates (ξ, η) may be obtained by making the value of η on a co-ordinate curve $\eta =$ constant jump discontinuously from positive to negative as it crosses the positive y_2 axis in the direction of y_1 decreasing. The range of values taken by the co-ordinates is then

$$-\infty < \eta < \infty, \quad 0 \leqslant \xi < \infty, \quad -\infty < z < \infty.$$

From (5.10) the only non-vanishing Christoffel symbols of the second kind are as

Sec. 5.2]   **Examples of orthogonal curvilinear co-ordinates**   133

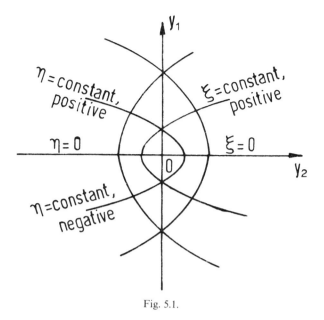

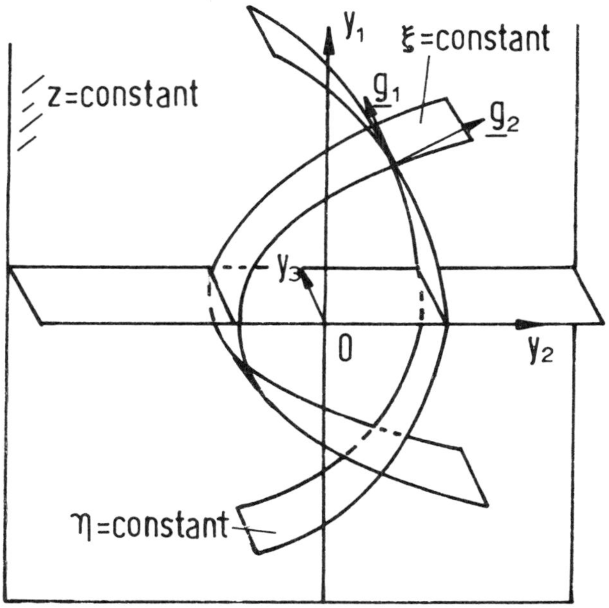

Fig. 5.1.

Fig. 5.2.

follows:

$$\Gamma^1_{11} = \frac{1}{h_1}\frac{\partial h_1}{\partial \xi} = \frac{1}{\sqrt{\xi^2+\eta^2}}\frac{\xi}{\sqrt{\xi^2+\eta^2}} = \frac{\xi}{\xi^2+\eta^2},$$

and, similarly,

$$\Gamma^2_{22} = \Gamma^1_{12} = \Gamma^1_{21} = \frac{\eta}{\xi^2+\eta^2}, \quad \Gamma^2_{21} = \Gamma^2_{12} = \frac{\xi}{\xi^2+\eta^2},$$

$$\Gamma^1_{22} = -\frac{\xi}{\xi^2+\eta^2}, \quad \Gamma^2_{11} = -\frac{\eta}{\xi^2+\eta^2}.$$

(5.31)

We can see from Fig. 5.2 that the covariant base vectors $\{\mathbf{g}_1, \mathbf{g}_2, \mathbf{g}_3\}$ form a right-handed set and can verify this by showing that the scalar triple product $(\mathbf{g}_1 \times \mathbf{g}_2)\cdot\mathbf{g}_3$ is everywhere positive.

5.2.4 Elliptic cylindrical co-ordinates

The transformation equations here are

$$\left.\begin{array}{l} y_1 = a\cosh u \cos v, \\ y_2 = a\sinh u \sin v \end{array}\right\}$$

(5.32)

(together with $y_3 = z$), where a is a positive constant. The new co-ordinates u, v generate co-ordinate curves which are ellipses ($u = $ constant)

$$\frac{y_1^2}{(a\cosh u)^2} + \frac{y_2^2}{(a\sinh u)^2} = 1$$

and hyperbolas ($v = $ constant)

$$\frac{y_1^2}{(a\cos v)^2} - \frac{y_2^2}{(a\sin v)^2} = 1.$$

These are illustrated in Fig. 5.3 and the co-ordinate surfaces (elliptic cylinders, hyperbolic cylinders, and planes $z = $ constant) in Fig. 5.4. By differentiating (5.32) with $x^1 = u, x^2 = v, x^3 = z$, we obtain covariant base vectors

$$\left.\begin{array}{l} \mathbf{g}_1 = [a\sinh u \cos v \quad a\cosh u \sin v \quad 0], \\ \mathbf{g}_2 = [-a\cosh u \sin v \quad a\sinh u \cos v \quad 0], \\ \mathbf{g}_3 = [0 \quad\quad\quad\quad\quad 0 \quad\quad\quad\quad 1], \end{array}\right\}$$

(5.33)

satisfying (5.1). Moreover

$$h_1^2 = h_2^2 = \mathbf{g}_1\cdot\mathbf{g}_1 = \mathbf{g}_2\cdot\mathbf{g}_2$$
$$= a^2\sinh^2 u \cos^2 v + a^2\cosh^2 u \sin^2 v$$
$$= a^2\sinh^2 u(1-\sin^2 v) + a^2(1+\sinh^2 u)\sin^2 v = a^2(\sinh^2 u + \sin^2 v).$$

Hence

$$h_1 = h_2 = a\sqrt{\sinh^2 u + \sin^2 v} \quad \text{and} \quad h_3 = 1.$$

(5.34)

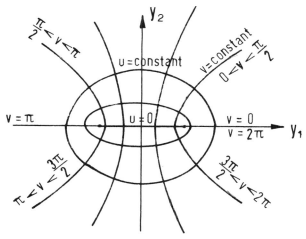

Fig. 5.3.

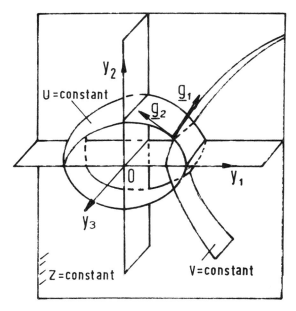

Fig. 5.4.

Now there are singularities where the scale factors vanish, which happens where $u = 0$ and $v = 0$ or $\pm \pi$, giving the points $(\pm a, 0)$ in the $Oy_1 y_2$ plane. We can ensure a one-to-one correspondence between the planar co-ordinates (y_1, y_2) and (u, v), except for points on the y_1 axis between the singular points, by allowing discontinuities in the values of v on the co-ordinate curves (hyperbolas) as they cross the y_1 axis, as indicated in Fig. 5.3. The range of values of u, v is then

$$0 \leq u < \infty, \qquad 0 \leq v < 2\pi, \qquad -\infty < z < \infty.$$

The non-vanishing components of the Christoffel symbols of the second kind are given by

$$\Gamma^1_{11} = \Gamma^2_{12} = \Gamma^2_{21} = -\Gamma^1_{22} = \frac{\sinh u \cosh u}{(\sinh^2 u + \sin^2 v)}$$

and

$$\Gamma^2_{22} = \Gamma^1_{12} = \Gamma^1_{21} = -\Gamma^2_{11} = \frac{\sin v \cos v}{(\sinh^2 u + \sin^2 v)}.$$

(5.35)

5.2.5 Bipolar co-ordinates

The transformation is to co-ordinates (ξ, η) in the $Oy_1 y_2$ plane, where

$$y_1 = \frac{a \sinh \eta}{\cosh \eta - \cos \xi},$$

$$y_2 = \frac{a \sin \xi}{\cosh \eta - \cos \xi}$$

(5.36)

and a is a positive constant. Noting the identity

$$\sinh^2 \eta + \sin^2 \xi \equiv \cosh^2 \eta - \cos^2 \xi,$$

we obtain

$$y_1^2 + y_2^2 = \frac{a^2(\sinh^2 \eta + \sin^2 \xi)}{(\cosh \eta - \cos \xi)^2} = \frac{a^2(\cosh^2 \eta - \cos^2 \xi)}{(\cosh \eta - \cos \xi)^2}$$

$$= \frac{a^2(\cosh \eta + \cosh \xi)}{\cosh \eta - \cos \xi} = a^2 \left(1 + \frac{2 \cos \xi}{\cosh \eta - \cos \xi}\right)$$

$$= a^2 \left(-1 + \frac{2 \cosh \eta}{\cosh \eta - \cos \xi}\right)$$

$$= a^2 + 2ay_2 \cot \xi = -a^2 + 2ay_1 \coth \eta.$$

Hence curves ξ = constant are the family of circles

$$y_1^2 + y_2^2 - 2ay_2 \cot \xi = a^2,$$

i.e.

$$y_1 + (y_2 - a \cot \xi)^2 = a^2 \operatorname{cosec}^2 \xi,$$

and curves η = constant are also circles

$$y_1^2 + y_2^2 - 2ay_1 \coth \eta + a^2 = 0,$$

i.e.

$$(y_1 - a \coth \eta)^2 + y_2^2 = a^2 \operatorname{cosech}^2 \eta.$$

These co-ordinate curves are illustrated in Fig. 5.5. Every circle in the family ξ = constant passes through the fixed points $(\pm a, 0)$, labelled A and B. The co-ordinate surfaces consist of two families of circular cylinders and planes z = constant.

The covariant base vectors are, by differentiating (5.36) and taking

$$x^1 = \xi, \quad x^2 = \eta, \quad x^3 = z,$$

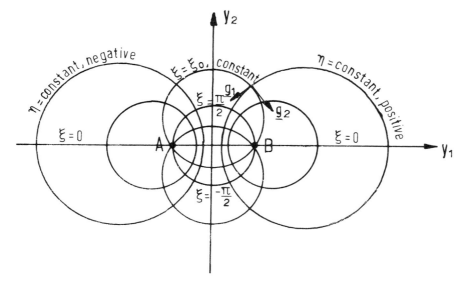

Fig. 5.5.

$$\mathbf{g}_1 = \begin{bmatrix} \dfrac{-a\sinh\eta\sin\xi}{(\cos\eta - \cos\xi)^2} & \dfrac{a(\cosh\eta\cos\xi - 1)}{(\cosh\eta - \cos\xi)^2} & 0 \end{bmatrix},$$

$$\mathbf{g}_2 = \begin{bmatrix} \dfrac{a(1 - \cosh\eta\cos\xi)}{(\cosh\eta - \cos\xi)^2} & \dfrac{-a\sinh\eta\sin\xi}{(\cosh\eta - \cos\xi)^2} & 0 \end{bmatrix}, \quad (5.37)$$

$$\mathbf{g}_3 = \begin{bmatrix} 0 & 0 & 1 \end{bmatrix},$$

and (5.1) is satisfied;

$$h_1^2 = h_2^2 = \frac{a^2 \sinh^2\eta \sin^2\xi}{(\cosh\eta - \cos\xi)^4} + \frac{a^2(\cosh\eta\cos\xi - 1)^2}{(\cosh\eta - \cos\xi)^4}$$

$$= \frac{a^2(\cosh\eta - \cos\xi)^2}{(\cosh\eta - \cos\xi)^4} = \frac{a^2}{(\cosh\eta - \cos\xi)^2}.$$

Hence

$$h_1 = h_2 = a(\cosh\eta - \cos\xi)^{-1}, \qquad h_3 = 1. \quad (5.38)$$

There are clearly singularities at A and B, where $\eta \to \pm\infty$ and h_1 and h_2 become zero. Each circle $\xi = $ constant passing through A and B is divided by the y_1 axis into two segments. If $\xi = \xi_0$ on the segment above the y_1 axis we may either take $\xi = \xi_0 - \pi$ on the segment below the y_1 axis, as indicated in Fig. 4.5, thus giving continuous variation in the ξ co-ordinate across the y_1 axis to the right of B and to the left of A, but involving a jump from $\xi = \pi$ to $\xi = -\pi$ as we cross the y_1 axis between A and B, or we may take $\xi = \xi_0 + \pi$ on the segment below the y_1 axis, giving continuity in ξ between A and B but a jump from $\xi = 0$ to $\xi = 2\pi$ as we cross the y_1 axis to the right of B or to the left of A. The range of values of ξ is either $-\pi < \xi \leq \pi$ or $0 \leq \xi < 2\pi$, while for η we have $-\infty < \eta < \infty$.

For the circular arcs in the upper half-plane, ξ may be interpreted in either case using simple geometry as the angle in the segment (i.e. the angle subtended by AB at any point on the circular arc). In the lower half-plane, ξ is also the angle in the segment, except that with the first choice of convention the angle must be taken with a negative sign and with the second choice the angle chosen must be greater than π.

The non-vanishing components of Γ^k_{ij} are given by

$$\left.\begin{aligned}\Gamma^1_{11} = \Gamma^2_{12} = \Gamma^2_{21} = -\Gamma^1_{22} &= -\frac{\sin\xi}{\cosh\eta - \cos\xi}, \\ \Gamma^2_{22} = \Gamma^1_{12} = \Gamma^1_{21} = -\Gamma^2_{11} &= -\frac{\sinh\eta}{\cosh\eta - \cos\xi}.\end{aligned}\right\} \quad (5.39)$$

In the following examples in sections 5.2.6–5.2.9 the patterns of co-ordinate curves shown in Figs. 5.1, 5.3 and 5.5 are rotated about one of the cartesian co-ordinate axes to generate new co-ordinate surfaces of revolution. It is clear that in each case the resulting curvilinear system of co-ordinates is orthogonal.

5.2.6 Parabolic co-ordinates

Here the pattern of curves in Fig. 5.1 is rotated about the y_2 axis to generate paraboloids of revolution ξ = constant and η = constant. The y_2 axis is relabelled the y_3 axis, and a new azimuthal angle ϕ is introduced as the third co-ordinate, specifying the angle between the plane $Oy_1 y_3$ and the plane containing a specified point in space and the y_3 axis. We then have the transformation equations

$$\left.\begin{aligned}y_1 &= \xi\eta\cos\phi, \\ y_2 &= \xi\eta\sin\phi, \\ y_3 &= \tfrac{1}{2}(\eta^2 - \xi^2),\end{aligned}\right\} \quad (5.40)$$

with $0 \leqslant \eta < \infty$, $0 \leqslant \xi < \infty$, $0 \leqslant \phi < 2\pi$, since the negative range of η shown in Fig. 5.1 is now not required. The co-ordinate surfaces are illustrated in Fig. 5.6. Any two paraboloids ξ = constant and η = constant intersect in a circle $y_1^2 + y_2^2 = \xi^2\eta^2$ of radius $\xi\eta$. This is a co-ordinate curve on which ϕ alone varies, and infinitesimal distance is measured by $\xi\eta\,d\phi$. Hence it is easy to see that, taking $x^1 = \xi$, $x^2 = \eta$, $x^3 = \phi$, the scale factors h_1 and h_2 are the same as for the parabolic cylindrical system, while $h_3 = \xi\eta$, i.e.

$$h_1 = h_2 = \sqrt{\xi^2 + \eta^2}, \qquad h_3 = \xi\eta. \quad (5.41)$$

Use of (5.10) now shows that (5.31) remains valid, except that we have additional non-zero components of Γ^k_{ij} as follows:

$$\left.\begin{aligned}\Gamma^3_{31} = \Gamma^3_{13} &= \frac{1}{\xi}, \quad \Gamma^3_{32} = \Gamma^3_{23} = \frac{1}{\eta}, \\ \Gamma^1_{33} &= \frac{-\xi\eta^2}{\xi^2 + \eta^2}, \quad \Gamma^2_{33} = \frac{-\xi^2\eta}{\xi^2 + \eta^2}.\end{aligned}\right\} \quad (5.42)$$

It may be observed from Fig. 5.6 that the covariant base vectors form a left-handed set, the scalar triple product $g_i \cdot (g_2 \times g_3)$ being negative (and equal to $-h_1 h_2 h_3$). This

Sec. 5.2]   **Examples of orthogonal curvilinear co-ordinates**   139

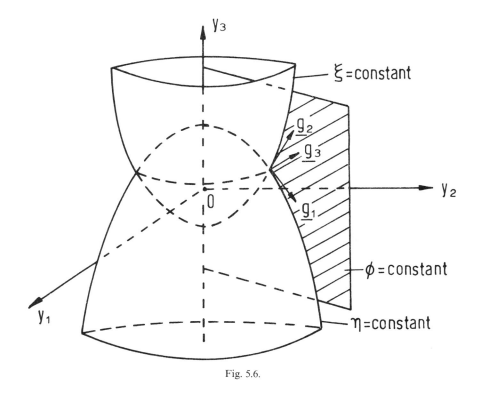

Fig. 5.6.

could have been avoided by a change in order of the curvilinear co-ordinates, e.g. taking $x^1 = \eta$, $x^2 = \xi$, $x^3 = \phi$.

The y_3 axis is singular for this set of co-ordinates since for points on this axis h_3 vanishes and the values of ϕ are unspecified (as for cylindrical polars).

5.2.7 Oblate spheroidal co-ordinates

The pattern of Fig. 5.3 is rotated about the y_2 axis, the ellipses $u = $ constant becoming co-ordinate surfaces $u = $ constant which are oblate spheroids (i.e. surfaces generated by revolving ellipses about their *minor* axis). The hyperbolas $v = $ constant become hyperboloids (of one sheet) of revolution. Relabelling the y_2 axis as y_3 and introducing the azimuthal angle ϕ we obtain the transformation equations

$$\left.\begin{aligned} y_1 &= a \cosh u \cos v \cos \phi, \\ y_2 &= a \cosh u \cos v \sin \phi, \\ y_3 &= a \sinh u \sin v. \end{aligned}\right\} \quad (5.43)$$

Separate branches of the hyperbolas in Fig. 5.3 become part of the same surface of revolution, and it is no longer necessary for v to take the range of values indicated there. A convenient range of values for the co-ordinates is given by

$$0 \leqslant u < \infty, \quad -\frac{\pi}{2} \leqslant v \leqslant \frac{\pi}{2}, \quad 0 \leqslant \phi < 2\pi.$$

Taking $x^1 = u, x^2 = v, x^3 = \phi$, we obtain the same scale factors h_1 and h_2 as in section 5.2.4 and the scale factor h_3 is given by $a \cosh u \cos v$, by the same argument as in section 5.2.6. Hence

$$h_1 = h_2 = a\sqrt{\sinh^2 u + \sin^2 v} \quad \text{and} \quad h_3 = a \cosh u \cos v. \tag{5.44}$$

The non-vanishing components of Γ^k_{ij} are given by (5.35), together with

$$\begin{aligned}\Gamma^3_{31} = \Gamma^3_{13} = \tanh u, & \qquad \Gamma^3_{32} = \Gamma^3_{23} = -\tanh v, \\ \Gamma^1_{33} = -\frac{\sinh u \cosh u \cos^2 v}{\sinh^2 u + \sin^2 v}, & \qquad \Gamma^2_{33} = \frac{\cosh^2 u \sin v \cos v}{\sinh^2 u + \sin^2 v}\end{aligned} \tag{5.45}$$

The covariant base vectors $\mathbf{g}_1, \mathbf{g}_2$ and $\mathbf{g}_3$ corresponding to (5.43) and our ordering of the co-ordinates again give a left-handed set, with $(\mathbf{g}_1 \times \mathbf{g}_2) \cdot \mathbf{g}_3 = -h_1 h_2 h_3$.

5.2.8 Prolate spheroidal co-ordinates

This case is similar to section 5.2.7, except that the pattern in Fig. 5.3 is rotated about the y_1 axis. Ellipses $u = $ constant become prolate spheroids (i.e. surfaces generated by revolving ellipses about their *major* axis), while hyperbolas $v = $ constant become hyperboloids (of two sheets) of revolution, each branch generating a different surface. We now relabel the y_1 axis as y_3, introduce the azimuthal angle ϕ, and the transformation equations are

$$\begin{aligned} y_1 &= a \sinh u \sin v \cos \phi, \\ y_2 &= a \sinh u \sin v \sin \phi, \\ y_3 &= a \cosh u \cos v. \end{aligned} \tag{5.46}$$

The co-ordinate surfaces are illustrated in Fig. 5.7. It is convenient to let v take constant values between 0 and $\pi/2$ on the hyperboloidal sheets lying above the $Oy_1 y_2$ plane and values between $\pi/2$ and π on those sheets below this plane, giving a range of co-ordinates

$$0 \leqslant u < \infty, \quad 0 \leqslant v \leqslant \pi, \quad 0 \leqslant \phi < 2\pi.$$

The scale factors are given, taking $x^1 = u, x^2 = v, x^3 = \phi$, by

$$h_1 = h_2 = a\sqrt{\sinh^2 u + \sin^2 v}, \quad h_3 = a \sinh u \sin v, \tag{5.47}$$

and the non-vanishing components of Γ^k_{ij} are given by (5.35), together with

$$\begin{aligned} \Gamma^3_{31} = \Gamma^3_{13} = \coth u, & \qquad \Gamma^3_{32} = \Gamma^3_{23} = \cot v, \\ \Gamma^1_{33} = -\frac{\sinh u \cosh u \sin^2 v}{\sinh^2 u + \sin^2 v}, & \qquad \Gamma^2_{33} = -\frac{\sinh^2 u \sin v \cos v}{\sinh^2 u + \sin^2 v}. \end{aligned} \tag{5.48}$$

Once again it may be seen from Fig. 5.7 and by explicit evaluation of $(\mathbf{g}_1 \times \mathbf{g}_2) \cdot \mathbf{g}_3$ that the covariant base vectors form a left-handed set.

5.2.9 Toroidal co-ordinates (1)

By rotating the pattern in Fig. 5.5 about Oy_2, relabelling y_2 as y_3 and introducing the azimuthal angle ϕ, we obtain the transformation equations

Examples of orthogonal curvilinear co-ordinates

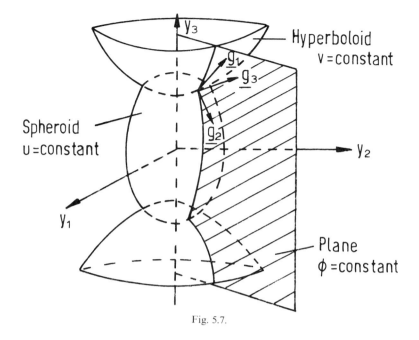

Fig. 5.7.

$$y_1 = \frac{a \sinh \eta \cos \phi}{\cosh \eta - \cos \xi},$$
$$y_2 = \frac{a \sinh \eta \sin \phi}{\cosh \eta - \cos \xi}, \quad (5.49)$$
$$y_3 = \frac{a \sin \xi}{\cosh \eta - \cos \xi}.$$

The co-ordinate curves $\xi = $ constant become spheres with centre on Oy_3, at a distance $\pm a \cot \xi$ from the origin and radius $a|\operatorname{cosec} \xi|$, while each co-ordinate curve $\eta = $ constant becomes a torus, or anchor ring, whose cross-section is a circle of radius $a \operatorname{cosech} \eta$ and whose axial circle has radius $a \coth \eta$. The third set of co-ordinate surfaces consists of planes $\phi = $ constant. There is a singular curve of points given by the circle $y_1^2 + y_2^2 = a^2$, $y_3 = 0$, on which η must take the value ∞ but ξ is unspecified. This circle lies on all the spheres $\xi = $ constant. As in Fig. 5.5, if $\xi = \xi_0$ on that part of a typical sphere which lies above the Oy_1y_2 plane, where $0 < \xi_0 < \pi$, we may put $\xi = \xi_0 - \pi$ (or $\xi_0 + \pi$) on that part which lies below the plane. This gives a range for the co-ordinates of

$$0 < \eta < \infty, \quad -\pi < \xi \leqslant \pi, \quad 0 \leqslant \phi < 2\pi$$

or

$$0 < \eta < \infty, \quad 0 \leqslant \xi < 2\pi, \quad 0 \leqslant \phi < 2\pi.$$

Putting $x^1 = \xi$, $x^2 = \eta$, $x^3 = \phi$, we obtain the scale factors

$$h_1 = h_2 = a(\cosh \eta - \cos \xi)^{-1}, \quad h_3 = a \sinh \eta (\cosh \eta - \cos \xi)^{-1}. \quad (5.50)$$

The non-vanishing components of Γ_{ij}^k are as given by (5.39), together with, after

142   Orthogonal curvilinear co-ordinates and physical components   [Ch. 5

some algebra,

$$\begin{aligned}
\Gamma^3_{31} &= \Gamma^3_{13} = -\sin\xi\,(\cosh\eta - \cos\xi)^{-1}, \\
\Gamma^3_{32} &= \Gamma^3_{23} = \operatorname{cosech}\eta\,(1 - \cosh\eta\cos\xi)(\cosh\eta - \cos\xi)^{-1}, \\
\Gamma^1_{33} &= \sinh^2\eta\sin\xi\,(\cosh\eta - \cos\xi)^{-1}, \\
\Gamma^2_{33} &= -\sinh\eta\,(1 - \cosh\eta\cos\xi)(\cosh\eta - \cos\xi)^{-1}.
\end{aligned} \qquad (5.51)$$

The set of covariant base vectors $\{\mathbf{g}_1, \mathbf{g}_2, \mathbf{g}_3\}$ is again left-handed.

5.2.10 Toroidal co-ordinates (2)

An alternative way of generating toroidal co-ordinates, although over only a bounded region of space, is to consider the rotation of circles of variable radius r in the Oy_1y_3 plane, with centre on the axis Oy_1 at a fixed distance R from O, about the axis Oy_3. Part of the torus produced is illustrated in Fig. 5.8, and the angular co-ordinates θ and ϕ determine the position of a point P on the toroidal surface. Clearly the equations relating these co-ordinates to the cartesian co-ordinates of P are

$$\left.\begin{aligned}
y_1 &= (R + r\cos\phi)\cos\theta, \\
y_2 &= (R + r\cos\phi)\sin\theta, \\
y_3 &= r\sin\phi.
\end{aligned}\right\} \qquad (5.52)$$

We have a set of co-ordinates r, θ, ϕ, with range

$$0 \leqslant r < R, \quad 0 \leqslant \theta < 2\pi, \quad 0 \leqslant \phi < 2\pi.$$

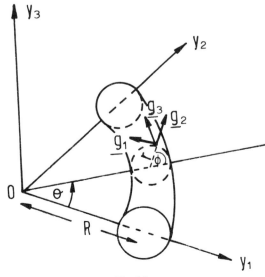

Fig. 5.8.

Sec. 5.2]   **Examples of orthogonal curvilinear co-ordinates**   143

Taking $x^1 = \phi, x^2 = r, x^3 = \theta$, we obtain by differentiation the base vectors

$$\begin{aligned}
\mathbf{g}_1 &= [-r\sin\phi\cos\theta & -r\sin\phi\sin\theta & r\cos\phi], \\
\mathbf{g}_2 &= [\cos\phi\cos\theta & \cos\phi\sin\theta & \sin\phi], \\
\mathbf{g}_3 &= [-(R + r\cos\phi)\sin\theta & (R + r\cos\phi)\cos\theta & 0],
\end{aligned} \quad (5.53)$$

which are mutually orthogonal, so that again we have a set of orthogonal curvilinear co-ordinates.

Moreover

$$h_1{}^2 = \mathbf{g}_1\cdot\mathbf{g}_1 = r^2\sin^2\phi\cos^2\theta + r^2\sin^2\phi\sin^2\theta + r^2\cos^2\phi = r^2,$$
$$h_2{}^2 = \mathbf{g}_2\cdot\mathbf{g}_2 = \cos^2\phi\cos^2\theta + \cos^2\phi\sin^2\theta + \sin^2\phi = 1$$

and $\quad h_3{}^2 = \mathbf{g}_3\cdot\mathbf{g}_3 = (R + r\cos\phi)^2\sin^2\theta + (R + r\cos\phi)^2\cos^2\theta = (R + r\cos\phi)^2.$

Hence the scale factors are

$$h_1 = r, \quad h_2 = 1, \quad h_3 = R + r\cos\phi. \quad (5.54)$$

From (5.6)–(5.10) we deduce that the only non-vanishing components of Γ_{ijk} and of Γ_{ij}^k are given by

$$\begin{aligned}
\Gamma_{121} &= \Gamma_{211} = -\Gamma_{112} = r, \\
\Gamma_{133} &= \Gamma_{313} = -\Gamma_{331} = -(R + r\cos\phi)r\sin\phi, \\
\Gamma_{233} &= \Gamma_{323} = -\Gamma_{332} = (R + r\cos\phi)\cos\phi
\end{aligned} \quad (5.55)$$

and

$$\begin{aligned}
\Gamma_{12}^1 &= \Gamma_{21}^1 = \frac{1}{r}, \quad \Gamma_{13}^3 = \Gamma_{31}^3 = \frac{-r\sin\phi}{R + r\cos\phi}, \quad \Gamma_{11}^2 = -r, \\
\Gamma_{23}^3 &= \Gamma_{32}^3 = \frac{\cos\phi}{R + r\cos\phi}, \quad \Gamma_{33}^1 = \frac{(R + r\cos\phi)\sin\phi}{r}, \\
\Gamma_{33}^2 &= -(R + r\cos\phi)\cos\phi.
\end{aligned} \quad (5.56)$$

5.2.11 Streamline co-ordinate systems

In the calculation of two-dimensional boundary layer flows of fluids over fixed convex surfaces a 'streamline co-ordinate system' is frequently used (see, for example, Jones and Watson 1963, Schlichting 1968). The situation is illustrated in Fig. 5.9. The surface has constant cross-section Γ in planes perpendicular to the axis Oz, and the co-ordinate s represents distance measured along Γ from some given point O. Normals to Γ are also shown, and the co-ordinate n represents distance measured along a normal from Γ. To complete a network of orthogonal curves we need to construct a family of curves which intersect the normals orthogonally and which include Γ. For a typical point P_1 in the plane the normal to Γ through P_1 intersects Γ at N, and the co-ordinates (s, n) of P_1 are just the distance from O to N along Γ and the distance P_1N, respectively.

An infinitesimal distance P_1P_2 in the plane is given by

$$(P_1P_2)^2 = (P_1P_3)^2 + (P_3P_2)^2,$$

since the network is orthogonal, where $P_3P_2 = \delta n$. If R is the radius of curvature at N,

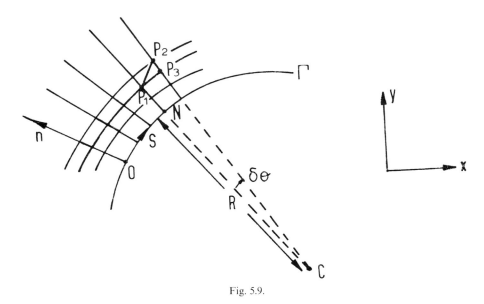

Fig. 5.9.

we have

$$P_1 P_3 = (R+n)\delta\theta = (R+n)\frac{\delta s}{R} = \left(1 + \frac{n}{R}\right)\delta s.$$

Hence distance is given by

$$(P_1 P_2)^2 = \left(1 + \frac{n}{R}\right)^2 (\delta s)^2 + (\delta n)^2.$$

If we take $x^1 = s$, $x^2 = n$, comparison with (5.2) shows that the scale factors here are given by

$$h_1 = 1 + \frac{n}{R}, \quad h_2 = 1, \tag{5.57}$$

where R, a function of s, is given by standard formulas. For example, if the cartesian equation of Γ is, in parametric form, $x = f(s)$ and $y = g(s)$, where f and g are given functions, then

$$R = [g'(s)f''(s) - f'(s)g''(s)]^{-1}. \tag{5.58}$$

Using (5.6)–(5.10) we may calculate the corresponding Christoffel symbols. The only non-vanishing components are given by

$$\left.\begin{array}{l} \Gamma_{111} = -\left(1 + \dfrac{n}{R}\right)\dfrac{n}{R^2}R', \quad \Gamma_{121} = \Gamma_{211} = -\Gamma_{112} = \left(1 + \dfrac{n}{R}\right)\dfrac{1}{R}, \\[2mm] \Gamma^1_{11} = \dfrac{-nR'}{R^2(1 + n/R)}, \quad \Gamma^1_{12} = \Gamma^1_{21} = \dfrac{1}{R(1 + n/R)}, \quad \Gamma^2_{11} = -\dfrac{1}{R}\left(1 + \dfrac{n}{R}\right). \end{array}\right\} \tag{5.59}$$

5.3 APPLICATION TO PARTICLE KINEMATICS

In newtonian mechanics it is often useful to be able to evaluate the components of the acceleration vector **a** of a particle with respect to non-cartesian co-ordinate systems. If a particle has cartesian co-ordinates (y_1, y_2, y_3) with respect to a rectangular cartesian system $Oy_1y_2y_3$, where each co-ordinate may be a function of time t, then the cartesian components of the acceleration vector with respect to the given reference system are given by $[\ddot{y}_1\ \ddot{y}_2\ \ddot{y}_3]$, where the double dot denotes the second derivative d^2/dt^2 with respect to t. We seek the generalization of this result to curvilinear co-ordinate systems.

If the curvilinear co-ordinates are x^i, the velocity vector **v** of a particle is given in terms of the position vector **r** by

$$\mathbf{v} = \frac{d\mathbf{r}}{dt} = \frac{\partial \mathbf{r}}{\partial x^i}\frac{dx^i}{dt} = \frac{\partial \mathbf{r}}{\partial x^i}\dot{x}^i,$$

using the chain rule for differentiation. In other words

$$\mathbf{v} = \dot{x}^i \mathbf{g}_i, \tag{5.60}$$

in terms of the covariant base vectors $\mathbf{g}_i$. Hence the contravariant components of **v** are

$$v^i = \dot{x}^i. \tag{5.61}$$

It follows from (5.13) that, if the curvilinear system is orthogonal, the physical components of **v** are given by

$$v_{(i)} = h_i \dot{x}^i \quad \text{(n.s.)}. \tag{5.62}$$

Differentiating (5.60) with respect to t gives

$$\mathbf{a} = \ddot{x}^i \mathbf{g}_i + \dot{x}^i \dot{\mathbf{g}}_i$$

$$= \ddot{x}^i \mathbf{g}_i + \dot{x}^i \frac{\partial \mathbf{g}_i}{\partial x^j}\dot{x}^j = \ddot{x}^i \mathbf{g}_i + \Gamma^k_{ij}\mathbf{g}_k \dot{x}^i \dot{x}^j$$

$$= (\ddot{x}^i + \Gamma^i_{jk}\dot{x}^j \dot{x}^k)\mathbf{g}_i.$$

Hence the contravariant components of acceleration are

$$a^i = \ddot{x}^i + \Gamma^i_{jk}\dot{x}^j \dot{x}^k, \tag{5.63}$$

and, if the system is orthogonal, the physical components of **a** are given by

$$a_{(i)} = h_i a^i \quad \text{(n.s.)}$$
$$= h_i \ddot{x}^i + h_i \Gamma^i_{jk}\dot{x}^j \dot{x}^k \quad \text{(n.s. on } i\text{)}, \tag{5.64}$$

where the summation convention applies only to the repeated indices j and k. From (5.55) the covariant components of **a** are given by

$$a_i = g_{il} a^l = g_{il}\ddot{x}^l + g_{il}\Gamma^l_{jk}\dot{x}^j \dot{x}^k$$
$$= g_{il}\ddot{x}^l + \Gamma_{jki}\dot{x}^j \dot{x}^k, \tag{5.65}$$

using (4.10).

In cylindrical polar co-ordinates, (5.64) shows that the physical components a_ρ, a_ϕ, a_z

of the acceleration vector are given by

$$\begin{aligned}
a_\rho &= \ddot{\rho} + \Gamma^1_{jk}\dot{x}^j\dot{x}^k = \ddot{\rho} + \Gamma^1_{22}\dot{x}^2\dot{x}^2 = \ddot{\rho} - \rho\dot{\phi}^2, \\
a_\phi &= \rho\ddot{\phi} + \rho\Gamma^2_{jk}\dot{x}^j\dot{x}^k = \rho\ddot{\phi} + \rho\Gamma^2_{12}\dot{x}^1\dot{x}^2 + \rho\Gamma^2_{21}\dot{x}^2\dot{x}^1 = \rho\ddot{\phi} + 2\dot{\rho}\dot{\phi} \\
a_z &= \ddot{z} + \Gamma^3_{jk}\dot{x}^j\dot{x}^k = \ddot{z}.
\end{aligned} \qquad (5.66)$$

For spherical polars we obtain physical components, using (4.15),

$$\begin{aligned}
a_r &= \ddot{r} + \Gamma^1_{jk}\dot{x}^j\dot{x}^k = \ddot{r} + \Gamma^1_{22}(\dot{x}^2)^2 + \Gamma^1_{33}(\dot{x}^3)^2 = \ddot{r} - r\dot{\theta}^2 - r\sin^2\theta\,\dot{\phi}^2, \\
a_\theta &= r\ddot{\theta} + r\Gamma^2_{jk}\dot{x}^j\dot{x}^k = r\ddot{\theta} + r[\Gamma^2_{12}\dot{x}^1\dot{x}^2 + \Gamma^2_{21}\dot{x}^2\dot{x}^1 + \Gamma^2_{33}(\dot{x}^3)^2] \\
&= r\ddot{\theta} + 2\dot{r}\dot{\theta} - r\sin\theta\cos\theta\,\dot{\phi}^2, \\
a_\phi &= r\sin\theta\,\ddot{\phi} + r\sin\theta\,\Gamma^3_{jk}\dot{x}^j\dot{x}^k \\
&= r\sin\theta\,\ddot{\phi} + r\sin\theta\,(\Gamma^3_{31}\dot{x}^3\dot{x}^1 + \Gamma^3_{13}\dot{x}^1\dot{x}^3 + \Gamma^3_{32}\dot{x}^3\dot{x}^2 + \Gamma^3_{23}\dot{x}^2\dot{x}^3) \\
&= r\sin\theta\,\ddot{\phi} + 2\sin\theta\,\dot{r}\dot{\phi} + 2r\cos\theta\,\dot{\theta}\dot{\phi}.
\end{aligned} \qquad (5.67)$$

As a final example, consider oblate spheroidal co-ordinates as in section 5.2.7. Using (5.35), (5.44) and (5.45), we obtain the physical components a_u, a_v and a_ϕ as follows:

$$\begin{aligned}
a_u &= a\sqrt{\sinh^2 u + \sin^2 v}\,\ddot{u} + \frac{a}{\sqrt{\sinh^2 u + \sin^2 v}}(\sinh u\cosh u\,\dot{u}^2 + 2\sin v\cos v\,\dot{u}\dot{v} \\
&\quad - \sinh u\cosh u\,\dot{v}^2 - \sinh u\cosh u\cos^2 v\,\dot{\phi}^2), \\
a_v &= a\sqrt{\sinh^2 u + \sin^2 v}\,\ddot{v} + \frac{a}{\sqrt{\sinh^2 u + \sin^2 v}}(-\sin v\cos v)\dot{u}^2 + 2\sinh u\cosh u\,\dot{u}\dot{v} \\
&\quad + \sin v\cos v\,\dot{v}^2 + \cosh^2 u\sin v\cos v\,\dot{\phi}^2), \\
a_\phi &= a\cosh u\cos v\,\ddot{\phi} + 2a\sinh u\cos v\,\dot{u}\dot{\phi} - 2a\cosh u\sin v\,\dot{v}\dot{\phi}.
\end{aligned} \qquad (5.68)$$

6

Applications in continuum mechanics

6.1 INTRODUCTION

In the following two chapters we apply tensor analysis to a number of expressions and equations frequently encountered in continuum mechanics, mainly in the areas of fluid mechanics and the theory of turbulence. Our approach is to take the form of these expressions with respect to an arbitrary rectangular cartesian co-ordinate system as known. (The reader should if necessary refer to standard textbooks for their derivation.) We consider a variety of examples in which we explicitly carry out the transformation of these expressions from a rectangular cartesian system to another co-ordinate system. The immediate objective is usually to produce valid generalized tensor expressions which reduce to the known form in the special case of rectangular cartesian co-ordinates. The generalized expressions may then be expressed in terms of any particular curvilinear co-ordinate system, or in terms of an arbitrary orthogonal curvilinear system. In principle, the transformations may be carried out in a straightforward way from any given co-ordinate system to any other, using the general tensor transformation laws.

The steps of the transformations are often shown in rather elaborate detail here, with the aim of clarifying points that may prove obscure to readers meeting these procedures for the first time.

6.2 APPLICATIONS

Example 1

Find expressions for the components of the strain-rate tensor in fluid mechanics (or, equivalently, the small strain tensor in solid mechanics) with respect to a general system of orthogonal curvilinear co-ordinates.

Method We start with the expression

$$\varepsilon_{ij} = \frac{1}{2}\left(\frac{\partial v_i}{\partial y_j} + \frac{\partial v_j}{\partial y_i}\right) \tag{6.1}$$

for the cartesian co-ordinates of the strain-rate tensor ε_{ij} with respect to a rectangular cartesian system $Oy_1y_2y_3$; **v** is the velocity vector. The following analysis is equally applicable to the small strain tensor

$$e_{ij} = \frac{1}{2}\left(\frac{\partial u_i}{\partial y_j} + \frac{\partial u_j}{\partial y_i}\right) \tag{6.2}$$

in solid mechanics, where **u** is the displacement vector.

The first step in transforming (6.1) is to note that the partial derivatives $\partial v_i/\partial y_j$ and $\partial v_j/\partial y_i$ must become the covariant derivatives $v_{i,j}$ and $v_{j,i}$. Secondly, the most straightforward attempt to write (6.1) in generalized form produces

$$\varepsilon_{ij} = \tfrac{1}{2}(v_{i,j} + v_{j,i}), \tag{6.3}$$

and this, interpreting ε_{ij} now as the covariant components of the strain-rate tensor, turns out to be a valid tensor identity, with the indices i and j each occurring once on either side of the identity and at the same levels in each term. It would not be admissible, for example, to write

$$\varepsilon^i_{.j} = \tfrac{1}{2}(v^i_{,j} + v^j_{,i})$$

for the mixed component $\varepsilon^i_{.j}$, since, for example, i does not appear at the same level in each term.

In fact the associated component $\varepsilon^i_{.j}$ may be determined from the valid expression (6.3) by the standard procedure of raising the first index, involving multiplication by g^{ij} (with summation over the repeated index, which must appear at different levels), i.e.

$$\varepsilon^i_{.j} = g^{ik}\varepsilon_{kj} = \tfrac{1}{2}g^{ik}(v_{k,j} + v_{j,k}),$$

which simplifies to

$$\varepsilon^i_{.j} = \tfrac{1}{2}(v^i_{,j} + g^{ik}v_{j,k}).$$

No further simplification is possible because of the covariant nature of the covariant derivative, which would render the expression

$$\varepsilon^i_{.j} = \tfrac{1}{2}(v^i_{,j} + v_j{}^{,i})$$

inadmissible, even though the indices appear at the same level in each term.

The associated contravariant, as well as the mixed, components of the strain-rate tensor may be obtained from (6.3) by raising the indices. We obtain

$$\varepsilon^{ij} = g^{ik}g^{jl}\varepsilon_{kl} = \tfrac{1}{2}g^{ik}g^{jl}(v_{k,l} + v_{l,k}) = \tfrac{1}{2}(g^{jl}v^i_{,l} + g^{ik}v^j_{,k}).$$

For a general curvilinear co-ordinate system (x^1, x^2, x^3), with Christoffel symbols of the second kind Γ^k_{ij}, it follows from (6.3), using (4.28), that

$$\varepsilon_{ij} = \frac{1}{2}\left(\frac{\partial v_i}{\partial x^j} - \Gamma^k_{ij}v_k + \frac{\partial v_j}{\partial x^i} - \Gamma^k_{ji}v_k\right) = \frac{1}{2}\left(\frac{\partial v_i}{\partial x^j} + \frac{\partial v_j}{\partial x^i} - 2\Gamma^k_{ij}v_k\right) \tag{6.4}$$

in view of (4.5).

If the curvilinear co-ordinate system is orthogonal, with scale factors h_1, h_2 and h_3, the physical components of the velocity vector and the strain-rate tensor are given by

$$\varepsilon_{(ij)} = \frac{1}{h_i h_j} \varepsilon_{ij} \quad \text{(n.s.)} \quad \text{and} \quad v_{(i)} = \frac{1}{h_i} v_i,$$

according to (5.14) and (5.16). Hence (6.4) gives

$$\varepsilon_{(ij)} = \frac{1}{2h_i h_j} \left(\frac{\partial}{\partial x^j}(h_i v_{(i)}) + \frac{\partial}{\partial x^i}(h_j v_{(j)}) - 2\sum_{k=1}^{3} \Gamma_{ij}^{k} h_k v_{(k)} \right) \quad \text{(n.s. on } i \text{ or } j\text{).} \quad (6.5)$$

Here we have explicitly written the summation sign for k to avoid confusion, the summation convention not being applicable. The components of Γ_{ij}^{k} are given by (5.10).

This solves the problem as posed. To take specific examples, we consider the following.

(a) Cylindrical polars With $x^1 = \rho$, $x^2 = \phi$, $x^3 = z$, and scale factors $h_1 = 1$, $h_2 = \rho$, $h_3 = 1$, the non-zero components of Γ_{ij}^{k} are given by (4.13). Denoting the physical components of strain rate by $\varepsilon_{\rho\rho}, \varepsilon_{\rho\phi}$, etc., we obtain firstly, taking $i = j = 1$ in (6.5),

$$\varepsilon_{(11)} = \varepsilon_{\rho\rho} = \frac{1}{2} \frac{1}{h_1^2} \left(\frac{\partial}{\partial x^1}(h_1 v_{(1)}) + \frac{\partial}{\partial x^1}(h_1 v_{(1)}) - 2\sum_{k=1}^{3} \Gamma_{11}^{k} h_k v_{(k)} \right)$$

$$= \frac{1}{2}\left(\frac{\partial v_\rho}{\partial \rho} + \frac{\partial v_\rho}{\partial \rho} - 2\Gamma_{11}^{1} h_1 v_\rho - 2\Gamma_{11}^{2} h_2 v_\phi - 2\Gamma_{11}^{3} h_3 v_z \right)$$

$$= \frac{1}{2}\left(2\frac{\partial v_\rho}{\partial \rho} - 0 - 0 - 0 \right) = \frac{\partial v_\rho}{\partial \rho}.$$

With $i=1, j=2$,

$$\varepsilon_{(12)} = \varepsilon_{\rho\phi} = \frac{1}{2h_1 h_2}\left(\frac{\partial}{\partial x^2}(h_1 v_{(1)}) + \frac{\partial}{\partial x^1}(h_2 v_{(2)}) - 2\sum_{k=1}^{3} \Gamma_{12}^{k} h_k v_{(k)} \right)$$

$$= \frac{1}{2\rho}\left(\frac{\partial v_\rho}{\partial \phi} + \frac{\partial}{\partial \rho}(\rho v_\phi) - 2\Gamma_{12}^{1} v_\rho - 2\Gamma_{12}^{2}\rho v_\phi - 2\Gamma_{12}^{3} v_z \right)$$

$$= \frac{1}{2\rho}\left[\frac{\partial v_\rho}{\partial \phi} + \left(\rho\frac{\partial v_\phi}{\partial \rho} + v_\phi \right) - 0 - 2\frac{1}{\rho}\rho v_\phi - 0 \right]$$

$$= \frac{1}{2}\left(\frac{1}{\rho}\frac{\partial v_\rho}{\partial \phi} + \frac{\partial v_\phi}{\partial \rho} - \frac{v_\phi}{\rho} \right).$$

Continuing in this way, we obtain the set of physical components

$$\varepsilon_{(ij)} = \begin{bmatrix} \dfrac{\partial v_\rho}{\partial \rho} & \dfrac{1}{2}\left(\dfrac{1}{\rho}\dfrac{\partial v_\rho}{\partial \phi} + \dfrac{\partial v_\phi}{\partial \rho} - \dfrac{v_\phi}{\rho} \right) & \dfrac{1}{2}\left(\dfrac{\partial v_\rho}{\partial z} + \dfrac{\partial v_z}{\partial \rho} \right) \\ \dfrac{1}{2}\left(\dfrac{1}{\rho}\dfrac{\partial v_\rho}{\partial \phi} + \dfrac{\partial v_\phi}{\partial \rho} - \dfrac{v_\phi}{\rho} \right) & \dfrac{1}{\rho}\left(\dfrac{\partial v_\phi}{\partial \phi} + v_\rho \right) & \dfrac{1}{2}\left(\dfrac{\partial v_\phi}{\partial z} + \dfrac{1}{\rho}\dfrac{\partial v_z}{\partial \phi} \right) \\ \dfrac{1}{2}\left(\dfrac{\partial v_\rho}{\partial z} + \dfrac{\partial v_z}{\partial \rho} \right) & \dfrac{1}{2}\left(\dfrac{\partial v_\phi}{\partial z} + \dfrac{1}{\rho}\dfrac{\partial v_z}{\partial \phi} \right) & \dfrac{\partial v_z}{\partial z} \end{bmatrix}. \quad (6.6)$$

150   Applications in continuum mechanics   [Ch. 6

(b) *Spherical polars* With $x^1 = r, x^2 = \theta, x^3 = \phi$, scale factors $h_1 = 1, h_2 = r$, $h_3 = r \sin \theta$, and Γ^k_{ij} given by (4.15), we obtain the following physical components, written $\varepsilon_{rr}, \varepsilon_{r\theta}$, etc:

$$\varepsilon_{(12)} = \varepsilon_{r\theta} = \frac{1}{2r}\left(\frac{\partial v_r}{\partial \theta} + \frac{\partial}{\partial r}(rv_\theta) - 2\sum_{k=1}^{3}\Gamma^k_{12}h_k v_{(k)}\right)$$

$$= \frac{1}{2r}\left(\frac{\partial v_r}{\partial \theta} + \frac{\partial}{\partial r}(rv_\theta) - 2\Gamma^2_{12} r v_\theta\right)$$

$$= \frac{1}{2r}\left(\frac{\partial v_r}{\partial \theta} + r\frac{\partial v_\theta}{\partial r} + v_\theta - 2v_\theta\right)$$

$$= \frac{1}{2}\left(\frac{1}{r}\frac{\partial v_r}{\partial \theta} + \frac{\partial v_\theta}{\partial r} - \frac{v_\theta}{r}\right),$$

$$\varepsilon_{(33)} = \varepsilon_{\phi\phi} = \frac{1}{2(r\sin\theta)^2}\left(2\frac{\partial}{\partial \phi}(r\sin\theta\, v_\phi) - 2\Gamma^1_{33} v_r - 2\Gamma^2_{33} r v_\theta\right)$$

$$= \frac{1}{(r\sin\theta)^2}\left(r\sin\theta \frac{\partial v_\phi}{\partial \phi} + r\sin^2\theta\, v_r + r\sin\theta\cos\theta\, v_\theta\right)$$

$$= \frac{1}{r\sin\theta}\frac{\partial v_\phi}{\partial \phi} + \frac{v_r}{r} + \frac{v_\theta}{r}\cot\theta, \text{ etc.}$$

The complete set of physical components is

$$\varepsilon_{(ij)} = \begin{bmatrix} \frac{\partial v_r}{\partial r} & \frac{1}{2}\left(\frac{\partial v_\theta}{\partial r} + \frac{1}{r}\frac{\partial v_r}{\partial \theta} - \frac{v_\theta}{r}\right) & \frac{1}{2}\left(\frac{1}{r\sin\theta}\frac{\partial v_r}{\partial \phi} + \frac{\partial v_\phi}{\partial r} - \frac{v_\phi}{r}\right) \\ \frac{1}{2}\left(\frac{\partial v_\theta}{\partial r} + \frac{1}{r}\frac{\partial v_r}{\partial \theta} - \frac{v_\theta}{r}\right) & \frac{1}{r}\left(\frac{\partial v_\theta}{\partial \theta} + v_r\right) & \frac{1}{2}\left(\frac{1}{r}\frac{\partial v_\phi}{\partial \theta} + \frac{1}{r\sin\theta}\frac{\partial v_\theta}{\partial \phi} - \frac{v_\phi \cot\theta}{r}\right) \\ \frac{1}{2}\left(\frac{1}{r\sin\theta}\frac{\partial v_r}{\partial \phi} + \frac{\partial v_\phi}{\partial r} - \frac{v_\phi}{r}\right) & \frac{1}{2}\left(\frac{1}{r}\frac{\partial v_\phi}{\partial \theta} + \frac{1}{r\sin\theta}\frac{\partial v_\theta}{\partial \phi} - \frac{v_\phi \cot\theta}{r}\right) & \frac{1}{r\sin\theta}\frac{\partial v_\phi}{\partial \phi} + \frac{v_r}{r} + \frac{v_\theta \cot\theta}{r} \end{bmatrix}$$
(6.7)

(c) *Toroidal co-ordinates* (2). With $x^1 = \phi, x^2 = r, x^3 = \theta$, and scale factors $h_1 = r$, $h_2 = 1, h_3 = R + r\cos\phi$, as in section 5.2.10, physical components of velocity v_ϕ, v_r, v_θ, and Γ^k_{ij} given by (5.56), we obtain from (6.5) the following physical components of strain rate:

$$\varepsilon_{\phi\phi} = \frac{1}{r^2}\left(\frac{\partial}{\partial \phi}(rv_\phi) + rv_r\right) = \frac{1}{r}\left(\frac{\partial v_\phi}{\partial \phi} + v_r\right),$$

$$\varepsilon_{rr} = \frac{1}{1}\left(\frac{\partial}{\partial r}v_r - 0\right) = \frac{\partial v_r}{\partial r},$$

$$\varepsilon_{\theta\theta} = \frac{1}{(R + r\cos\phi)^2}\left(\frac{\partial}{\partial \theta}[(R + r\cos\phi)v_\theta] - \Gamma^1_{33}h_1 v_\phi - \Gamma^2_{33}h_2 v_r\right)$$

$$= \frac{1}{(R + r\cos\phi)}\left(\frac{\partial v_\theta}{\partial \theta} - v_\phi \sin\phi + v_r \cos\phi\right), \quad (6.8)$$

$$\varepsilon_{\phi r} = \varepsilon_{r\phi} = \frac{1}{2r}\left(\frac{\partial}{\partial r}(rv_\phi) + \frac{\partial v_r}{\partial \phi} - 2\Gamma^1_{12}h_1 v_\phi\right) = \frac{1}{2}\left(\frac{\partial v_\phi}{\partial r} + \frac{1}{r}\frac{\partial v_r}{\partial \phi} - \frac{v_\phi}{r}\right),$$

$$\varepsilon_{r\theta} = \varepsilon_{\theta r} = \frac{1}{2(R+r\cos\phi)}\left(\frac{\partial v_r}{\partial \theta} + \frac{\partial}{\partial r}[(R+r\cos\phi)v_\theta] - 2\Gamma^3_{23}h_3 v_\theta\right)$$

$$= \frac{1}{2}\left(\frac{1}{R+r\cos\phi}\frac{\partial v_r}{\partial \theta} + \frac{\partial v_\theta}{\partial r} - \frac{\cos\phi}{R+r\cos\phi}v_\theta\right),$$

$$\varepsilon_{\theta\phi} = \varepsilon_{\phi\theta} = \frac{1}{2r(R+r\cos\phi)}\left(\frac{\partial}{\partial \theta}(rv_\phi) + \frac{\partial}{\partial \phi}[(R+r\cos\phi)v_\theta] - 2\Gamma^3_{13}h_3 v_\theta\right)$$

$$= \frac{1}{2}\left(\frac{1}{R+r\cos\phi}\frac{\partial v_\phi}{\partial \theta} + \frac{1}{r}\frac{\partial v_\theta}{\partial \phi} + \frac{\sin\phi}{R+r\cos\phi}v_\theta\right).$$

The identity (6.5) may be expressed in an alternative form by making explicit use of the components (5.10) of Γ^k_{ij}. The reader should verify the following general expressions for the physical components $\varepsilon_{(ij)}$ in an orthogonal curvilinear co-ordinate system:

$$\begin{aligned}
\varepsilon_{(11)} &= \frac{1}{h_1}\frac{\partial v_{(1)}}{\partial x^1} + \frac{v_{(2)}}{h_1 h_2}\frac{\partial h_1}{\partial x^2} + \frac{v_{(3)}}{h_1 h_3}\frac{\partial h_1}{\partial x^3},\\
\varepsilon_{(22)} &= \frac{1}{h_2}\frac{\partial v_{(2)}}{\partial x^2} + \frac{v_{(1)}}{h_1 h_2}\frac{\partial h_2}{\partial x^1} + \frac{v_{(3)}}{h_2 h_3}\frac{\partial h_2}{\partial x^3},\\
\varepsilon_{(33)} &= \frac{1}{h_3}\frac{\partial v_{(3)}}{\partial x^3} + \frac{v_{(1)}}{h_1 h_3}\frac{\partial h_3}{\partial x^1} + \frac{v_{(2)}}{h_2 h_3}\frac{\partial h_3}{\partial x^2},\\
\varepsilon_{(12)} &= \frac{1}{2}\left[\frac{h_1}{h_2}\frac{\partial}{\partial x^2}\left(\frac{v_{(1)}}{h_1}\right) + \frac{h_2}{h_1}\frac{\partial}{\partial x^1}\left(\frac{v_{(2)}}{h_2}\right)\right],\\
\varepsilon_{(23)} &= \frac{1}{2}\left[\frac{h_2}{h_3}\frac{\partial}{\partial x^3}\left(\frac{v_{(2)}}{h_2}\right) + \frac{h_3}{h_2}\frac{\partial}{\partial x^2}\left(\frac{v_{(3)}}{h_3}\right)\right],\\
\varepsilon_{(31)} &= \frac{1}{2}\left[\frac{h_3}{h_1}\frac{\partial}{\partial x^1}\left(\frac{v_{(3)}}{h_3}\right) + \frac{h_1}{h_3}\frac{\partial}{\partial x^3}\left(\frac{v_{(1)}}{h_1}\right)\right].
\end{aligned} \qquad (6.9)$$

(Note that once the expressions for $\varepsilon_{(11)}$ and $\varepsilon_{(12)}$ have been obtained the remaining ones can be immediately written down by cyclic permutation of the indices.)

Example 2

Find expressions for the components of Green strain (a finite strain tensor) in cylindrical polar co-ordinates.

Method

When the strains in a solid body (for example, made of a rubber-like material) are no longer small, it becomes necessary to use a finite strain tensor rather than (6.2).

A common choice is the (symmetric) Green strain tensor, with cartesian components

$$E_{ij} = \frac{1}{2}\left(\frac{\partial u_i}{\partial y_j} + \frac{\partial u_j}{\partial y_i} + \frac{\partial u_k}{\partial y_i}\frac{\partial u_k}{\partial y_j}\right) \qquad (6.10)$$

in terms of the displacement vector **u**. Typical components are, with $y_1 = x$, $y_2 = y$, $y_3 = z$, $u_1 = u$, $u_2 = v$ and $u_3 = w$,

$$E_{11} = \frac{\partial u}{\partial x} + \frac{1}{2}\left[\left(\frac{\partial u}{\partial x}\right)^2 + \left(\frac{\partial v}{\partial x}\right)^2 + \left(\frac{\partial w}{\partial x}\right)^2\right]$$

and
$$E_{12} = \frac{1}{2}\left(\frac{\partial u}{\partial y} + \frac{\partial v}{\partial x}\right) + \frac{1}{2}\left(\frac{\partial u}{\partial x}\frac{\partial u}{\partial y} + \frac{\partial v}{\partial x}\frac{\partial v}{\partial y} + \frac{\partial w}{\partial x}\frac{\partial w}{\partial y}\right).$$

Following Example 1, it is straightforward to generalize (6.10) to arbitrary co-ordinate systems by expressing the covariant component of Green strain as

$$E_{ij} = \tfrac{1}{2}(u_{i,j} + u_{j,i} + u_{k,i}u^k{}_{,j}). \qquad (6.11)$$

In the last term the repeated index k must appear at different levels to produce a valid tensor expression. Equally valid would be

$$E_{ij} = \tfrac{1}{2}(u_{i,j} + u_{j,i} + u^k{}_{,i}u_{k,j}).$$

The associated components of E_{ij} may be obtained from (6.11). For example, the contravariant components are

$$E^{ij} = g^{ik}g^{jl}E_{kl} = \tfrac{1}{2}g^{ik}g^{jl}(u_{k,l} + u_{l,k} + u_{m,k}u^m{}_{,l})$$

(introducing a new dummy suffix m, since k and l are already in use), which reduces to

$$E^{ij} = \tfrac{1}{2}(g^{jl}u^i{}_{,l} + g^{ik}u^j{}_{,l} + g^{ik}g^{jl}u_{m,k}u^m{}_{,l}).$$

Note that (6.11) differs from (6.2) only in the extra term $\tfrac{1}{2}u_{k,i}u^k{}_{,j}$, which may be written as

$$\frac{1}{2}\left(\frac{\partial u_k}{\partial x^i} - \Gamma^l_{ki}u_l\right)\left(\frac{\partial u^k}{\partial x^j} + \Gamma^k_{mj}u^m\right)$$

$$= \frac{1}{2}\left(\frac{\partial u_k}{\partial x^i}\frac{\partial u^k}{\partial x^j} - \Gamma^l_{ki}u_l\frac{\partial u^k}{\partial x^j} + \Gamma^k_{mj}u^m\frac{\partial u_k}{\partial x^i} - \Gamma^l_{ki}\Gamma^k_{mj}u_l u^m\right),$$

using (4.26) and (4.28). The last term involves a triple summation over the indices k, l and m, and care must be taken when evaluating this term in particular co-ordinate systems to make sure that no terms are inadvertently omitted. An alternative method is to return to the expression $\tfrac{1}{2}u_{k,i}u^k{}_{,j}$ and write out the components of $u_{k,i}$ and $u^k{}_{,j}$ as separate matrix arrays. Then the components of $u_{k,i}u^k{}_{,j}$ can be obtained by matrix multiplication according to the rule (2.2).

For cylindrical polar co-ordinates, for example, we have by (4.13), (4.26) and (4.28)

Sec. 6.2] **Applications** 153

the matrix arrays

$$\begin{bmatrix} u_{1,1} & u_{2,1} & u_{3,1} \\ u_{1,2} & u_{2,2} & u_{3,2} \\ u_{1,3} & u_{2,3} & u_{3,3} \end{bmatrix} = \begin{bmatrix} \dfrac{\partial u_1}{\partial \rho} & \dfrac{\partial u_2}{\partial \rho} - \dfrac{1}{\rho} u_2 & \dfrac{\partial u_3}{\partial \rho} \\ \dfrac{\partial u_1}{\partial \phi} - \dfrac{1}{\rho} u_2 & \dfrac{\partial u_2}{\partial \phi} + u_1 & \dfrac{\partial u_3}{\partial \phi} \\ \dfrac{\partial u_1}{\partial z} & \dfrac{\partial u_2}{\partial z} & \dfrac{\partial u_3}{\partial z} \end{bmatrix}$$

$$= \begin{bmatrix} \dfrac{\partial u_\rho}{\partial \rho} & \rho \dfrac{\partial u_\phi}{\partial \rho} & \dfrac{\partial u_z}{\partial \rho} \\ \dfrac{\partial u_\rho}{\partial \phi} - u_\phi & \rho \dfrac{\partial u_\phi}{\partial \phi} + \rho u_\rho & \dfrac{\partial u_z}{\partial \phi} \\ \dfrac{\partial u_\rho}{\partial z} & \rho \dfrac{\partial u_\phi}{\partial z} & \dfrac{\partial u_z}{\partial z} \end{bmatrix}, \qquad (6.12)$$

and

$$\begin{bmatrix} u^1{}_{,1} & u^1{}_{,2} & u^1{}_{,3} \\ u^2{}_{,1} & u^2{}_{,2} & u^2{}_{,3} \\ u^3{}_{,1} & u^3{}_{,2} & u^3{}_{,3} \end{bmatrix} = \begin{bmatrix} \dfrac{\partial u^1}{\partial \rho} & \dfrac{\partial u^1}{\partial \phi} - \rho u^2 & \dfrac{\partial u^1}{\partial z} \\ \dfrac{\partial u^2}{\partial \rho} + \dfrac{1}{\rho} u^2 & \dfrac{\partial u^2}{\partial \phi} + \dfrac{1}{\rho} u^1 & \dfrac{\partial u^2}{\partial z} \\ \dfrac{\partial u^3}{\partial \rho} & \dfrac{\partial u^3}{\partial \phi} & \dfrac{\partial u^3}{\partial z} \end{bmatrix}$$

$$= \begin{bmatrix} \dfrac{\partial u_\rho}{\partial \rho} & \dfrac{\partial u_\rho}{\partial \phi} - u_\phi & \dfrac{\partial u_\rho}{\partial z} \\ \dfrac{1}{\rho} \dfrac{\partial u_\phi}{\partial \rho} & \dfrac{1}{\rho} \dfrac{\partial u_\phi}{\partial \phi} + \dfrac{u_\rho}{\rho} & \dfrac{1}{\rho} \dfrac{\partial u_\phi}{\partial z} \\ \dfrac{\partial u_z}{\partial \rho} & \dfrac{\partial u_z}{\partial \phi} & \dfrac{\partial u_z}{\partial z} \end{bmatrix}, \qquad (6.13)$$

making use of (5.13) and (5.14). (Note that it is possible to obtain (6.13) directly from (6.12) by using the identity $u^k{}_{,j} = g^{ik} u_{i,j}$ and premultiplying the transpose of the matrix (6.12) by the matrix corresponding to g^{ij}.)

The components $u_{k,i} u^k{}_{,j}$ are obtained by direct matrix multiplication of (6.12) and (6.13). If the physical components are required, the final step is to recognize that $\tfrac{1}{2} u_{k,i} u^k{}_{,j}$ is a covariant second-order tensor, and hence, according to (5.16) its physical components are obtained by multiplying each $i-j$ component by the corresponding factor $h_i h_j$. The resulting array is added to the matrix (6.6) to give the physical components of Green strain. These components are

$$E_{\rho\rho} = \frac{\partial u_\rho}{\partial \rho} + \frac{1}{2}\left[\left(\frac{\partial u_\rho}{\partial \rho}\right)^2 + \left(\frac{\partial u_\phi}{\partial \rho}\right)^2 + \left(\frac{\partial u_z}{\partial \rho}\right)^2\right],$$

$$E_{\phi\phi} = \frac{1}{\rho}\left(\frac{\partial u_\phi}{\partial \phi} + u_\rho\right) + \frac{1}{2\rho^2}\left[\left(\frac{\partial u_\rho}{\partial \phi} - u_\phi\right)^2 + \left(\frac{\partial u_\phi}{\partial \phi} + u_\rho\right)^2 + \left(\frac{\partial u_z}{\partial \phi}\right)^2\right],$$

$$E_{zz} = \frac{\partial u_z}{\partial z} + \frac{1}{2}\left[\left(\frac{\partial u_\rho}{\partial z}\right)^2 + \left(\frac{\partial u_\phi}{\partial z}\right)^2 + \left(\frac{\partial u_z}{\partial z}\right)^2\right],$$

$$E_{\rho\phi} = \frac{1}{2}\left(\frac{1}{\rho}\frac{\partial u_\rho}{\partial \phi} + \frac{\partial u_\phi}{\partial \rho} - \frac{u_\phi}{\rho}\right) + \frac{1}{2\rho}\left[\frac{\partial u_\rho}{\partial \rho}\left(\frac{\partial u_\rho}{\partial \phi} - u_\phi\right)\right.$$
$$\left. + \frac{\partial u_\phi}{\partial \rho}\left(\frac{\partial u_\phi}{\partial \phi} + u_\rho\right) + \frac{\partial u_z}{\partial \rho}\frac{\partial u_z}{\partial \phi}\right]$$

$$E_{\phi z} = \frac{1}{2}\left(\frac{\partial u_\phi}{\partial z} + \frac{1}{\rho}\frac{\partial u_z}{\partial \phi}\right) + \frac{1}{2\rho}\left[\frac{\partial u_\rho}{\partial z}\left(\frac{\partial u_\rho}{\partial \phi} - u_\phi\right) + \frac{\partial u_\phi}{\partial z}\left(\frac{\partial u_\phi}{\partial \phi} + u_\rho\right) + \frac{\partial u_z}{\partial \phi}\frac{\partial u_z}{\partial z}\right],$$

$$E_{\rho z} = \frac{1}{2}\left(\frac{\partial u_\rho}{\partial z} + \frac{\partial u_z}{\partial \rho}\right) + \frac{1}{2}\left(\frac{\partial u_\rho}{\partial \rho}\frac{\partial u_\rho}{\partial z} + \frac{\partial u_\phi}{\partial \rho}\frac{\partial u_\phi}{\partial z} + \frac{\partial u_z}{\partial \rho}\frac{\partial u_z}{\partial z}\right).$$

(6.14)

Example 3

Transform the constitutive equation for

(a) the classical newtonian fluid and
(b) the isotropic linear elastic solid, from a rectangular cartesian co-ordinate system to a generalized co-ordinate system.

Method

(a) Classical newtonian fluid A constitutive equation for a particular material relates the stress tensor (e.g. the Cauchy stress σ_{ij}) at a point in the material to the deformation or rate of deformation currently taking place at that point. For the classical newtonian fluid the equation is

$$\sigma_{ij} = -p\delta_{ij} + 2\mu\varepsilon_{ij} - \tfrac{2}{3}\mu \operatorname{div} \mathbf{v}\, \delta_{ij} \tag{6.15}$$

in rectangular cartesians, where p is the fluid pressure, ε_{ij} is the strain rate, $\mathbf{v}$ is the velocity and μ is the coefficient of viscosity. It is often assumed that the fluid is incompressible, in which case $\operatorname{div} \mathbf{v} = 0$, and the last term disappears.

With respect to an arbitrary set of generalized co-ordinates, (6.15) may immediately be regarded as the covariant form of the constitutive equation, except that δ_{ij} must be replaced by the metric tensor g_{ij} throughout, since g_{ij} is the unique covariant tensor which reduces to δ_{ij} when the system is rectangular cartesian. Thus we have

$$\sigma_{ij} = -pg_{ij} + 2\mu\varepsilon_{ij} - \tfrac{2}{3}\mu \operatorname{div} \mathbf{v}\, g_{ij}. \tag{6.16}$$

A convenient expression for $\operatorname{div} \mathbf{v}$ has been given in (4.34). We also have

$$\operatorname{div} \mathbf{v} = v^k{}_{,k} = \varepsilon^k{}_{,k} = g^{kl}\varepsilon_{kl}.$$

For an arbitrary set of *orthogonal* curvilinear co-ordinates, it follows from (5.5) and (5.16) that the physical components of stress $\sigma_{(ij)}$ are given in terms of the physical components of strain rate $\varepsilon_{(ij)}$ by

$$\sigma_{(ij)} = -p\delta_{ij} + 2\mu\varepsilon_{(ij)} - \tfrac{2}{3}\mu \operatorname{div} \mathbf{v}\, \delta_{ij}, \tag{6.17}$$

where $\varepsilon_{(ij)}$ may be determined from (6.5) in terms of the physical components of $\mathbf{v}$. Moreover, div $\mathbf{v}$ may be evaluated with the help of (5.19).

As an example we consider spherical polar co-ordinates, in which the physical components of σ_{ij} are written σ_{rr}, $\sigma_{r\theta}$, etc. Using (5.26) and (6.7), we obtain

$$\left.\begin{aligned}
\sigma_{rr} &= -p + 2\mu\frac{\partial v_r}{\partial r} - \tfrac{2}{3}\mu \operatorname{div} \mathbf{v} \quad \text{(putting } i=j=1 \text{ in (6.17))}, \\
\sigma_{\theta\theta} &= -p + \frac{2\mu}{r}\left(\frac{\partial v_\theta}{\partial \theta} + v_r\right) - \tfrac{2}{3}\mu \operatorname{div} \mathbf{v}, \\
\sigma_{\phi\phi} &= -p + \frac{2\mu}{r}\left(\frac{1}{\sin\theta}\frac{\partial v_\phi}{\partial \phi} + v_r + v_\theta \cot\theta\right) - \tfrac{2}{3}\mu \operatorname{div} \mathbf{v}, \\
\sigma_{r\theta} &= \mu\left(\frac{\partial v_\theta}{\partial r} + \frac{1}{r}\frac{\partial v_r}{\partial \theta} - \frac{v_\theta}{r}\right) \quad \text{(putting } i=1, j=2 \text{ in (6.17))}, \\
\sigma_{\theta\phi} &= \mu\left(\frac{1}{r}\frac{\partial v_\phi}{\partial \theta} + \frac{1}{r\sin\theta}\frac{\partial v_\theta}{\partial \phi} - \frac{v_\phi}{r}\cot\theta\right), \\
\sigma_{r\phi} &= \mu\left(\frac{1}{r\sin\theta}\frac{\partial v_r}{\partial \phi} + \frac{\partial v_\phi}{\partial r} - \frac{v_\phi}{r}\right),
\end{aligned}\right\} \tag{6.18}$$

where

$$\operatorname{div} \mathbf{v} = \frac{1}{r^2}\frac{\partial}{\partial r}(r^2 v_r) + \frac{1}{r\sin\theta}\frac{\partial}{\partial \theta}(v_\theta \sin\theta) + \frac{1}{r\sin\theta}\frac{\partial v_\phi}{\partial \phi}.$$

(b) Isotropic linear elastic solid The isotropic linear elastic constitutive equation for a solid undergoing small strain can be written

$$\sigma_{ij} = \lambda \delta_{ij} e_{kk} + 2G e_{ij} \tag{6.19}$$

in rectangular cartesians, where e_{ij} is the small strain tensor, and λ and G are physical properties of the material called the Lamé constants. The equation generalizes immediately to the covariant form

$$\sigma_{ij} = \lambda g_{ij} e^k{}_{.k} + 2G e_{ij} \tag{6.20}$$

$$= \lambda g_{ij} u^k{}_{,k} + G(u_{i,j} + u_{j,i}) \tag{6.21}$$

in terms of the displacement $\mathbf{u}$.

For orthogonal curvilinear co-ordinate systems we may deduce the relationship between physical components

$$\sigma_{(ij)} = \lambda \delta_{ij} \operatorname{div} \mathbf{u} + 2G e_{(ij)}, \tag{6.22}$$

where, again, $e_{(ij)}$ may be determined from (6.5) or (6.9) expressed in terms of $\mathbf{u}$ instead of $\mathbf{v}$, and div $\mathbf{u}$ may be found from (5.19). In spherical polar co-ordinates, for example, this gives

$$\begin{aligned}
\sigma_{rr} &= \lambda \operatorname{div} \mathbf{u} + 2G \frac{\partial u_r}{\partial r}, \\
\sigma_{\theta\theta} &= \lambda \operatorname{div} \mathbf{u} + \frac{2G}{r}\left(\frac{\partial u_\theta}{\partial \theta} + u_r\right), \\
\sigma_{\phi\phi} &= \lambda \operatorname{div} \mathbf{u} + \frac{2G}{r}\left(\frac{1}{\sin\theta}\frac{\partial u_\phi}{\partial \phi} + u_r + u_\theta \cot\theta\right), \\
\sigma_{r\theta} &= G\left(\frac{\partial u_\theta}{\partial r} + \frac{1}{r}\frac{\partial u_r}{\partial \theta} - \frac{u_\theta}{r}\right), \\
\sigma_{\theta\phi} &= G\left(\frac{1}{r}\frac{\partial u_\phi}{\partial \theta} + \frac{1}{r\sin\theta}\frac{\partial u_\theta}{\partial \phi} - \frac{u_\phi \cot\theta}{r}\right), \\
\sigma_{r\phi} &= G\left(\frac{1}{r\sin\theta}\frac{\partial u_r}{\partial \phi} + \frac{\partial u_\phi}{\partial r} - \frac{u_\phi}{r}\right),
\end{aligned} \qquad (6.23)$$

where $\operatorname{div} \mathbf{u}$ is given by (5.26).

Note that in general we also have

$$\operatorname{div} \mathbf{u} = e_{(11)} + e_{(22)} + e_{(33)},$$

so that (6.22) may be written

$$\begin{aligned}
\sigma_{(11)} &= \lambda \operatorname{div} \mathbf{u} + 2Ge_{(11)} = \lambda(e_{(11)} + e_{(22)} + e_{(33)}) + 2Ge_{(11)} \\
&= (\lambda + 2G)e_{(11)} + \lambda e_{(22)} + \lambda e_{(33)},
\end{aligned}$$

and two other equations obtainable by cyclic permutation, as well as

$$\sigma_{(12)} = 2Ge_{(12)}, \text{ etc.}$$

Example 4

Write the continuity equation in fluid dynamics (relating the density ρ and velocity field $\mathbf{v}$ in a fluid) in terms of

(a) spherical polar co-ordinates and
(b) parabolic cylindrical co-ordinates.

Method

If the fluid is *incompressible*, the density ρ at a point moving with the fluid does not change with time, and the continuity equation is

$$\operatorname{div} \mathbf{v} = 0,$$

which in rectangular cartesian co-ordinates can be expressed as

$$\frac{\partial v_i}{\partial y_i} = \frac{\partial v_1}{\partial y_1} + \frac{\partial v_2}{\partial y_2} + \frac{\partial v_3}{\partial y_3} = 0. \qquad (6.24)$$

Sec. 6.2]   **Applications**   157

With respect to an arbitrary co-ordinate system, this becomes

$$v^i{}_{,i} = 0, \qquad (6.25)$$

where the repeated index i must occur once in the upper position and once in the lower position, in order that the relevant tensor properties apply. In this case these properties are that $v^i{}_{,i}$ represents a scalar on a general basis, but $v_{i,i}$ does not. The continuity equation is a scalar identity, and cannot be represented generally by $v_{i,i} = 0$. It is possible to use the covariant derivative of the *covariant* component of velocity if required, but this involves writing (6.25) as the valid tensor identity

$$g^{ik} v_{k,i} = 0.$$

For an orthogonal curvilinear co-ordinate system, (6.25) can be expressed in terms of (5.19), i.e.

$$\sum_{i=1}^{3} \frac{1}{h_1 h_2 h_3} \frac{\partial}{\partial x^i} \left(\frac{h_1 h_2 h_3}{h_i} v_{(i)} \right) = 0,$$

using physical components of **v**. For example, in the toroidal co-ordinate system (2), with $x^1 = \phi$, $x^2 = r$, $x^3 = \theta$, and physical components v_ϕ, v_r, v_θ, we obtain

$$\frac{1}{r(R + r\cos\phi)} \left[\frac{\partial}{\partial \phi}((R + r\cos\phi)v_\phi) + \frac{\partial}{\partial r}[(R + r\cos\phi)rv_r] + \frac{\partial}{\partial \theta}(rv_\theta) \right]$$

$$= \frac{1}{r} \frac{\partial v_\phi}{\partial \phi} + \frac{\partial v_r}{\partial r} + \frac{1}{R + r\cos\phi} \frac{\partial v_\theta}{\partial \theta} - \frac{v_\phi \sin\phi}{R + r\cos\phi} + \frac{R + 2r\cos\phi}{r(R + r\cos\phi)} v_r = 0. \qquad (6.26)$$

For *compressible* fluids the continuity equation is

$$\frac{D\rho}{Dt} + \rho \operatorname{div} \mathbf{v} = \frac{\partial \rho}{\partial t} + v_i \frac{\partial \rho}{\partial y_i} + \rho \frac{\partial v_i}{\partial y_i} = 0 \qquad (6.27)$$

in rectangular cartesians, where D/Dt represents the usual 'material derivative', i.e. the rate of change associated with a given particle of the fluid. This generalizes to

$$\frac{\partial \rho}{\partial t} + v^i \rho_{,i} + \rho v^i{}_{,i} = 0, \qquad (6.28)$$

for arbitrary co-ordinate systems, where, since the repeated index i must appear at different levels, the contravariant component v^i must accompany the covariant derivative $\rho_{,i}$ in the second term. Since ρ is a scalar, $\rho_{,i}$ is just the same as the partial derivative $\partial \rho / \partial x^i$.

Taking account of the relation between contravariant and physical components given by (5.13), and again using (5.19) to evaluate the last term, we can express (6.28) in the following orthogonal curvilinear co-ordinate systems:

(a) Spherical polars   In this system we have

$$\frac{\partial \rho}{\partial t} + v_r \frac{\partial \rho}{\partial r} + \frac{v_\theta}{r} \frac{\partial \rho}{\partial \theta} + \frac{v_\phi}{r \sin\theta} \frac{\partial \rho}{\partial \phi} + \frac{\rho}{r^2} \frac{\partial}{\partial r}(r^2 v_r)$$

$$+ \frac{\rho}{r \sin\theta} \frac{\partial}{\partial \theta}(v_\theta \sin\theta) + \frac{1}{r \sin\theta} \frac{\partial v_\phi}{\partial \phi} = 0,$$

i.e.
$$\frac{\partial \rho}{\partial t} + \frac{1}{r^2}\frac{\partial}{\partial r}(\rho r^2 v_r) + \frac{1}{r\sin\theta}\frac{\partial}{\partial \theta}(\rho v_\theta \sin\theta) + \frac{1}{r\sin\theta}\frac{\partial}{\partial \phi}(\rho v_\phi) = 0. \tag{6.29}$$

(b) Parabolic cylindrical co-ordinates   With scale factors given by (5.30) and physical components of **v** denoted by v_ξ, v_η, v_z, we obtain

$$\frac{\partial \rho}{\partial t} + \frac{v_\xi}{\sqrt{\xi^2+\eta^2}}\frac{\partial \rho}{\partial \xi} + \frac{v_\eta}{\sqrt{\xi^2+\eta^2}}\frac{\partial \rho}{\partial \eta} + v_z \frac{\partial \rho}{\partial z}$$
$$+ \frac{\rho}{\xi^2+\eta^2}\left(\frac{\partial}{\partial \xi}(\sqrt{\xi^2+\eta^2}\, v_\xi) + \frac{\partial}{\partial \eta}(\sqrt{\xi^2+\eta^2}\, v_\eta) + \frac{\partial}{\partial z}[(\xi^2+\eta^2)v_z]\right) = 0,$$

i.e.
$$\frac{\partial \rho}{\partial t} + \frac{1}{\xi^2+\eta^2}\left(\frac{\partial}{\partial \xi}(\sqrt{\xi^2+\eta^2}\, \rho v_\xi) + \frac{\partial}{\partial \eta}(\sqrt{\xi^2+\eta^2}\, \rho v_\eta)\right) + \frac{\partial}{\partial z}(\rho v_z) = 0. \tag{6.30}$$

Example 5

The expression $\partial T_{ij}/\partial y_j$ in rectangular cartesians, where T_{ij} is a symmetric CT2, arises in certain equations of theoretical physics and engineering. For example, if T_{ij} is the Cauchy stress σ_{ij} in a continuum at rest, the equilibrium equations with body forces neglected are

$$\frac{\partial \sigma_{ij}}{\partial y_j} = 0, \qquad i = 1, 2, 3. \tag{6.31}$$

Transform $\partial T_{ij}/\partial y_j$ to generalized tensor form, and then express it in physical components with reference to an arbitrary orthogonal curvilinear co-ordinate system.

Method

First, observe that $\partial T_{ij}/\partial y_j$ is a contracted form of the CT3 appearing in problem 3.2.20, and so constitutes a CT1 (a vector). A valid generalized tensor expression is $T^{ij}{}_{,j}$; with i at the upper level, this gives the contravariant components of the vector. Note that the valid third-order tensor $T_{ij,k}$ must not be contracted to give $T_{ij,j}$, which will not transform as a covariant vector since the j do not appear at different levels. The associated covariant components of the contravariant $T^{ij}{}_{,j}$, however, are obtained by merely lowering the free index i to get $T_i{}^j{}_{,j}$, which involves the mixed components $T_i{}^j$.

From (4.40) we know that
$$T^{ij}{}_{,j} = \frac{\partial T^{ij}}{\partial x^j} + \Gamma^i_{lj} T^{lj} + \Gamma^j_{lj} T^{il}$$
$$= \frac{\partial T^{ij}}{\partial x^j} + \Gamma^i_{lj} T^{lj} + \Gamma^l_{jl} T^{ij},$$

interchanging the repeated indices j and l in the last term. Using (4.23) with (4.5), we can write

$$T^{ij}{}_{,j} = \frac{\partial T^{ij}}{\partial x^j} + \Gamma^i_{lj} T^{lj} + \frac{\partial}{\partial x^j}(\log \sqrt{g}) T^{ij}$$

$$= \frac{1}{\sqrt{g}} \frac{\partial}{\partial x^j}(\sqrt{g} T^{ij}) + \Gamma^i_{lj} T^{lj}. \tag{6.32}$$

Now consider a set of orthogonal curvilinear co-ordinates with scale factors h_1, h_2, h_3, and evaluate (6.32) for the case $i=1$, bearing in mind that, from (5.11), $\sqrt{g} = h_1 h_2 h_3$. We obtain

$$T^{1j}{}_{,j} = \frac{1}{h_1 h_2 h_3} \frac{\partial}{\partial x^j}(h_1 h_2 h_3 T^{1j}) + \Gamma^1_{lj} T^{lj},$$

which may be expressed in terms of physical components $T_{(ij)}$, according to (5.15), as

$$\frac{1}{h_1 h_2 h_3}\left[\frac{\partial}{\partial x^1}\left(\frac{h_2 h_3}{h_1} T_{(11)}\right) + \frac{\partial}{\partial x^2}(h_3 T_{(12)}) + \frac{\partial}{\partial x^3}(h_2 T_{(13)})\right]$$

$$+ \frac{\Gamma^1_{11}}{h_1{}^2} T_{(11)} + \frac{\Gamma^1_{12} T_{(12)}}{h_1 h_2} + \frac{\Gamma^1_{13} T_{(13)}}{h_1 h_3} + \frac{\Gamma^1_{21} T_{(21)}}{h_2 h_1} + \frac{\Gamma^1_{22} T_{(22)}}{h_2{}^2} + \frac{\Gamma^1_{23} T_{(23)}}{h_2 h_3}$$

$$+ \frac{\Gamma^1_{31} T_{(31)}}{h_3 h_1} + \frac{\Gamma^1_{32} T_{(32)}}{h_3 h_2} + \frac{\Gamma^1_{33} T_{(33)}}{h_3{}^2}$$

$$= \frac{1}{h_1 h_2 h_3}\left[\frac{\partial}{\partial x^1}\left(\frac{h_2 h_3}{h_1} T_{(11)}\right) + \frac{\partial}{\partial x^2}(h_3 T_{(12)}) + \frac{\partial}{\partial x^3}(h_2 T_{(13)})\right]$$

$$+ \frac{1}{h_1{}^3} \frac{\partial h_1}{\partial x_1} T_{(11)} + \frac{2}{h_1{}^2 h_2} \frac{\partial h_1}{\partial x^2} T_{(12)} + \frac{2}{h_1{}^2 h_3} \frac{\partial h_1}{\partial x^3} T_{(13)}$$

$$- \frac{1}{h_1{}^2 h_2} \frac{\partial h_2}{\partial x^1} T_{(22)} - \frac{1}{h_1{}^2 h_3} \frac{\partial h_3}{\partial x^1} T_{(33)},$$

using (5.10) and the symmetry of T_{ij} (and, consequently, of $T_{(ij)}$). A little further manipulation, using product rules for differentiation, reduces this to

$$\frac{1}{h_1{}^2 h_2 h_3}\left(\frac{\partial}{\partial x^1}(h_2 h_3 T_{(11)}) + \frac{1}{h_1}\frac{\partial}{\partial x^2}(h_1{}^2 h_3 T_{(12)}) + \frac{1}{h_1}\frac{\partial}{\partial x^3}(h_1{}^2 h_2 T_{(13)})\right)$$

$$- \frac{1}{h_1{}^2 h_2} \frac{\partial h_2}{\partial x^1} T_{(22)} - \frac{1}{h_1{}^2 h_3} \frac{\partial h_3}{\partial x^1} T_{(33)}.$$

Recalling that $T^{ij}{}_{,j}$ is a contravariant vector, and using (5.13), we can now write down the physical component corresponding to $i=1$ by multiplying the above expression by h_1. The terms corresponding to $i=2$ and 3 follow by cyclic permutation of the indices. Hence the components are as follows: for $i=1$,

$$\frac{1}{h_1h_2h_3}\left(\frac{\partial}{\partial x^1}(h_2h_3T_{(11)}) + \frac{1}{h_1}\frac{\partial}{\partial x^2}(h_1{}^2h_3T_{(12)}) + \frac{1}{h_1}\frac{\partial}{\partial x^3}(h_1{}^2h_2T_{(13)})\right.$$
$$\left. - h_3\frac{\partial h_2}{\partial x^1}T_{(22)} - h_2\frac{\partial h_3}{\partial x^1}T_{(33)}\right);$$

for $i = 2$,

$$\frac{1}{h_1h_2h_3}\left(\frac{\partial}{\partial x^2}(h_3h_1T_{(22)}) + \frac{1}{h_2}\frac{\partial}{\partial x^3}(h_2{}^2h_1T_{(23)}) + \frac{1}{h_2}\frac{\partial}{\partial x^1}(h_2{}^2h_3T_{(12)})\right.$$
$$\left. - h_1\frac{\partial h_3}{\partial x^2}T_{(33)} - h_3\frac{\partial h_1}{\partial x^2}T_{(11)}\right);$$

for $i = 3$,

$$\frac{1}{h_1h_2h_3}\left(\frac{\partial}{\partial x^3}(h_1h_2T_{(33)}) + \frac{1}{h_3}\frac{\partial}{\partial x^1}(h_3{}^2h_2T_{(13)}) + \frac{1}{h_3}\frac{\partial}{\partial x^2}(h_3{}^2h_1T_{(23)})\right.$$
$$\left. - h_2\frac{\partial h_1}{\partial x^3}T_{(11)} - h_1\frac{\partial h_2}{\partial x^3}T_{(22)}\right).$$

(6.33)

This solves the problem as posed, but for specific examples we now write down the equilibrium equations (6.31) in the following co-ordinate systems.

(a) Cylindrical polars We obtain

$$\frac{1}{\rho}\left(\frac{\partial}{\partial\rho}(\rho\sigma_{\rho\rho}) + \frac{\partial\sigma_{\rho\phi}}{\partial\phi} + \frac{\partial}{\partial z}(\rho\sigma_{\rho z}) - \sigma_{\phi\phi}\right) = 0,$$
$$\frac{1}{\rho}\left(\frac{\partial\sigma_{\phi\phi}}{\partial\phi} + \frac{1}{\rho}\frac{\partial}{\partial z}(\rho^2\sigma_{\phi z}) + \frac{1}{\rho}\frac{\partial}{\partial\rho}(\rho^2\sigma_{\rho\phi})\right) = 0,$$
$$\frac{1}{\rho}\left(\frac{\partial}{\partial z}(\rho\sigma_{zz}) + \frac{\partial}{\partial\rho}(\rho\sigma_{\rho z}) + \frac{\partial\sigma_{\phi z}}{\partial\phi}\right) = 0.$$

(6.34)

(b) Spherical polars We find that

$$\frac{1}{r^2\sin\theta}\left(\frac{\partial}{\partial r}(r^2\sigma_{rr}\sin\theta) + \frac{\partial}{\partial\theta}(r\sigma_{r\theta}\sin\theta) + \frac{\partial}{\partial\phi}(r\sigma_{r\phi}) - r\sin\theta(\sigma_{\theta\theta} + \sigma_{\phi\phi})\right) = 0,$$
$$\frac{1}{r^2\sin\theta}\left(\frac{\partial}{\partial\theta}(r\sigma_{\theta\theta}\sin\theta) + \frac{1}{r}\frac{\partial}{\partial\phi}(r^2\sigma_{\theta\phi}) + \frac{1}{r}\frac{\partial}{\partial r}(r^3\sigma_{r\theta}\sin\theta) - r\sigma_{\phi\phi}\cos\theta\right) = 0,$$
$$\frac{1}{r^2\sin\theta}\left(\frac{\partial}{\partial\phi}(r\sigma_{\phi\phi}) + \frac{1}{r\sin\theta}\frac{\partial}{\partial r}(r^3\sin^2\theta\sigma_{r\phi}) + \frac{1}{r\sin\theta}\frac{\partial}{\partial\theta}(r^2\sin^2\theta\sigma_{\theta\phi})\right) = 0.$$

(6.35)

Example 6

In continuum mechanics the acceleration vector $D\mathbf{v}/Dt$, the material derivative of the velocity field, is a basic feature of the equations of motion. Determine the physical

Sec. 6.2] **Applications** 161

components of D**v**/D*t* with respect to an arbitrary set of ('eulerian') orthogonal curvilinear co-ordinates.

Method

In rectangular cartesians,

$$\frac{Dv_i}{Dt} = \frac{\partial v_i}{\partial t} + v_j \frac{\partial v_i}{\partial y_j}, \qquad (6.36)$$

where the velocity field **v** is expressed in terms of eulerian (spatial) variables y_1, y_2, y_3, and time t. This identity is often expressed in vector form as

$$\frac{D\mathbf{v}}{Dt} = \frac{\partial \mathbf{v}}{\partial t} + (\mathbf{v}\cdot\text{grad})\mathbf{v}.$$

It is important to realize, however, that we cannot proceed immediately to the required physical components in an orthogonal curvilinear co-ordinate system by substituting the corresponding operator expressions for 'grad' directly into this expression. For example, in cylindrical polars the physical components of 'grad' are the operators.

$$\left[\frac{\partial}{\partial \rho} \quad \frac{1}{\rho}\frac{\partial}{\partial \phi} \quad \frac{\partial}{\partial z}\right]$$

and the differential operator (**v**·grad) is given by

$$\mathbf{v}\cdot\text{grad} = v_\rho \frac{\partial}{\partial \rho} + \frac{v_\phi}{\rho}\frac{\partial}{\partial \phi} + v_z \frac{\partial}{\partial z},$$

but the physical ρ component of (**v**·grad)**v** cannot be obtained by applying this operator to v_ρ to produce the expression

$$v_\rho \frac{\partial v_\rho}{\partial \rho} + \frac{v_\phi}{\rho}\frac{\partial v_\rho}{\partial \phi} + v_z \frac{\partial v_\rho}{\partial z}. \qquad (6.37)$$

The reason for this may be seen by considering the generalized form of (6.36). In covariant form this may be written straightforwardly as

$$\left(\frac{Dv}{Dt}\right)_i = \frac{\partial v_i}{\partial t} + v^j v_{i,j}. \qquad (6.38)$$

Now, from (4.28), (6.38) is equivalent to

$$\left(\frac{D\mathbf{v}}{Dt}\right)_i = \frac{\partial v_i}{\partial t} + v^j \frac{\partial v_i}{\partial x^j} - v^j \Gamma^k_{ij} v_k. \qquad (6.39)$$

For $i = 1$ in cylindrical polars the middle term on the RHS of (6.39) corresponds to the expression (6.37), but the final term also needs to be taken into account. With $i = 1$ in (6.39) we obtain in general

$$\frac{\partial v_1}{\partial t} + v^j \frac{\partial v_1}{\partial x^j} - v^j \Gamma^k_{1j} v_k$$

$$= \frac{\partial v_1}{\partial t} + v^1 \frac{\partial v_1}{\partial x^1} + v^2 \frac{\partial v_1}{\partial x^2} + v^3 \frac{\partial v_1}{\partial x^3}$$
$$- \Gamma^1_{11} v^1 v_1 - \Gamma^1_{12} v^2 v_1 - \Gamma^1_{13} v^3 v_1 - \Gamma^2_{11} v^1 v_2 - \Gamma^2_{12} v^2 v_2 - \Gamma^2_{13} v^3 v_2$$
$$- \Gamma^3_{11} v^1 v_3 - \Gamma^3_{12} v^2 v_3 - \Gamma^3_{13} v^3 v_3, \tag{6.40}$$

performing the summations over j and k in the final term. For cylindrical polars this reduces to

$$\frac{\partial v_\rho}{\partial t} + v_\rho \frac{\partial v_\rho}{\partial \rho} + \frac{v_\phi}{\rho} \frac{\partial v_\rho}{\partial \phi} + v_z \frac{\partial v_\rho}{\partial z} - \frac{v_\phi^2}{\rho},$$

since Γ^2_{12} supplies the only non-zero term in the summation.

For $i = 2$, however, the physical ϕ component is obtained from the covariant component by dividing by ρ, according to (5.14), and we get

$$\frac{\partial v_\phi}{\partial t} + \frac{v_\rho}{\rho} \frac{\partial}{\partial \rho}(\rho v_\phi) + \frac{v_\phi}{\rho^2} \frac{\partial}{\partial \phi}(\rho v_\phi) + \frac{v_z}{\rho} \frac{\partial}{\partial z}(\rho v_\phi) - \frac{1}{\rho}\left[\Gamma^2_{21} v_\rho (\rho v_\phi) + \Gamma^1_{22} v_\rho \left(\frac{v_\phi}{\rho}\right)\right]$$
$$= \frac{\partial v_\phi}{\partial t} + v_\rho \frac{\partial v_\phi}{\partial \rho} + \frac{v_\phi}{\rho} \frac{\partial v_\phi}{\partial \phi} + v_z \frac{\partial v_\phi}{\partial z} + \frac{v_\rho v_\phi}{\rho},$$

and similarly for $i = 3$. To sum up, the physical components of $D\mathbf{v}/Dt$ in cylindrical polar co-ordinates are as follows:

$$\left. \begin{aligned} &\frac{\partial v_\rho}{\partial t} + v_\rho \frac{\partial v_\rho}{\partial \rho} + \frac{v_\phi}{\rho} \frac{\partial v_\rho}{\partial \phi} + v_z \frac{\partial v_\rho}{\partial z} - \frac{v_\phi^2}{\rho}, \\ &\frac{\partial v_\phi}{\partial t} + v_\rho \frac{\partial v_\phi}{\partial \rho} + \frac{v_\phi}{\rho} \frac{\partial v_\phi}{\partial \phi} + v_z \frac{\partial v_\phi}{\partial z} + \frac{1}{\rho} v_\rho v_\phi, \\ &\frac{\partial v_z}{\partial t} + v_\rho \frac{\partial v_z}{\partial \rho} + \frac{v_\phi}{\rho} \frac{\partial v_z}{\partial \phi} + v_z \frac{\partial v_z}{\partial z}. \end{aligned} \right\} \tag{6.41}$$

Reverting to the more general form (6.40) and substituting (5.10), we obtain after a little manipulation the physical components corresponding to $i = 1$:

$$\frac{\partial v_{(1)}}{\partial t} + \frac{1}{h_1} v_{(1)} \frac{\partial v_{(1)}}{\partial x^1} + \frac{1}{h_2} v_{(2)} \frac{\partial v_{(1)}}{\partial x^2} + \frac{1}{h_3} v_{(3)} \frac{\partial v_{(1)}}{\partial x^3}$$
$$+ \frac{1}{h_1 h_2} \frac{\partial h_1}{\partial x^2} v_{(1)} v_{(2)} + \frac{1}{h_1 h_3} \frac{\partial h_1}{\partial x^3} v_{(1)} v_{(3)} - \frac{1}{h_1 h_2} \frac{\partial h_2}{\partial x^1} v_{(2)}^2 - \frac{1}{h_1 h_3} \frac{\partial h_3}{\partial x^1} v_{(3)}^2. \tag{6.42}$$

The physical components corresponding to $i = 2$ and 3 may be written immediately by cyclic permutation of the indices. As a further example we have for spherical polar co-ordinates the physical components

$$\left. \begin{aligned} &\frac{\partial v_r}{\partial t} + v_r \frac{\partial v_r}{\partial r} + \frac{v_\theta}{r} \frac{\partial v_r}{\partial \theta} + \frac{v_\phi}{r \sin \theta} \frac{\partial v_r}{\partial \phi} - \frac{v_\theta^2}{r} - \frac{v_\phi^2}{r}, \\ &\frac{\partial v_\theta}{\partial t} + v_r \frac{\partial v_\theta}{\partial r} + \frac{v_\theta}{r} \frac{\partial v_\theta}{\partial \theta} + \frac{v_\phi}{r \sin \theta} \frac{\partial v_\theta}{\partial \phi} + \frac{v_r v_\theta}{r} - \frac{v_\phi^2}{r} \cot \theta, \end{aligned} \right\} \tag{6.43}$$

$$\frac{\partial v_\phi}{\partial t} + v_r \frac{\partial v_\phi}{\partial r} + \frac{v_\theta}{r}\frac{\partial v_\phi}{\partial \theta} + \frac{v_\phi}{r\sin\theta}\frac{\partial v_\phi}{\partial \phi} + \frac{v_r v_\phi}{r} + \frac{v_\theta v_\phi}{r}\cot\theta.\Bigg]$$

Example 7

Transform the equations of motion for a continuum in rectangular cartesians

$$\frac{\partial \sigma_{ij}}{\partial y_j} + \rho g_i = \rho \frac{Dv_i}{Dt}, \qquad (6.44)$$

where **g** is the body force per unit mass, to a general co-ordinate system.

Method

Example 5 above suggests the generalization of (6.44) to contravariant form

$$\sigma^{ij}{}_{,j} + \rho g^i = \rho \frac{Dv^i}{Dt} \qquad (6.45)$$

(although Example 6 concentrated on the covariant form of D**v**/Dt). This completes the transformation.

Using (6.32) and (4.26) we may write (6.45) as

$$\frac{1}{\sqrt{g}}\frac{\partial}{\partial x^j}(\sqrt{g}\,\sigma^{ij}) + \Gamma^i_{kj}\sigma^{kj} + \rho g^i = \rho\left(\frac{\partial v^i}{\partial t} + v^j v^i{}_{,j}\right)$$

$$= \rho\left(\frac{\partial v^i}{\partial t} + v^j \frac{\partial v^i}{\partial x^j} + v^j \Gamma^i_{kj} v^k\right). \qquad (6.46)$$

With orthogonal curvilinear co-ordinates we may proceed immediately to the expression of (6.45) in physical components if we know the physical components of the individual terms. For example, in spherical polars we have, by (6.35) and (6.43),

$$\left.\begin{aligned}
&\frac{1}{r^2}\frac{\partial}{\partial r}(r^2 \sigma_{rr}) + \frac{1}{r\sin\theta}\frac{\partial}{\partial \theta}(\sigma_{r\theta}\sin\theta) + \frac{1}{r\sin\theta}\frac{\partial \sigma_{r\phi}}{\partial \phi} - \frac{1}{r}(\sigma_{\theta\theta}+\sigma_{\phi\phi}) + \rho g_r\\
&= \rho\left(\frac{\partial v_r}{\partial t} + v_r\frac{\partial v_r}{\partial r} + \frac{v_\theta}{r}\frac{\partial v_r}{\partial \theta} + \frac{v_\phi}{r\sin\theta}\frac{\partial v_r}{\partial \phi} - \frac{v_\theta^2}{r} - \frac{v_\phi^2}{r}\right),\\
&\frac{1}{r\sin\theta}\frac{\partial}{\partial \theta}(\sigma_{\theta\theta}\sin\theta) + \frac{1}{r\sin\theta}\frac{\partial \sigma_{\theta\phi}}{\partial \phi} + \frac{1}{r^3}\frac{\partial}{\partial r}(r^3\sigma_{r\theta}) - \frac{1}{r}\sigma_{\phi\phi}\cot\theta + \rho g_\theta\\
&= \rho\left(\frac{\partial v_\theta}{\partial t} + v_r\frac{\partial v_\theta}{\partial r} + \frac{v_\theta}{r}\frac{\partial v_\theta}{\partial \theta} + \frac{v_\phi}{r\sin\theta}\frac{\partial v_\theta}{\partial \phi} + \frac{v_r v_\theta}{r} - \frac{v_\phi^2}{r}\cot\theta\right),\\
&\frac{1}{r\sin\theta}\frac{\partial \sigma_{\phi\phi}}{\partial \phi} + \frac{1}{r^3}\frac{\partial}{\partial r}(r^3\sigma_{r\phi}) + \frac{1}{r\sin^2\theta}\frac{\partial}{\partial \theta}(\sigma_{\theta\phi}\sin^2\theta) + \rho g_\phi\\
&= \rho\left(\frac{\partial v_\phi}{\partial t} + v_r\frac{\partial v_\phi}{\partial r} + \frac{v_\theta}{r}\frac{\partial v_\phi}{\partial \theta} + \frac{v_\phi}{r\sin\theta}\frac{\partial v_\phi}{\partial \phi} + \frac{v_r v_\phi}{r} + \frac{v_\theta v_\phi}{r}\cot\theta\right)
\end{aligned}\right\} \quad (6.47)$$

where g_r, g_θ, g_ϕ are the physical components of **g**.

Example 8

Transform Navier's equation in rectangular cartesians

$$(\lambda + G)\frac{\partial^2 u_j}{\partial y_i \partial y_j} + G\frac{\partial^2 u_i}{\partial y_j \partial y_j} + \rho g_i = \rho \frac{\partial^2 u_i}{\partial t^2} \qquad (6.48)$$

for the motion of an isotropic linear elastic solid undergoing small strains, where $\mathbf{u} = \mathbf{u}(y_1, y_2, y_3, t)$ is the displacement field.

Method

Navier's equation may be deduced from (6.44), using (6.19) and (6.2) and neglecting terms on the RHS which involve products of terms containing $\mathbf{u}$. Indeed, in orthogonal curvilinear co-ordinate systems we may obtain relations between the physical components of σ_{ij} and e_{ij}, and e_{ij} and $\mathbf{u}$, and substitute directly into the appropriate equations of motion, thus giving the physical components of Navier's equation directly. For example, with spherical polars the appropriate equations are (6.23). These give expressions for σ_{rr}, etc., may be substituted directly into (6.47). The RHSs of (6.47) are approximated by $\rho \partial^2 u_r/\partial t^2$, $\rho \partial^2 u_\theta/\partial t^2$ and $\rho \partial^2 u_\phi/\partial t^2$, respectively.

However, it is instructive to start from the cartesian form (6.48). The first term generalizes to $(\lambda + G)u^j{}_{,ij}$; we check that the repeated index j appears at different levels and observe that the free index takes the lower position corresponding to covariant differentiation. The second term, however, cannot be expressed immediately in corresponding covariant form as $\mu u_{i,jj}$, since the repeated j indices must not appear at the same level. A tempting possibility here arises from observing that in rectangular cartesians the second term in (6.48) is

$$G\frac{\partial^2 u_i}{\partial y_j \partial y_j} = G\left(\frac{\partial^2 u_i}{\partial y_1{}^2} + \frac{\partial^2 u_i}{\partial y_2{}^2} + \frac{\partial^2 u_i}{\partial y_3{}^2}\right) = G\nabla^2 u_i$$

in terms of the Laplacian operator ∇^2. It might then be expected that for orthogonal curvilinear co-ordinates the physical components of the second term could be written down directly by applying the operator ∇^2 as it appears in (5.20) to the physical components of $\mathbf{u}$. However, for reasons similar to those given in Example 6 when we discussed the operator $(\mathbf{v} \cdot \text{grad})$, this procedure will give an incorrect result.

In a valid generalized tensor expression which reduces to $\partial^2 u_i/(\partial y_j \partial y_j)$ in rectangular cartesians, the repeated j indices must appear at different levels. One possibility is to consider the third-order covariant tensor $u_{i,jk}$ and to raise the last index in the usual way by multiplying by g^{ij}. This gives $u_{i,jk}g^{jk}$; we check that this is a valid tensor which reduces to the required form in rectangular cartesians, since g^{jk} reduces to δ_{jk}.

Thus we have the generalized (covariant) form

$$(\lambda + G)u^j{}_{,ij} + Gu_{i,jk}g^{jk} + \rho g_i = \frac{\partial^2 u_i}{\partial t^2}. \qquad (6.49)$$

This completes the transformation, and in principle we can apply (6.49) to any co-ordinate system, making use of the expression (4.52) for the second covariant derivative $u^i{}_{,jk}$ and a similar lengthy expression for $u_{i,jk}$. We pursue this method in Example 9. An alternative approach is to return to cartesians and observe that the first term of (6.48) is just

$$(\lambda + G)\frac{\partial}{\partial y_i}\left(\frac{\partial u_j}{\partial y_j}\right) = (\lambda + G)[\text{grad}(\text{div } \mathbf{u})]_i.$$

Moreover, there is a convenient identity in cartesians

$$\text{curl curl } \mathbf{u} \equiv \text{grad div } \mathbf{u} - \nabla^2 \mathbf{u}, \qquad (6.50)$$

which may be established using the permutation identity (1.17):

$$(\text{curl curl } \mathbf{u})_i = e_{ijk}\frac{\partial}{\partial y_j}(\text{curl } \mathbf{u})_k$$

$$= e_{ijk}\frac{\partial}{\partial y_j}\left(e_{klm}\frac{\partial u_m}{\partial y_l}\right) = e_{ijk}e_{klm}\frac{\partial^2 u_m}{\partial y_j \partial y_l}$$

$$= (\delta_{il}\delta_{jm} - \delta_{im}\delta_{jl})\frac{\partial^2 u_m}{\partial y_j \partial y_l} = \frac{\partial^2 u_j}{\partial y_j \partial y_i} - \frac{\partial^2 u_i}{\partial y_j \partial y_j}$$

$$= \frac{\partial}{\partial y_i}(\text{div } \mathbf{u}) - \nabla^2 u_i = (\text{grad div } \mathbf{u})_i - (\nabla^2 \mathbf{u})_i.$$

This shows that $\nabla^2 \mathbf{u}$, which as we indicated above cannot be evaluated using the laplacian operator in a general orthogonal curvilinear system, is given by

$$\nabla^2 \mathbf{u} = \text{grad div } \mathbf{u} - \text{curl curl } \mathbf{u},$$

and now the RHS is expressed in terms of standard operators. Navier's equation becomes

$$(\lambda + G)\text{ grad div } \mathbf{u} + G\nabla^2 \mathbf{u} + \rho \mathbf{g}$$

$$= (\lambda + 2G)\text{ grad div } \mathbf{u} - G\text{ curl curl } \mathbf{u} + \rho \mathbf{g} = \rho\frac{\partial^2 \mathbf{u}}{\partial t^2} \qquad (6.51)$$

in invariant vector form. In any orthogonal curvilinear co-ordinate system, (6.51) may be expressed directly in terms of physical components.

For example, in spherical polars, curl $\mathbf{u}$ is given by (5.27), and curl curl $\mathbf{u}$ has physical components

$$\left.\begin{aligned}&\frac{1}{r\sin\theta}\left\{\frac{\partial}{\partial\theta}\left[\left(\frac{1}{r}\frac{\partial}{\partial r}(ru_\theta) - \frac{1}{r}\frac{\partial u_r}{\partial\theta}\right)\sin\theta\right] - \frac{\partial}{\partial\phi}\left(\frac{1}{r\sin\theta}\frac{\partial u_r}{\partial\phi} - \frac{1}{r}\frac{\partial}{\partial r}(ru_\phi)\right)\right\},\\&\frac{1}{r\sin\theta}\frac{\partial}{\partial\phi}\left[\frac{1}{r\sin\theta}\left(\frac{\partial}{\partial\theta}(u_\phi\sin\theta) - \frac{\partial u_\theta}{\partial\phi}\right)\right] - \frac{1}{r}\frac{\partial}{\partial r}\left(\frac{\partial}{\partial r}(ru_\theta) - \frac{\partial u_r}{\partial\theta}\right),\\&\frac{1}{r}\frac{\partial}{\partial r}\left(\frac{1}{\sin\theta}\frac{\partial u_r}{\partial\phi} - \frac{\partial}{\partial r}(ru_\phi)\right) - \frac{1}{r}\frac{\partial}{\partial\theta}\left[\frac{1}{r\sin\theta}\left(\frac{\partial}{\partial\theta}(u_\phi\sin\theta) - \frac{\partial u_\theta}{\partial\phi}\right)\right].\end{aligned}\right\} \quad (6.52)$$

Hence the physical r component of (6.51) may be written as

$$(\lambda + 2G)\frac{\partial}{\partial r}\left(\frac{1}{r^2}\frac{\partial}{\partial r}(r^2 u_r) + \frac{1}{r\sin\theta}\frac{\partial}{\partial\theta}(\sin\theta\, u_\theta) + \frac{1}{r\sin\theta}\frac{\partial u_\phi}{\partial\phi}\right)$$

$$-\frac{G}{r\sin\theta}\left\{\frac{\partial}{\partial\theta}\left[\left(\frac{1}{r}\frac{\partial}{\partial r}(ru_\theta)-\frac{1}{r}\frac{\partial u_r}{\partial\theta}\right)\sin\theta\right]-\frac{\partial}{\partial\phi}\left(\frac{1}{r\sin\theta}\frac{\partial u_r}{\partial\phi}-\frac{1}{r}\frac{\partial}{\partial r}(ru_\phi)\right)\right\}+\rho g_r$$

$$=\rho\frac{\partial^2 u_r}{\partial t^2},$$

and the other two component equations similarly.

Example 9

Transform the Navier–Stokes equations for an incompressible viscous fluid in rectangular cartesians

$$\frac{Dv_i}{Dt}=g_i-\frac{1}{\rho}\frac{\partial p}{\partial y_i}+v\frac{\partial^2 v_i}{\partial y_j\,\partial y_j}, \qquad (6.53)$$

where v is the kinematic viscosity of the fluid ($v=\mu/\rho$), to a general system of co-ordinates, and express them with respect to spherical polar co-ordinates.

Method

We observe first that (6.53) is the equation of motion of the classical newtonian fluid satisfying the constraint of incompressibility. (It is assumed here that the viscosity μ is a constant.) In rectangular cartesians, (6.53) may be derived by substituting the constitutive equation (6.15) for σ_{ij}, with the last term in div $\mathbf{v}$ omitted, into the general equation of motion (6.44) and expressing ε_{ij} in terms of $\mathbf{v}$ by (6.1).

Equations (6.53) transform immediately to the covariant form

$$\frac{Dv_i}{Dt}=g_i-\frac{1}{\rho}p_{,i}+vv_{i,jk}g^{jk}. \qquad (6.54)$$

The equivalent contravariant form in this case is obtained merely by raising the index i throughout to get

$$\frac{Dv^i}{Dt}=g^i-\frac{1}{\rho}g^{ij}p_{,j}+vv^i{}_{,jk}g^{jk}. \qquad (6.55)$$

We now proceed to evaluate (6.54) directly in terms of spherical polar co-ordinates. Since the only non-zero components of g^{ij} are $g^{11}=1, g^{22}=r^{-2}, g^{33}=r^{-2}\sin^{-2}\theta$, we obtain

$$\frac{Dv_i}{Dt}=g_i-\frac{1}{\rho}p_{,i}+vv_{i,11}+\frac{v}{r^2}v_{i,22}+\frac{v}{r^2\sin^2\theta}v_{i,33}, \qquad (6.56)$$

carrying out the summation over j and k in the last term.

Now we use (4.41) with $v_{i,j}$ in place of T_{ij} to obtain expressions for the components of $v_{i,jk}$. For example, we need to evaluate

$$v_{i,11}=(v_{i,1})_{,1}=\frac{\partial}{\partial x^1}(v_{i,1})-\Gamma^l_{i1}v_{l,1}-\Gamma^l_{11}v_{i,l}$$

$$=\frac{\partial}{\partial x^1}(v_{i,1})-\Gamma^2_{i1}v_{2,1}-\Gamma^3_{i1}v_{3,1}$$

Sec. 6.2] **Applications** 167

by (4.15), since $\Gamma^l_{11} = 0$ for $l = 1, 2, 3$. Hence

$$v_{1,11} = \frac{\partial}{\partial x^1}(v_{1,1}),$$

$$v_{2,11} = \frac{\partial}{\partial x^1}(v_{2,1}) - \frac{1}{r}v_{2,1}$$

and

$$v_{3,11} = \frac{\partial}{\partial x^1}(v_{3,1}) - \frac{1}{r}v_{3,1},$$

and we can use (4.28) to evaluate the components $v_{i,j}$. This gives

$$v_{1,1} = \frac{\partial v_1}{\partial x^1} - \Gamma^k_{11}v_k = \frac{\partial v_1}{\partial x^1} - \Gamma^1_{11}v_1 - \Gamma^2_{11}v_2 - \Gamma^3_{11}v_3 = \frac{\partial v_1}{\partial x^1},$$

$$v_{2,1} = \frac{\partial v_2}{\partial x^1} - \Gamma^k_{21}v_k = \frac{\partial v_2}{\partial x^1} - \Gamma^2_{21}v_2 = \frac{\partial v_2}{\partial x^1} - \frac{1}{r}v_2$$

and

$$v_{3,1} = \frac{\partial v_3}{\partial x^1} - \Gamma^k_{31}v_k = \frac{\partial v_3}{\partial x^1} - \Gamma^3_{31}v_3 = \frac{\partial v_3}{\partial x^1} - \frac{1}{r}v_3.$$

Hence

$$v_{1,11} = \frac{\partial}{\partial x^1}\left(\frac{\partial v_1}{\partial x^1}\right) = \frac{\partial^2 v_1}{\partial (x^1)^2} = \frac{\partial^2 v_r}{\partial r^2},$$

$$v_{2,11} = \frac{\partial}{\partial x^1}\left(\frac{\partial v_2}{\partial x^1} - \frac{1}{r}v_2\right) - \frac{1}{r}\left(\frac{\partial v_2}{\partial x^1} - \frac{1}{r}v_2\right) = \frac{\partial}{\partial r}\left(\frac{\partial}{\partial r}(rv_\theta) - v_\theta\right) - \frac{1}{r}\left(\frac{\partial}{\partial r}(rv_\theta) - v_\theta\right)$$

$$= \frac{\partial}{\partial r}\left(r\frac{\partial v_\theta}{\partial r}\right) - \frac{\partial v_\theta}{\partial r} = r\frac{\partial^2 v_\theta}{\partial r^2},$$

$$v_{3,11} = \frac{\partial}{\partial x^1}\left(\frac{\partial v_3}{\partial x^1} - \frac{1}{r}v_3\right) - \frac{1}{r}\left(\frac{\partial v_3}{\partial x^1} - \frac{1}{r}v_3\right)$$

$$= \frac{\partial}{\partial r}\left(\frac{\partial}{\partial r}(r\sin\theta\, v_\phi) - \sin\theta\, v_\phi\right) - \frac{1}{r}\left(\frac{\partial}{\partial r}(r\sin\theta\, v_\phi) - \sin\theta\, v_\phi\right) = r\sin\theta\frac{\partial^2 v_\phi}{\partial r^2}.$$

Here we have introduced physical components v_r, v_θ and v_ϕ.

The other relevant second covariant components of (6.56) may be determined similarly. We get

$$v_{1,22} = \frac{\partial^2 v_r}{\partial \theta^2} + r\frac{\partial v_r}{\partial r} - v_r - 2\frac{\partial v_\theta}{\partial \theta},$$

$$v_{2,22} = r\frac{\partial^2 v_\theta}{\partial \theta^2} + 2r\frac{\partial v_r}{\partial \theta} + r^2\frac{\partial v_\theta}{\partial r} - rv_\theta,$$

$$v_{3,22} = r\sin\theta\frac{\partial^2 v_\phi}{\partial \theta^2} + r^2\sin\theta\frac{\partial v_\phi}{\partial r},$$

$$v_{1,33} = \frac{\partial^2 v_r}{\partial \phi^2} + r\sin^2\theta\frac{\partial v_r}{\partial r} + \sin\theta\cos\theta\frac{\partial v_r}{\partial \theta} - \sin^2\theta\, v_r$$

168   Applications in continuum mechanics   [Ch. 6

$$-2\sin\theta\cos\theta\, v_\theta - 2\sin\theta\frac{\partial v_\phi}{\partial \phi},$$

$$v_{2,33} = r\frac{\partial^2 v_\theta}{\partial \phi^2} - 2r\cos\theta\frac{\partial v_\phi}{\partial \phi} - r\cos^2\theta\, v_\theta + r^2\sin^2\theta\frac{\partial v_\theta}{\partial r} + r\sin\theta\cos\theta\frac{\partial v_\theta}{\partial \theta},$$

and $\quad v_{3,33} = r\sin\theta\frac{\partial^2 v_\phi}{\partial \phi^2} + 2r\sin^2\theta\frac{\partial v_r}{\partial \phi} + 2r\sin\theta\cos\theta\frac{\partial v_\theta}{\partial \phi} + r^2\sin^3\theta\frac{\partial v_\phi}{\partial r}$

$$+ r\sin^2\theta\cos\theta\frac{\partial v_\phi}{\partial \theta} - r\sin\theta\, v_\phi.$$

We are now in a position to write the Navier–Stokes equations in spherical polar co-ordinates by putting $i = 1, 2, 3$ in turn into (6.56). We can use the physical components of acceleration previously obtained in (6.43), and we must remember to divide covariant components by the corresponding scale factors to get physical components. The result is

$$\left.\begin{aligned}
&\frac{\partial v_r}{\partial t} + v_r\frac{\partial v_r}{\partial r} + \frac{v_\theta}{r}\frac{\partial v_r}{\partial \theta} + \frac{v_\phi}{r\sin\theta}\frac{\partial v_r}{\partial \phi} - \frac{1}{r}(v_\theta^2 + v_\phi^2) \\
&= g_r - \frac{1}{\rho}\frac{\partial p}{\partial r} \\
&\quad + \nu\left(\frac{\partial^2 v_r}{\partial r^2} + \frac{2}{r}\frac{\partial v_r}{\partial r} - \frac{2v_r}{r^2} + \frac{1}{r^2}\frac{\partial^2 v_r}{\partial \theta^2} + \frac{\cot\theta}{r^2}\frac{\partial v_r}{\partial \theta} + \frac{1}{r^2\sin^2\theta}\frac{\partial^2 v_r}{\partial \phi^2}\right. \\
&\qquad \left. - \frac{2}{r^2}\frac{\partial v_\theta}{\partial \theta} - \frac{2\cot\theta}{r^2}v_\theta - \frac{2}{r^2\sin\theta}\frac{\partial v_\phi}{\partial \phi}\right), \\
&\frac{\partial v_\theta}{\partial t} + v_r\frac{\partial v_\theta}{\partial r} + \frac{v_\theta}{r}\frac{\partial v_\theta}{\partial \theta} + \frac{v_\phi}{r\sin\theta}\frac{\partial v_\theta}{\partial \phi} + \frac{v_r v_\theta}{r} - \frac{v_\phi^2}{r}\cot\theta \\
&= g_\theta - \frac{1}{\rho}\frac{1}{r}\frac{\partial p}{\partial \theta} \\
&\quad + \nu\left(\frac{\partial^2 v_\theta}{\partial r^2} + \frac{1}{r^2}\frac{\partial^2 v_\theta}{\partial \theta^2} + \frac{1}{r^2\sin^2\theta}\frac{\partial^2 v_\theta}{\partial \phi^2} + \frac{2}{r}\frac{\partial v_\theta}{\partial r} + \frac{1}{r^2}\cot\theta\frac{\partial v_\theta}{\partial \theta}\right. \\
&\qquad \left. - \frac{1}{r^2\sin^2\theta}v_\theta + \frac{2}{r^2}\frac{\partial v_r}{\partial \theta} - \frac{2\cos\theta}{r^2\sin^2\theta}\frac{\partial v_\phi}{\partial \phi}\right), \\
&\frac{\partial v_\phi}{\partial t} + v_r\frac{\partial v_\phi}{\partial r} + \frac{v_\theta}{r}\frac{\partial v_\phi}{\partial \theta} + \frac{v_\phi}{r\sin\theta}\frac{\partial v_\phi}{\partial \phi} + \frac{v_r v_\phi}{r} + \frac{v_\theta v_\phi}{r}\cot\theta \\
&= g_\phi - \frac{1}{\rho}\frac{1}{r\sin\theta}\frac{\partial p}{\partial \phi} \\
&\quad + \nu\left(\frac{\partial^2 v_\phi}{\partial r^2} + \frac{2}{r}\frac{\partial v_\phi}{\partial r} + \frac{1}{r^2}\frac{\partial^2 v_\phi}{\partial \theta^2} + \frac{1}{r^2}\cot\theta\frac{\partial v_\phi}{\partial \theta} + \frac{1}{r^2\sin^2\theta}\frac{\partial^2 v_\phi}{\partial \phi^2}\right. \\
&\qquad \left. - \frac{1}{r^2\sin^2\theta}v_\phi + \frac{2}{r^2\sin\theta}\frac{\partial v_r}{\partial \phi} + \frac{2\cos\theta}{r^2\sin^2\theta}\frac{\partial v_\theta}{\partial \phi}\right).
\end{aligned}\right\} \quad (6.57)$$

The alternative approach followed in Example 8 is to express (6.53) in the co-ordinate-free form, using (6.50),

$$\frac{D\mathbf{v}}{Dt} = \mathbf{g} - \frac{1}{\rho}\operatorname{grad} p + v(\operatorname{grad}\operatorname{div}\mathbf{v} - \operatorname{curl}\operatorname{curl}\mathbf{v}) \tag{6.58}$$

and use physical components from the outset. For incompressible fluids, (6.58) becomes

$$\frac{D\mathbf{v}}{Dt} = \mathbf{g} - \frac{1}{\rho}\operatorname{grad} p - v\operatorname{curl}\operatorname{curl}\mathbf{v}. \tag{6.59}$$

6.3 TRANSPORT EQUATIONS

6.3.1 The energy equations

In applying the first law of thermodynamics (conservation of energy) to a fixed control volume of space occupied by a continuous material, the rate at which the total energy contained in the volume is increasing is equated to the sum of the following terms:

(i) the rate at which work is being done on the volume by external forces;
(ii) the rate at which heat is being added to the volume.

This leads to the following equation (in rectangular cartesian coordinates) for E, the total energy per unit volume; the equation must hold at each point of the continuum:

$$\frac{\partial E}{\partial t} + \frac{\partial}{\partial y_i}(Ev_i) = \frac{\partial Q}{\partial t} - \frac{\partial q_i}{\partial y_i} + \frac{\partial}{\partial y_j}(\sigma_{ij}v_i) + \rho g_i v_i. \tag{6.60}$$

Here $\mathbf{v}$ is the velocity vector at each point of the continuum, $\mathbf{q}$ is the heat flux vector, $\partial Q/\partial t$ is the rate of heating produced by external agencies (e.g. radiation) per unit volume, σ_{ij} is the Cauchy stress tensor and $\mathbf{g}$ is the body force per unit mass.

Individual terms in (6.60) may be interpreted as follows: $\partial E/\partial t$ is the local rate of increase in total energy per unit volume; $(\partial/\partial y_i)(Ev_i)$ is the rate of increase in total energy per unit volume due to convection; $(\partial/\partial y_i)(\sigma_{ij}v_j)$ is the rate of working on a small element locally due to hydrostatic (pressure) forces and viscous forces; $\rho g_i v_i$ is the rate of working of the body forces.

For heat transfer by conduction, $\mathbf{q}$ is given in terms of temperature T and thermal conductivity κ by

$$\mathbf{q} = -\kappa\operatorname{grad} T. \tag{6.61}$$

When the continuum is a classical newtonian fluid with viscosity μ, we can express σ_{ij}, according to (6.15), as

$$\sigma_{ij} = -p\delta_{ij} + \tau_{ij}, \tag{6.62}$$

where $p = -\frac{1}{3}\sigma_{kk}$ (in cartesians) and τ_{ij} can be regarded as giving the viscous stress components;

$$\tau_{ij} = 2\mu(\varepsilon_{ij} - \tfrac{1}{3}\operatorname{div}\mathbf{v}\,\delta_{ij}). \tag{6.63}$$

We then have

$$\sigma_{ij}v_j = (-p\delta_{ij} + \tau_{ij})v_j = -pv_i + \tau_{ij}v_j.$$

It follows that (6.60) may be expressed, using the obvious notation for rectangular cartesians, as

$$0 = \frac{\partial E}{\partial t} - \frac{\partial Q}{\partial t} - \rho(g_x u + g_y v + g_z w)$$

$$+ \frac{\partial}{\partial x}(Eu + pu - u\tau_{xx} - v\tau_{xy} - w\tau_{xz} + q_x)$$

$$+ \frac{\partial}{\partial y}(Ev + pv - u\tau_{xy} - v\tau_{yy} - w\tau_{yz} + q_y)$$

$$+ \frac{\partial}{\partial z}(Ew + pw - u\tau_{xz} - v\tau_{yz} - w\tau_{zz} + q_z). \tag{6.64}$$

In this form of the energy equation all terms apart from those involving body force (which are often negligible) appear under a partial derivative. This has certain advantages in numerical computations when the solution domain is discretized. Integration of the terms involving spatial derivatives $\partial/\partial y_j$ yields terms which depend only on boundary values. Since such terms cancel in pairs at cell boundaries in the interior of the solution domain, the conservation property of the transport equation is retained in each cell and is also satisfied in the whole solution domain, thus giving good numerical accuracy. We may say that (6.64) is the energy transport equation in **strong conservation form**.

Recognizing that the energy equation is a *scalar* equation, we also have

$$\frac{\partial E}{\partial t} + (Ev^i)_{,i} = \frac{\partial Q}{\partial t} - q^i{}_{,i} + (\sigma^{ij}v_i)_{,j} + \rho g^i v_i \tag{6.65}$$

with respect to generalized co-ordinates.

The energy function E may be expressed as

$$E = \rho(e + \tfrac{1}{2}v^2), \tag{6.66}$$

where e is the internal energy per unit mass and v is the magnitude of **v**. With the help of the continuity equation (6.27) and the equation of motion (6.44), we may deduce from (6.60) the **internal energy transport equation**

$$\rho \frac{De}{Dt} = \frac{\partial Q}{\partial t} - \frac{\partial q_i}{\partial y_i} + \sigma_{ij}\frac{\partial v_i}{\partial y_j}$$

$$= \frac{\partial Q}{\partial t} - \frac{\partial q_i}{\partial y_i} - p\frac{\partial v_i}{\partial y_i} + \tau_{ij}\frac{\partial v_i}{\partial y_j}. \tag{6.67}$$

The term $\tau_{ij}\partial v_i/\partial y_j$ is often referred to as the **dissipation function**, and we shall represent it by Φ. By (6.63) we have, in rectangular cartesians,

$$\Phi = 2\mu[\varepsilon_{ij} - \tfrac{1}{3}(\text{div }\mathbf{v})\delta_{ij}]v_{i,j}$$
$$= 2\mu[\varepsilon_{ij}\varepsilon_{ij} - \tfrac{1}{3}(\text{div }\mathbf{v})^2]$$
$$= 2\mu\left[\left(\frac{\partial u}{\partial x}\right)^2 + \left(\frac{\partial v}{\partial y}\right)^2 + \left(\frac{\partial w}{\partial z}\right)^2 + \frac{1}{2}\left(\frac{\partial u}{\partial y} + \frac{\partial v}{\partial x}\right)^2 + \frac{1}{2}\left(\frac{\partial v}{\partial z} + \frac{\partial w}{\partial y}\right)^2\right.$$

$$+\frac{1}{2}\left(\frac{\partial u}{\partial z}+\frac{\partial w}{\partial x}\right)^2 - \frac{1}{3}\left(\frac{\partial u}{\partial x}+\frac{\partial v}{\partial y}+\frac{\partial w}{\partial z}\right)^2 \Bigg] \qquad (6.68)$$

using the obvious notation; this represents the heat equivalent of the rate at which mechanical energy is expended in the deformation of the fluid due to viscous forces. In generalized co-ordinates,

$$\Phi = \tau^{ij}v_{i,j} = \tau^{ij}\varepsilon_{ij}$$
$$= 2\mu(\varepsilon^{ij} - \tfrac{1}{3}v^k{}_{,k}g^{ij})\varepsilon_{ij} = 2\mu[\varepsilon^{ij}\varepsilon_{ij} - \tfrac{1}{3}(v^k{}_{,k})^2]. \qquad (6.69)$$

If we introduce the enthalpy h, defined by

$$h = e + \frac{p}{\rho}, \qquad (6.70)$$

we can deduce from (6.67) the **enthalpy transport equation**

$$\rho\frac{Dh}{Dt} = \frac{Dp}{Dt} + \frac{\partial Q}{\partial t} - \frac{\partial q_i}{\partial y_i} + \Phi. \qquad (6.71)$$

For incompressible fluids satisfying (6.24), with constant thermal conductivity κ, the transport equation (6.67) becomes, using (6.61),

$$\rho\frac{De}{Dt} = \frac{\partial Q}{\partial t} + \kappa\nabla^2 T + \Phi. \qquad (6.72)$$

6.3.2 Momentum transport

Multiplying the continuity equation (6.27) through by v_i and adding to the equation of motion (6.44) in the form

$$\rho\frac{\partial v_i}{\partial t} + \rho v_j \frac{\partial v_i}{\partial y_j} = \frac{\partial \sigma_{ij}}{\partial y_j} + \rho g_i$$

yield the equation

$$\frac{\partial}{\partial t}(\rho v_i) + \frac{\partial}{\partial y_j}(\rho v_i v_j) = \frac{\partial \sigma_{ij}}{\partial y_j} + \rho g_i$$
$$= -\frac{\partial p}{\partial y_i} + \frac{\partial \tau_{ij}}{\partial y_j} + \rho g_i \qquad (6.73)$$

for a classical newtonian fluid.

This equation can be regarded as governing the transport of *momentum*, since $\rho\mathbf{v}$ is momentum per unit volume; hence it is called the **momentum equation**. It is, of course, a vector equation, and its three components in rectangular cartesians may be written out in full as

$$\left.\begin{aligned}\frac{\partial}{\partial t}(\rho u) + \frac{\partial}{\partial x}(\rho u^2 + p - \tau_{xx}) + \frac{\partial}{\partial y}(\rho uv - \tau_{xy}) + \frac{\partial}{\partial z}(\rho uw - \tau_{xz}) &= \rho g_x, \\ \frac{\partial}{\partial t}(\rho v) + \frac{\partial}{\partial x}(\rho uv - \tau_{xy}) + \frac{\partial}{\partial y}(\rho v^2 + p - \tau_{yy}) + \frac{\partial}{\partial z}(\rho vw - \tau_{yz}) &= \rho g_y,\end{aligned}\right\} \quad (6.74)$$

$$\frac{\partial}{\partial t}(\rho w) + \frac{\partial}{\partial x}(\rho u w - \tau_{xz}) + \frac{\partial}{\partial y}(\rho v w - \tau_{yz}) + \frac{\partial}{\partial z}(\rho w^2 + p - \tau_{zz}) = \rho g_z.$$

Since vector quantities are involved, (6.73) may be expressed in either covariant or contravariant form when referred to generalized co-ordinates. The contravariant form is

$$\frac{\partial}{\partial t}(\rho v^i) + (\rho v^i v^j)_{,j} = \sigma^{ij}{}_{,j} + \rho g^i$$

$$= -g^{ij} p_{,j} + \tau^{ij}{}_{,j} + \rho g^i. \qquad (6.75)$$

Note that the contravariant metric tensor g^{ij} has to be introduced here to preserve the consistent tensorial character of each term, thus ensuring that each term remains a contravariant vector in the index i.

For a classical newtonian fluid, (6.63) generalizes to

$$\tau^{ij} = 2\mu(\varepsilon^{ij} - \tfrac{1}{3} v^k{}_{,k} g^{ij}),$$

and (6.75) becomes

$$\frac{\partial}{\partial t}(\rho v^i) + (\rho v^i v^j)_{,j} = -g^{ij} p_{,j} + 2\mu(\varepsilon^{ij} - \tfrac{1}{3} v^k{}_{,k} g^{ij})_{,j} + \rho g^i.$$

When the fluid is also *incompressible*, with $\mathrm{div}\, \mathbf{v} = v^k{}_{,k} = 0$, this reduces to

$$\frac{\partial v^i}{\partial t} + (v^i v^j)_{,j} = -\frac{g^{ij}}{\rho} p_{,j} + 2\nu \varepsilon^{ij}{}_{,j} + g^i$$

$$= -\frac{g^{ij}}{\rho} p_{,j} + 2\nu [\tfrac{1}{2}(g^{jl} v^i{}_{,l} + g^{ik} v^j{}_{,k})]_{,j} + g^i$$

$$= -\frac{g^{ij}}{\rho} p_{,j} + \nu g^{jl} v^i{}_{,jl} + g^i, \qquad (6.76)$$

making use of (4.45). This equation is equivalent to (6.55).

Example

Express the momentum equation (6.76) in terms of the two-dimensional streamline co-ordinate system (see section 5.2.11).

Solution

If we denote the physical components of velocity in the s and n directions by U and V, respectively, we have, by (5.13) and (5.57),

$$v^1 = \frac{1}{h_1} U = \left(1 + \frac{n}{R}\right)^{-1} U$$

and

$$v^2 = \frac{1}{h_2} V = V.$$

Sec. 6.3] Transport equations 173

Moreover, (6.76) is in contravariant form so that, if we want the equations of momentum transport in physical components, we shall have to multiply the resulting equations by h_i, according to (5.13).

We can make use of the result (6.32), with $T^{ij} = v^i v^j$, to transform the term $(v^i v^j)_{,j}$, noting that $\sqrt{g} = h_1$ in this example. The case $i = 1$ gives

$$(v^1 v^j)_{,j} = \frac{1}{h_1} \frac{\partial}{\partial x^j} (h_1 v^1 v^j) + \Gamma^1_{lj} v^l v^j$$

$$= \frac{1}{h_1} \frac{\partial}{\partial x^1} [h_1 (v^1)^2] + \frac{1}{h_1} \frac{\partial}{\partial x^2} (h_1 v^1 v^2) + \Gamma^1_{11} (v^1)^2$$

$$+ \Gamma^1_{12} v^1 v^2 + \Gamma^1_{21} v^2 v^1 \qquad \text{by (5.59)}$$

$$= \frac{1}{h_1} \frac{\partial}{\partial s} \left(\frac{U^2}{h_1} \right) + \frac{1}{h_1} \frac{\partial}{\partial n} (UV) - \frac{nR'}{R^2 h_1^3} U^2 + \frac{2}{Rh_1^2} UV$$

$$= \frac{1}{h_1^2} \frac{\partial}{\partial s} (U^2) - \frac{1}{h_1^3} \frac{\partial h_1}{\partial s} U^2 + \frac{1}{h_1} \frac{\partial}{\partial n} (UV) + \frac{1}{h_1^3} \frac{\partial h_1}{\partial s} U^2 + \frac{2UV}{Rh_1^2}. \qquad (6.77)$$

Multiplying by h_1 to get the physical component of the LHS of (6.76) in the s direction, we obtain

$$\frac{\partial U}{\partial t} + \frac{1}{h_1} \frac{\partial}{\partial s} (U^2) + \frac{\partial}{\partial n} (UV) + \frac{2UV}{Rh_1}. \qquad (6.78)$$

A similar calculation gives, for the case $i = 2$,

$$\frac{\partial V}{\partial t} + \frac{1}{h_1} \frac{\partial}{\partial s} (UV) + \frac{\partial}{\partial n} (V^2) + \frac{V^2 - U^2}{Rh_1}, \qquad (6.79)$$

which represents the physical component of the LHS of (6.76) in the n direction.

The term $g^{jl} v^i_{,jl}$ in (6.76) may be transformed by the method used in Example 9 to transform the last term of (6.54). Here we have $g^{11} = h_1^{-2}, g^{22} = 1$, and hence

$$g^{jl} v^i_{,jl} = g^{11} v^i_{,11} + g^{22} v^i_{,22} = h_1^{-2} v^i_{,11} + v^i_{,22}.$$

Now by (4.26) we obtain

$$v^1_{,1} = \frac{\partial v^1}{\partial s} + \Gamma^1_{k1} v^k$$

$$= \frac{\partial}{\partial s} \left(\frac{U}{h_1} \right) + \frac{1}{h_1^2} \frac{\partial h}{\partial s} U + \frac{1}{Rh_1} V$$

$$= \frac{1}{h_1} \frac{\partial U}{\partial s} + \frac{1}{Rh_1} V,$$

and, similarly,

$$v^1_{,2} = \frac{1}{h_1} \frac{\partial U}{\partial n}, \qquad v^2_{,1} = \frac{\partial V}{\partial s} - \frac{U}{R}, \qquad v^2_{,2} = \frac{\partial V}{\partial n}.$$

Then, using (4.42) with $T^i{}_{,j} = v^i{}_{,j}$, we find

$$v^i{}_{,11} = \frac{\partial}{\partial s}(v^i{}_{,1}) + \Gamma^i_{l1}v^l{}_{,1} - \Gamma^l_{11}v^i{}_{,l}.$$

and

$$v^i{}_{,22} = \frac{\partial}{\partial s}(v^i{}_{,2}) + \Gamma^i_{l2}v^l{}_{,2} - \Gamma^l_{22}v^i{}_{,l}.$$

For the case $i = 1$ we obtain

$$g^{jl}v^1{}_{,jl} = \frac{1}{h_1{}^2}v^1{}_{,11} + v^1{}_{,22}$$

$$= \frac{1}{h_1{}^2}\left(\frac{\partial}{\partial s}(v^1{}_{,1}) + \Gamma^1_{l1}v^l{}_{,1} - \Gamma^l_{11}y^1{}_{,l}\right) + \left(\frac{\partial}{\partial n}(v^1{}_{,2}) + \Gamma^1_{l2}v^l{}_{,2} - \Gamma^l_{22}v^1{}_{,l}\right)$$

$$= \frac{1}{h_1{}^2}\left\{\frac{\partial}{\partial s}\left[\frac{1}{h_1}\left(\frac{\partial U}{\partial s} + \frac{V}{R}\right)\right] + \frac{1}{Rh_1}\left(\frac{\partial V}{\partial s} - \frac{U}{R}\right) + \frac{h_1}{R}\left(\frac{1}{h_1}\frac{\partial U}{\partial n}\right)\right\}$$

$$+ \frac{\partial}{\partial n}\left(\frac{1}{h_1}\frac{\partial U}{\partial n}\right) + \frac{1}{Rh_1}\left(\frac{1}{h_1}\frac{\partial U}{\partial n}\right)\bigg\}$$

$$= \frac{1}{h_1{}^2}\frac{\partial}{\partial s}\left[\frac{1}{h_1}\left(\frac{\partial U}{\partial s} + \frac{V}{R}\right)\right] + \frac{1}{h_1}\frac{\partial^2 U}{\partial n^2} + \frac{1}{Rh_1{}^2}\frac{\partial U}{\partial n} + \frac{1}{Rh_1{}^3}\left(\frac{\partial V}{\partial s} - \frac{U}{R}\right).$$

Further simplification arises from using the incompressibility conditions div $\mathbf{v} = 0$, which, by (4.34), may be written

$$\frac{\partial U}{\partial s} + \frac{\partial}{\partial n}(h_1 V) = 0,$$

i.e.

$$\frac{\partial U}{\partial s} + \frac{V}{R} = -h_1\frac{\partial V}{\partial n}. \tag{6.80}$$

The above ($i = 1$) component becomes, after some manipulation,

$$\frac{1}{h_1}\left[-\frac{1}{h_1}\frac{\partial^2 V}{\partial s\, \partial n} + \frac{\partial^2 U}{\partial n^2} + \frac{1}{Rh_1}\frac{\partial U}{\partial n} + \frac{1}{Rh_1{}^2}\left(\frac{\partial V}{\partial s} - \frac{U}{R}\right)\right]$$

$$= \frac{1}{h_1}\frac{\partial}{\partial n}\left[\frac{1}{h_1}\left(\frac{\partial}{\partial n}(h_1 U) - \frac{\partial V}{\partial s}\right)\right].$$

Multiplying this contravariant component by h_1 to obtain the physical component, and substituting into (6.76) together with the transformed term (6.78), we may obtain the following equations for the physical component of momentum transport in the s direction:

$$\frac{\partial U}{\partial t} + \frac{1}{h_1}\frac{\partial}{\partial s}(U^2) + \frac{\partial}{\partial n}(UV) + \frac{2UV}{Rh_1}$$

$$= -\frac{1}{h_1\rho}\frac{\partial p}{\partial s} + v\frac{\partial}{\partial n}\left[\frac{1}{h_1}\left(\frac{\partial}{\partial n}(h_1 U) - \frac{\partial V}{\partial s}\right)\right] + g_{(s)} \tag{6.81}$$

where $g_{(s)}$ is the physical component of body force in the s direction (see Jones and Watson 1963).

In a similar way the component $i = 2$ of (6.76) yields the physical component of momentum transport in the n direction (after substituting a number of times for $\partial V/\partial n$ from (6.80)):

$$\frac{\partial V}{\partial t} + \frac{1}{h_1}\frac{\partial}{\partial s}(UV) + \frac{\partial}{\partial n}(V^2) + \frac{V^2 - U^2}{Rh_1}$$
$$= -\frac{1}{\rho}\frac{\partial p}{\partial n} - \frac{v}{h_1}\frac{\partial}{\partial s}\left[\frac{1}{h_1}\left(\frac{\partial}{\partial n}(h_1 U) - \frac{\partial V}{\partial s}\right)\right] + g_{(n)} \qquad (6.82)$$

where $g_{(n)}$ is the physical component of body force in the n direction.

Note that, because of the extra terms generated by the Christoffel symbol, we lose the strong conservation property of the momentum equation (6.74) expressed in rectangular cartesian co-ordinates.

6.3.3 Alternative representation of the Navier–Stokes equations

With the continuity equation (6.27) expressed in rectangular cartesian form

$$\frac{\partial \rho}{\partial t} + \frac{\partial}{\partial x}(\rho u) + \frac{\partial}{\partial y}(\rho v) + \frac{\partial}{\partial z}(\rho w) = 0; \qquad (6.83)$$

equations (6.83), (6.74) and (6.64) have a similar form, representing the transport of mass, momentum and energy, respectively. When body forces $\mathbf{g}$ and external heat addition $\partial Q/\partial t$ are neglected, these five equations may be written in vector form as

$$\frac{\partial \mathbf{Q}}{\partial t} + \frac{\partial \mathbf{E}}{\partial x} + \frac{\partial \mathbf{F}}{\partial y} + \frac{\partial \mathbf{G}}{\partial z} = 0, \qquad (6.84)$$

where $\mathbf{Q}, \mathbf{E}, \mathbf{F}$ and $\mathbf{G}$ are 5×1 column vectors given by (Anderson et al. 1984)

$$\mathbf{Q} = \begin{bmatrix} \rho \\ \rho u \\ \rho v \\ \rho w \\ E \end{bmatrix}, \quad \mathbf{E} = \begin{bmatrix} \rho u \\ \rho u^2 + p - \tau_{xx} \\ \rho u v - \tau_{xy} \\ \rho u w - \tau_{xz} \\ (E + p)u - u\tau_{xx} - v\tau_{xy} - w\tau_{xz} + q_x \end{bmatrix},$$

$$\mathbf{F} = \begin{bmatrix} \rho v \\ \rho u v - \tau_{xy} \\ \rho v^2 + p - \tau_{yy} \\ \rho v w - \tau_{yz} \\ (E + p)v - u\tau_{xy} - v\tau_{yy} - w\tau_{yz} + q_y \end{bmatrix}, \qquad (6.85)$$

$$\mathbf{G} = \begin{bmatrix} \rho w \\ \rho u w - \tau_{xz} \\ \rho v w - \tau_{yz} \\ \rho w^2 + p - \tau_{zz} \\ (E + p)w - u\tau_{xz} - v\tau_{yz} - w\tau_{zz} + q_z \end{bmatrix}.$$

The set of equations (6.84), together with (6.63), which may be written as

$$\begin{aligned}
\tau_{xx} &= \tfrac{2}{3}\mu\left(2\frac{\partial u}{\partial x} - \frac{\partial v}{\partial y} - \frac{\partial w}{\partial z}\right), \\
\tau_{yy} &= \tfrac{2}{3}\mu\left(2\frac{\partial v}{\partial y} - \frac{\partial u}{\partial x} - \frac{\partial w}{\partial z}\right), \\
\tau_{zz} &= \tfrac{2}{3}\mu\left(2\frac{\partial w}{\partial z} - \frac{\partial u}{\partial x} - \frac{\partial v}{\partial y}\right), \\
\tau_{xy} &= \mu\left(\frac{\partial u}{\partial y} + \frac{\partial v}{\partial x}\right), \\
\tau_{yz} &= \mu\left(\frac{\partial v}{\partial z} + \frac{\partial w}{\partial y}\right), \\
\tau_{xz} &= \mu\left(\frac{\partial u}{\partial z} + \frac{\partial w}{\partial x}\right),
\end{aligned} \qquad (6.86)$$

are a convenient representation of the Navier–Stokes equations for compressible unsteady, viscous fluid flows (with body force and external heat addition neglected). Note that, although we have referred to (6.84) as being written in *vector* form, the term 'vector' does not here have the sense of 'physical vector' but stands for a convenient shorthand by which five equations (in rectangular cartesians) are compressed to a single one. Thus we cannot use a single tensor transformation rule to transform (6.84) directly to a 5×1 vector form referred to a general co-ordinate system.

It may be instructive to show how (6.84) may be transformed in a general two-dimensional case (with $\partial \mathbf{G}/\partial z = 0$) from rectangular cartesians (x, y) to (not necessarily orthogonal) curvilinear co-ordinates (ξ, η), by initially using just the chain rule for partial differentiation to give

$$\frac{\partial \mathbf{Q}}{\partial t} + \frac{\partial \mathbf{E}}{\partial \xi}\frac{\partial \xi}{\partial x} + \frac{\partial \mathbf{E}}{\partial \eta}\frac{\partial \eta}{\partial x} + \frac{\partial \mathbf{F}}{\partial \xi}\frac{\partial \xi}{\partial y} + \frac{\partial \mathbf{F}}{\partial \eta}\frac{\partial \eta}{\partial y} = 0, \qquad (6.87)$$

where each term is now a 4×1 column vector.

Note that, with the notation of section 3.3.1,

$$L = \begin{bmatrix} \dfrac{\partial x}{\partial \xi} & \dfrac{\partial y}{\partial \xi} \\ \dfrac{\partial x}{\partial \eta} & \dfrac{\partial y}{\partial \eta} \end{bmatrix}, \quad M = \begin{bmatrix} \dfrac{\partial \xi}{\partial x} & \dfrac{\partial \xi}{\partial y} \\ \dfrac{\partial \eta}{\partial x} & \dfrac{\partial \eta}{\partial y} \end{bmatrix},$$

and $L = (M^{\mathrm{T}})^{-1}$. Thus

$$\frac{\partial x}{\partial \xi} = \frac{1}{J^*}\frac{\partial \eta}{\partial y}, \quad \frac{\partial x}{\partial \eta} = -\frac{1}{J^*}\frac{\partial \xi}{\partial y}, \quad \frac{\partial y}{\partial \xi} = -\frac{1}{J^*}\frac{\partial \eta}{\partial x}, \quad \frac{\partial y}{\partial \eta} = \frac{1}{J^*}\frac{\partial \xi}{\partial x}, \qquad (6.88)$$

where

$$J^* = \det M = \frac{\partial \xi}{\partial x}\frac{\partial \eta}{\partial y} - \frac{\partial \xi}{\partial y}\frac{\partial \eta}{\partial x}. \qquad (6.89)$$

Sec. 6.3]   **Transport equations**   177

Dividing (6.87) by J^* then gives

$$\frac{\partial}{\partial t}\left(\frac{\mathbf{Q}}{J^*}\right) + \frac{1}{J^*}\frac{\partial \mathbf{E}}{\partial \xi}\frac{\partial \xi}{\partial x} + \frac{1}{J^*}\frac{\partial \mathbf{F}}{\partial \xi}\frac{\partial \xi}{\partial y} + \frac{1}{J^*}\frac{\partial \mathbf{E}}{\partial \eta}\frac{\partial \eta}{\partial x} + \frac{1}{J^*}\frac{\partial \mathbf{F}}{\partial \eta}\frac{\partial \eta}{\partial y}$$

$$= \frac{\partial}{\partial t}\left(\frac{\mathbf{Q}}{J^*}\right) + \frac{\partial}{\partial \xi}\left(\frac{\mathbf{E}}{J^*}\frac{\partial \xi}{\partial x}\right) + \frac{\partial}{\partial \xi}\left(\frac{\mathbf{F}}{J^*}\frac{\partial \xi}{\partial y}\right) - \mathbf{E}\frac{\partial}{\partial \xi}\left(\frac{1}{J^*}\frac{\partial \xi}{\partial x}\right) - \mathbf{F}\frac{\partial}{\partial \xi}\left(\frac{1}{J^*}\frac{\partial \xi}{\partial y}\right)$$

$$+ \frac{\partial}{\partial \eta}\left(\frac{\mathbf{E}}{J^*}\frac{\partial \eta}{\partial x}\right) + \frac{\partial}{\partial \eta}\left(\frac{\mathbf{F}}{J^*}\frac{\partial \eta}{\partial y}\right) - \mathbf{E}\frac{\partial}{\partial \eta}\left(\frac{1}{J^*}\frac{\partial \eta}{\partial x}\right) - \mathbf{F}\frac{\partial}{\partial \eta}\left(\frac{1}{J^*}\frac{\partial \eta}{\partial y}\right)$$

$$= \frac{\partial}{\partial t}\left(\frac{\mathbf{Q}}{J^*}\right) + \frac{\partial}{\partial \xi}\left(\frac{\mathbf{E}\,\partial \xi/\partial x + \mathbf{F}\,\partial \xi/\partial y}{J^*}\right) - \mathbf{E}\frac{\partial}{\partial \xi}\left(\frac{\partial y}{\partial \eta}\right) + \mathbf{F}\frac{\partial}{\partial \xi}\left(\frac{\partial x}{\partial \eta}\right)$$

$$+ \frac{\partial}{\partial \eta}\left(\frac{\mathbf{E}\,\partial \eta/\partial x + \mathbf{F}\,\partial \eta/\partial y}{J^*}\right) + \mathbf{E}\frac{\partial}{\partial \eta}\left(\frac{\partial y}{\partial \xi}\right) - \mathbf{F}\frac{\partial}{\partial \eta}\left(\frac{\partial x}{\partial \xi}\right) \qquad \text{by (6.88).}$$

Hence, retaining the strong conservation form of (6.84),

$$\frac{\partial}{\partial t}\left(\frac{\mathbf{Q}}{J^*}\right) + \frac{\partial}{\partial \xi}\left(\frac{\mathbf{E}\,\partial \xi/\partial x + \mathbf{F}\,\partial \xi/\partial y}{J^*}\right) + \frac{\partial}{\partial \eta}\left(\frac{\mathbf{E}\,\partial \eta/\partial x + \mathbf{F}\,\partial \eta/\partial y}{J^*}\right) = \mathbf{0},$$

i.e.

$$\left.\begin{aligned}\frac{\partial}{\partial t}\hat{\mathbf{Q}} + \frac{\partial}{\partial \xi}\hat{\mathbf{E}} + \frac{\partial}{\partial \eta}\hat{\mathbf{F}} &= 0,\\ \hat{\mathbf{Q}} &= (J^*)^{-1}\mathbf{Q},\\ \hat{\mathbf{E}} &= \frac{1}{J^*}\left(\mathbf{E}\frac{\partial \xi}{\partial x} + \mathbf{F}\frac{\partial \xi}{\partial y}\right),\\ \hat{\mathbf{F}} &= \frac{1}{J^*}\left(\mathbf{E}\frac{\partial \eta}{\partial x} + \mathbf{F}\frac{\partial \eta}{\partial y}\right).\end{aligned}\right\} \qquad (6.90)$$

where

and

The components of $\hat{\mathbf{E}}$ and $\hat{\mathbf{F}}$ may be simplified slightly if the contravariant components of velocity U, V in the ξ, η directions are introduced. By (3.47), we have

$$\left.\begin{aligned}U &= \frac{\partial \xi}{\partial x}u + \frac{\partial \xi}{\partial y}v\\ V &= \frac{\partial \eta}{\partial x}u + \frac{\partial \eta}{\partial y}v.\end{aligned}\right\} \qquad (6.91)$$

and

We obtain the 4×1 column vectors

$$\hat{\mathbf{E}} = \frac{1}{J^*}\begin{bmatrix}\rho U \\ \rho u U + p\frac{\partial \xi}{\partial x} - \left(\tau_{xx}\frac{\partial \xi}{\partial x} + \tau_{xy}\frac{\partial \xi}{\partial y}\right) \\ \rho v U + p\frac{\partial \xi}{\partial y} - \left(\tau_{xy}\frac{\partial \xi}{\partial x} + \tau_{yy}\frac{\partial \xi}{\partial y}\right) \\ (E + p)U - \left[u\tau_{xx}\frac{\partial \xi}{\partial x} + \tau_{xy}\left(v\frac{\partial \xi}{\partial x} + u\frac{\partial \xi}{\partial y}\right) + v\tau_{yy}\frac{\partial \xi}{\partial y}\right] - \left(q_x\frac{\partial \xi}{\partial x} + q_y\frac{\partial \xi}{\partial y}\right)\end{bmatrix}$$

and

$$\hat{\mathbf{F}} = \frac{1}{J^*} \begin{bmatrix} \rho V \\ \rho u V + p\dfrac{\partial \eta}{\partial x} - \left(\tau_{xx}\dfrac{\partial \eta}{\partial x} + \tau_{xy}\dfrac{\partial \eta}{\partial y}\right) \\ \rho v V + p\dfrac{\partial \eta}{\partial y} - \left(\tau_{xy}\dfrac{\partial \eta}{\partial x} + \tau_{yy}\dfrac{\partial \eta}{\partial y}\right) \\ (E+p)V - \left[u\tau_{xx}\dfrac{\partial \eta}{\partial x} + \tau_{xy}\left(v\dfrac{\partial \eta}{\partial x} + u\dfrac{\partial \eta}{\partial y}\right) + v\tau_{yy}\dfrac{\partial \eta}{\partial y}\right] - \left(q_x\dfrac{\partial \eta}{\partial x} + q_y\dfrac{\partial \eta}{\partial y}\right) \end{bmatrix}$$

(6.92)

The viscous stresses τ_{ij} are given by, from (6.86) and the chain rule,

$$\begin{aligned} \tau_{xx} &= \tfrac{2}{3}\mu\left[2\left(\dfrac{\partial u}{\partial \xi}\dfrac{\partial \xi}{\partial x} + \dfrac{\partial u}{\partial \eta}\dfrac{\partial \eta}{\partial x}\right) - \left(\dfrac{\partial v}{\partial \xi}\dfrac{\partial \xi}{\partial y} + \dfrac{\partial v}{\partial \eta}\dfrac{\partial \eta}{\partial y}\right)\right], \\ \tau_{xy} &= \mu\left(\dfrac{\partial u}{\partial \xi}\dfrac{\partial \xi}{\partial y} + \dfrac{\partial u}{\partial \eta}\dfrac{\partial \eta}{\partial y} + \dfrac{\partial v}{\partial \xi}\dfrac{\partial \xi}{\partial y} + \dfrac{\partial v}{\partial \eta}\dfrac{\partial \eta}{\partial y}\right), \\ \tau_{yy} &= \tfrac{2}{3}\mu\left[2\left(\dfrac{\partial v}{\partial \xi}\dfrac{\partial \xi}{\partial y} + \dfrac{\partial v}{\partial \eta}\dfrac{\partial \eta}{\partial y}\right) - \left(\dfrac{\partial u}{\partial \xi}\dfrac{\partial \xi}{\partial x} + \dfrac{\partial u}{\partial \eta}\dfrac{\partial \eta}{\partial x}\right)\right]. \end{aligned}$$

(6.93)

and

In flows over a rigid body a 'thin-layer approximation' may be used. If we take the ξ co-ordinate to run parallel to the surface of the body, this approximation entails the neglect of all derivatives with respect to ξ occurring in viscous terms, while retaining those with respect to η. A corresponding numerical approach would involve the close spacing of 'streamwise' grid lines in a boundary layer. If we separate the viscous terms from (6.90), the equations become

$$\frac{\partial \hat{\mathbf{Q}}}{\partial t} + \frac{\partial \tilde{\mathbf{E}}}{\partial \xi} + \frac{\partial \tilde{\mathbf{F}}}{\partial \eta} = \frac{\partial \mathbf{R}}{\partial \eta},$$

(6.94)

where

$$\tilde{\mathbf{E}} = \frac{1}{J^*}\begin{bmatrix} \rho U \\ \rho u U + p\dfrac{\partial \xi}{\partial x} \\ \rho v U + p\dfrac{\partial \xi}{\partial y} \\ (E+p)U - \left(q_x\dfrac{\partial \xi}{\partial x} + q_y\dfrac{\partial \xi}{\partial y}\right) \end{bmatrix}, \quad \tilde{\mathbf{F}} = \frac{1}{J^*}\begin{bmatrix} \rho V \\ \rho u V + p\dfrac{\partial \eta}{\partial x} \\ \rho v V + p\dfrac{\partial \eta}{\partial y} \\ (E+p)V - \left(q_x\dfrac{\partial \eta}{\partial x} + q_y\dfrac{\partial \eta}{\partial y}\right) \end{bmatrix}$$

and

Transport equations

$$\mathbf{R} = \frac{1}{J^*} \begin{bmatrix} 0 \\ \mu\left[\left(\frac{\partial \eta}{\partial x}\right)^2 + \left(\frac{\partial \eta}{\partial y}\right)^2\right]\frac{\partial u}{\partial \eta} + \frac{\mu}{3}\frac{\partial \eta}{\partial x}\left(\frac{\partial u}{\partial \eta}\frac{\partial \eta}{\partial x} + \frac{\partial v}{\partial \eta}\frac{\partial \eta}{\partial y}\right) \\ \mu\left[\left(\frac{\partial \eta}{\partial x}\right)^2 + \left(\frac{\partial \eta}{\partial y}\right)^2\right]\frac{\partial v}{\partial \eta} + \frac{\mu}{3}\frac{\partial \eta}{\partial y}\left(\frac{\partial u}{\partial \eta}\frac{\partial \eta}{\partial x} + \frac{\partial v}{\partial \eta}\frac{\partial \eta}{\partial y}\right) \\ \frac{\mu}{2}\left[\left(\frac{\partial \eta}{\partial x}\right)^2 + \left(\frac{\partial \eta}{\partial y}\right)^2\right]\frac{\partial}{\partial \eta}(u^2 + v^2) + \frac{\mu}{3}\frac{\partial \eta}{\partial x}\frac{\partial \eta}{\partial y}\frac{\partial}{\partial \eta}(uv) \\ + \frac{\mu}{6}\left[\left(\frac{\partial \eta}{\partial x}\right)^2\frac{\partial}{\partial \eta}(u^2) + \left(\frac{\partial \eta}{\partial y}\right)^2\frac{\partial}{\partial \eta}(v^2)\right] \end{bmatrix}.$$

Note that we have not attempted to eliminate completely the cartesian components of velocity and heat flux from (6.94). It would be possible to complete the transformation of (6.84) using (6.91) and the analogous equations for $\mathbf{q}$.

We shall not carry out the *tensor* transformation of the various components of (6.84) for the general two-dimensional case. This would require the evaluation of the metric tensor g_{ij}, which by (3.38) has the 2×2 matrix representation

$$\mathbf{G} = \mathbf{L}\mathbf{L}^T = \begin{bmatrix} \left(\frac{\partial x}{\partial \xi}\right)^2 + \left(\frac{\partial y}{\partial \xi}\right)^2 & \frac{\partial x}{\partial \xi}\frac{\partial x}{\partial \eta} + \frac{\partial y}{\partial \xi}\frac{\partial y}{\partial \eta} \\ \frac{\partial x}{\partial \xi}\frac{\partial x}{\partial \eta} + \frac{\partial y}{\partial \xi}\frac{\partial y}{\partial \eta} & \left(\frac{\partial x}{\partial \eta}\right)^2 + \left(\frac{\partial y}{\partial \eta}\right)^2 \end{bmatrix}, \qquad (6.95)$$

followed by the evaluation of the Christoffel symbols, using, for example, (4.17). This gives

$$\Gamma_{111} = \frac{1}{2}\frac{\partial g_{11}}{\partial x^1} = \frac{\partial x}{\partial \xi}\frac{\partial^2 x}{\partial \xi^2} + \frac{\partial y}{\partial \xi}\frac{\partial^2 y}{\partial \xi^2},$$

$$\Gamma_{112} = \frac{1}{2}\left(2\frac{\partial g_{12}}{\partial x^1} - \frac{\partial g_{11}}{\partial x^2}\right) = \frac{\partial x}{\partial \eta}\frac{\partial^2 x}{\partial \xi^2} + \frac{\partial y}{\partial \eta}\frac{\partial^2 y}{\partial \xi^2}, \quad \text{etc.}$$

We would then be in a position to express the scalar equations (6.25) and (6.65) together with the vector equation (6.76) with respect to the (ξ, η) co-ordinates.

7

Turbulence equations

7.1 INTRODUCTION

Turbulence is a characteristic of certain fluid flows rather than a property of fluids such as viscosity. It may be regarded as a continuum phenomenon insofar as the smallest length scale in turbulent flows, the Kolmogorov microscale, is generally much larger than the molecular length scales. Although turbulence is rather difficult to define, various typical features may be listed, such as irregularities of flow, diffusivity of mass, momentum and heat, high Reynolds number, three-dimensionality, vorticity fluctuation, and dissipation of energy (Hinze 1959).

The continuum view of turbulence leads one to expect that the Navier–Stokes equations for the flow of a viscous fluid should still be applicable, together with the continuity equation. The statistical approach to these equations with which we shall be concerned here uses the concept of time averaging, introduced by Reynolds in 1894. The rules of time averaging give rise to good approximations for fluid flows in which the fluctuations are sufficiently numerous and distributed at random. The turbulent flows considered here are assumed to have steady mean properties.

However, time averaging the equations of fluid flow always generates equations in which there are more unknown quantities than equations. This problem is generally referred to as the **closure problem** of turbulence. The solutions of the equations are indeterminate unless further relations between the unknowns are postulated to close the set of equations. This procedure is akin to the formulation of constitutive equations for materials and is called **turbulence modelling**. The degree of complexity of the particular turbulence model one will choose or formulate in a given problem will depend on the degree of accuracy required in the results and the cost of computing the solutions.

In this chapter we consider a selection of the more important equations which arise in this approach to turbulence, our primary objective being to illustrate their

7.2 EXAMPLES

Example 1

Transform the time-averaged Navier–Stokes equations from a rectangular cartesian co-ordinate system to cylindrical polar co-ordinates.

Method

The time-averaged Navier–Stokes equations for incompressible fluid flow are given in rectangular cartesian co-ordinates (neglecting body forces) by

$$U_j \frac{\partial U_i}{\partial y_j} = -\frac{1}{\rho}\frac{\partial p}{\partial y_i} + v\frac{\partial^2 U_i}{\partial y_j \partial y_j} - \frac{\partial}{\partial y_j}(\overline{u_i u_j}), \tag{7.1}$$

where the velocity vector **v** has been decomposed into the sum of mean and fluctuating velocities **U** and **u**, with components U_i and u_i, respectively; the over-bar denotes time averaging. Equations (7.1) are commonly known as the **Reynolds equations**. The essential difference between (7.1) and (6.53) is the absence of the unsteady term $\partial v_i/\partial t$ in (7.1) and the extra term $\partial(\overline{u_i u_j})/\partial y_j$ arising from time averaging. This extra term corresponds to the gradient of 'Reynolds stress' $-\rho\overline{u_i u_j}$.

The time-averaged equation (7.1) is still a *vector* equation (in the *i* component) and can be represented either covariantly or contravariantly with respect to generalized co-ordinates. The contravariant form is

$$U^j U^i{}_{,j} = -\frac{1}{\rho} g^{ij} p_{,j} + v g^{jl} U^i{}_{,jl} - \overline{u^i u^j}{}_{,j}. \tag{7.2}$$

With the notation (r, θ, z) for cylindrical polar co-ordinates instead of (ρ, ϕ, z), the physical components of **U** are given by

$$U_r = U^1, \quad U_\theta = rU^2, \quad U_z = U^3,$$

and the three contravariant components of the LHS of (7.2) may be obtained by multiplying the matrix (6.13) into the column vector of contravariant components U^j. Thus

$$\begin{bmatrix} \dfrac{\partial U_r}{\partial r} & \dfrac{\partial U_r}{\partial \theta} - U_\theta & \dfrac{\partial U_r}{\partial z} \\ \dfrac{1}{r}\dfrac{\partial U_\theta}{\partial r} & \dfrac{1}{r}\dfrac{\partial U_\theta}{\partial \theta} + \dfrac{U_r}{r} & \dfrac{1}{r}\dfrac{\partial U_\theta}{\partial z} \\ \dfrac{\partial U_z}{\partial r} & \dfrac{\partial U_z}{\partial \theta} & \dfrac{\partial U_z}{\partial z} \end{bmatrix} \begin{bmatrix} U_r \\ \dfrac{1}{r}U_\theta \\ U_z \end{bmatrix}$$

$$= \begin{bmatrix} U_r \dfrac{\partial U_r}{\partial r} + \dfrac{1}{r} U_\theta \dfrac{\partial U_r}{\partial \theta} - \dfrac{1}{r} U_\theta^2 + U_z \dfrac{\partial U_r}{\partial z} \\[6pt] \dfrac{1}{r} U_r \dfrac{\partial U_\theta}{\partial r} + \dfrac{1}{r^2} U_\theta \dfrac{\partial U_\theta}{\partial \theta} + \dfrac{1}{r^2} U_r U_\theta + \dfrac{1}{r} U_z \dfrac{\partial U_\theta}{\partial z} \\[6pt] U_r \dfrac{\partial U_z}{\partial r} + \dfrac{1}{r} U_\theta \dfrac{\partial U_z}{\partial \theta} + U_z \dfrac{\partial U_z}{\partial \theta} \end{bmatrix}. \qquad (7.3)$$

Note that, after multiplying each row by the appropriate scale factor to obtain the physical components, this result is identical with (6.41), apart from the time-derivative term; (6.41) was derived using the covariant form of the acceleration vector $D\mathbf{v}/Dt$.

The middle term of the RHS of (7.2) can be shown, by the method used in Chapter 6, Example 9, for spherical polar co-ordinates, to yield the set of contravariant components:

$$\nu \begin{bmatrix} \dfrac{\partial^2 U_r}{\partial r^2} + \dfrac{1}{r^2} \dfrac{\partial^2 U_r}{\partial \theta^2} + \dfrac{\partial^2 U_r}{\partial z^2} + \dfrac{1}{r} \dfrac{\partial U_r}{\partial r} - \dfrac{2}{r^2} \dfrac{\partial U_\theta}{\partial \theta} - \dfrac{U_r}{r^2} \\[6pt] \dfrac{1}{r} \dfrac{\partial^2 U_\theta}{\partial r^2} + \dfrac{1}{r^3} \dfrac{\partial^2 U_\theta}{\partial \theta^2} + \dfrac{1}{r} \dfrac{\partial^2 U_\theta}{\partial z^2} + \dfrac{1}{r^2} \dfrac{\partial U_\theta}{\partial r} + \dfrac{2}{r^3} \dfrac{\partial U_r}{\partial \theta} - \dfrac{U_\theta}{r^3} \\[6pt] \dfrac{\partial^2 U_z}{\partial r^2} + \dfrac{1}{r^2} \dfrac{\partial^2 U_z}{\partial \theta^2} + \dfrac{\partial^2 U_z}{\partial z^2} + \dfrac{1}{r} \dfrac{\partial U_z}{\partial r} \end{bmatrix}. \qquad (7.4)$$

Finally, the last term in (7.2) may be evaluated using (6.32) with $T^{ij} = \overline{u^i u^j}$. In cylindrical polars

$$\overline{u^i u^j}_{,j} = \frac{1}{r} \frac{\partial}{\partial x^j} (r \overline{u^i u^j}) + \Gamma^i_{lj} \overline{u^l u^j}$$

$$= \frac{1}{r} \frac{\partial}{\partial x^j} (r \overline{u^i u^j}) + \Gamma^i_{12} \overline{u^1 u^2} + \Gamma^i_{21} \overline{u^2 u^1} + \Gamma^i_{22} \overline{(u^2)^2},$$

using (4.13). Also, by (5.15),

$$\overline{u^i u^j} = \frac{1}{h_i h_j} \overline{u_{(i)} u_{(j)}} \quad \text{(n.s.)}$$

in terms of physical components.

Thus for $i = 1$ we obtain

$$\overline{u^1 u^j}_{,j} = \frac{1}{r} \frac{\partial}{\partial x^j} (r \overline{u^1 u^j}) + \Gamma^1_{22} \overline{(u^2)^2}$$

$$= \frac{1}{r} \left(\frac{\partial}{\partial r} (r \overline{(u^1)^2}) + \frac{\partial}{\partial \theta} (r \overline{u^1 u^2}) + \frac{\partial}{\partial z} (r \overline{u^1 u^3}) \right) - r \overline{(u^2)^2}$$

$$= \frac{1}{r} \left(\frac{\partial}{\partial r} (r \overline{u_r^2}) + \frac{\partial}{\partial \theta} (\overline{u_r u_\theta}) + \frac{\partial}{\partial z} (r \overline{u_r u_z}) \right) - \frac{1}{r} \overline{u_\theta^2}$$

$$= \frac{\partial}{\partial r} (\overline{u_r^2}) + \frac{1}{r} \frac{\partial}{\partial \theta} (\overline{u_r u_\theta}) + \frac{\partial}{\partial z} (\overline{u_r u_z}) + \frac{\overline{u_r^2} - \overline{u_\theta^2}}{r}.$$

Similarly, with $i = 2$ we have the contravariant component

$$\overline{u^2 u^j}_{,j} = \frac{1}{r}\frac{\partial}{\partial r}(\overline{u_r u_\theta}) + \frac{1}{r^2}\frac{\partial}{\partial \theta}(\overline{u_\theta^2}) + \frac{1}{r}\frac{\partial}{\partial z}(\overline{u_\theta u_z}) + \frac{2}{r^2}\overline{u_r u_\theta},$$

and, for $i = 3$,

$$\overline{u^3 u^j}_{,j} = \frac{\partial}{\partial r}(\overline{u_r u_z}) + \frac{1}{r}\frac{\partial}{\partial \theta}(\overline{u_\theta u_z}) + \frac{\partial}{\partial z}(\overline{u_z^2}) + \frac{\overline{u_r u_z}}{r}.$$

Substituting these contravariant components, together with (7.3) and (7.4), into (7.2) and multiplying the $i = 2$ component by the scale factor r to give the physical component in the θ direction, we obtain the three equations:

$$\left.\begin{aligned}
& U_r \frac{\partial U_r}{\partial r} + \frac{1}{r} U_\theta \frac{\partial U_r}{\partial \theta} - \frac{1}{r} U_\theta^2 + U_z \frac{\partial U_r}{\partial z} \\
& = -\frac{1}{\rho}\frac{\partial p}{\partial r} + \nu\left(\frac{\partial^2 U_r}{\partial r^2} + \frac{1}{r^2}\frac{\partial^2 U_r}{\partial \theta^2} + \frac{\partial^2 U_r}{\partial z^2} + \frac{1}{r}\frac{\partial U_r}{\partial r} - \frac{2}{r^2}\frac{\partial U_\theta}{\partial \theta} - \frac{U_r}{r^2}\right) \\
& \quad - \left(\frac{\partial}{\partial r}(\overline{u_r^2}) + \frac{1}{r}\frac{\partial}{\partial \theta}(\overline{u_r u_\theta}) + \frac{\partial}{\partial z}(\overline{u_r u_z}) + \frac{\overline{u_r^2} - \overline{u_\theta^2}}{r}\right), \\
\\
& U_r \frac{\partial U_\theta}{\partial r} + \frac{1}{r} U_\theta \frac{\partial U_\theta}{\partial \theta} + \frac{U_r U_\theta}{r} + U_z \frac{\partial U_\theta}{\partial z} \\
& = -\frac{1}{\rho r}\frac{\partial p}{\partial \theta} + \nu\left(\frac{\partial^2 U_\theta}{\partial r^2} + \frac{1}{r^2}\frac{\partial^2 U_\theta}{\partial \theta^2} + \frac{\partial^2 U_\theta}{\partial z^2} + \frac{1}{r}\frac{\partial U_\theta}{\partial r} + \frac{2}{r^2}\frac{\partial U_r}{\partial \theta} - \frac{U_\theta}{r^2}\right) \\
& \quad - \left(\frac{\partial}{\partial r}(\overline{u_r u_\theta}) + \frac{1}{r}\frac{\partial}{\partial \theta}(\overline{u_\theta^2}) + \frac{\partial}{\partial z}(\overline{u_\theta u_z}) + \frac{2}{r}\overline{u_r u_\theta}\right),
\end{aligned}\right.$$

and

$$\left.\begin{aligned}
& U_r \frac{\partial U_z}{\partial r} + \frac{1}{r} U_\theta \frac{\partial U_z}{\partial \theta} + U_z \frac{\partial U_z}{\partial z} \\
& = -\frac{1}{\rho}\frac{\partial p}{\partial z} + \nu\left(\frac{\partial^2 U_z}{\partial r^2} + \frac{1}{r^2}\frac{\partial^2 U_z}{\partial \theta^2} + \frac{\partial^2 U_z}{\partial z^2} + \frac{1}{r}\frac{\partial U_z}{\partial r}\right) \\
& \quad - \left(\frac{\partial}{\partial r}(\overline{u_r u_z}) + \frac{1}{r}\frac{\partial}{\partial \theta}(\overline{u_\theta u_z}) + \frac{\partial}{\partial z}(\overline{u_z^2}) + \frac{\overline{u_r u_z}}{r}\right).
\end{aligned}\right\} \quad (7.5)$$

Example 2

Transform the Reynolds stress transport equation from rectangular cartesian co-ordinates to cylindrical polar co-ordinates.

Method

The transport equation for Reynolds stress in rectangular cartesians with buoyancy forces neglected is given by

$$U_k \frac{\partial}{\partial y_k} \overline{u_i u_i} = \underbrace{-\frac{\partial}{\partial y_k}\left(-v\frac{\partial}{\partial y_k}(\overline{u_i u_j}) + \overline{u_k u_i u_j} + \frac{1}{\rho}\overline{(\delta_{jk}u_i + \delta_{ik}u_j)p'}\right)}_{\text{diffusion}}$$

$$\underbrace{-\left(\overline{u_i u_k}\frac{\partial U_j}{\partial y_k} + \overline{u_j u_k}\frac{\partial U_i}{\partial y_k}\right)}_{\text{production}} + \underbrace{\frac{1}{\rho}\overline{p'\left(\frac{\partial u_i}{\partial y_j} + \frac{\partial u_j}{\partial y_i}\right)}}_{\text{redistribution}} - \underbrace{2v\overline{\frac{\partial u_i}{\partial y_k}\frac{\partial u_j}{\partial y_k}}}_{\text{dissipation}}, \quad (7.6)$$

where the conventional description of the function of various terms, or groups of terms, is indicated; p' is the fluctuation in static pressure. This equation arises frequently in 'second-order modelling' in the theory of turbulence. In particular, the modelling of the 'pressure–strain correlation' (the redistribution term above) has been the subject of intensive research.

Note that the equation is of second-order tensor character, with nine component equations (corresponding to two free indices i and j), only six of which, however, are independent because of the symmetry of each term indicated in (7.6). Thus, in expressing (7.6) with respect to a set of generalized co-ordinates, we have a choice of three representations, namely contravariant, covariant or mixed. (The two mixed representations are equivalent because of the symmetry.)

Choosing the contravariant representation yields

$$\underbrace{U^k \overline{u^i u^j}_{,k}}_{\text{convection}} = \underbrace{-\left(-vg^{kl}(\overline{u^i u^j})_{,kl} + (\overline{u^k u^i u^j})_{,k} + \frac{1}{\rho}[(\overline{p'u^i})_{,k}g^{jk} + (\overline{p'u^j})_{,k}g^{ik}]\right)}_{\text{diffusion}}$$

$$\underbrace{-\overline{u^i u^k}\, U^j_{,k} - \overline{u^j u^k}\, U^i_{,k}}_{\text{production}} + \underbrace{\frac{1}{\rho}(\overline{p'u^i_{,k}}g^{jk} + \overline{p'u^j_{,k}}g^{ik})}_{\text{redistribution}} - \underbrace{2v\overline{u^i_{,l}u^j_{,k}}g^{kl}}_{\text{dissipation}}. \quad (7.7)$$

This is the unique valid generalized contravariant second-order tensor form which reduces to (7.6) in rectangular cartesians.

Because of the length of the algebra involved, we evaluate (7.7) in cylindrical polar co-ordinates only for the two cases $i = j = 1$ and $i = 1, j = 2$. Since $h_1 = 1$ and $h_2 = r$ in cylindrical polars, the contravariant component with $i = j = 1$ is also the (1–1) physical component, according to (5.15), but the contravariant component for $i = 1$, $j = 2$, must be multiplied by r to give the corresponding physical component.

The named terms in (7.7) are each associated with second-order tensors, and we can represent the equation symbolically as

$$C^{ij} = D^{ij} + P^{ij} + \Phi^{ij} - \mathscr{E}^{ij}.$$

We proceed to evaluate each of these terms in cylindrical polars.

(a) Transformation of convection term According to (4.40), the convection term is

$$C^{ij} = U^k \overline{u^i u^j}_{,k} = U^k\left(\frac{\partial}{\partial x^k}(\overline{u^i u^j}) + \Gamma^i_{lk}\overline{u^l u^j} + \Gamma^j_{lk}\overline{u^i u^l}\right).$$

Examples

First consider the case $i = j = 1$. We obtain

$$U^k\left(\frac{\partial}{\partial x^k}(\overline{u^1 u^1}) + \Gamma^1_{lk}\overline{u^l u^1} + \Gamma^1_{lk}\overline{u^1 u^l}\right),$$

with summation understood over k and l but, because of (4.13), only the values $l = 2$ and $k = 2$ make a non-zero contribution to the last two terms. Thus we have

$$C^{11} = U^1\frac{\partial}{\partial x^1}(\overline{(u^1)^2}) + U^2\frac{\partial}{\partial x^2}(\overline{(u^1)^2}) + U^3\frac{\partial}{\partial x^3}(\overline{(u^1)^2}) + U^2\Gamma^1_{22}\overline{u^2 u^1} + U^2\Gamma^1_{22}\overline{u^1 u^2}$$

$$= U_r\frac{\partial}{\partial r}\overline{u_r^2} + \frac{U_\theta}{r}\frac{\partial}{\partial \theta}\overline{u_r^2} + U_z\frac{\partial}{\partial z}\overline{u_r^2} - 2\frac{U_\theta}{r}\overline{u_r u_\theta}, \tag{7.8}$$

since $U^1 = U_r$, $U^2 = U_\theta/r$, etc., by (5.13).

For $i = 1$, $j = 2$, the convection term is

$$C^{12} = U^k\left(\frac{\partial}{\partial x^k}(\overline{u^1 u^2}) + \Gamma^1_{lk}\overline{u^l u^2} + \Gamma^2_{lk}\overline{u^1 u^l}\right)$$

$$= U^1\frac{\partial}{\partial x^1}(\overline{u^1 u^2}) + U^2\frac{\partial}{\partial x^2}(\overline{u^1 u^2}) + U^3\frac{\partial}{\partial x^3}(\overline{u^1 u^2})$$

$$+ U^2\Gamma^1_{22}\overline{(u^2)^2} + U^1\Gamma^2_{21}\overline{u^1 u^2} + U^2\Gamma^2_{12}\overline{(u^1)^2},$$

summing over the non-zero terms, i.e.

$$U_r\frac{\partial}{\partial r}\left(\frac{1}{r}\overline{u_r u_\theta}\right) + \frac{U_\theta}{r}\frac{\partial}{\partial \theta}\left(\frac{1}{r}\overline{u_r u_\theta}\right) + U_z\frac{\partial}{\partial z}\left(\frac{1}{r}\overline{u_r u_\theta}\right) - \frac{U_\theta}{r^2}\overline{u_\theta^2} + \frac{U_r}{r^2}\overline{u_r u_\theta} + \frac{U_\theta}{r^2}\overline{u_r^2}. \tag{7.9}$$

This contravariant 1–2 component has to be multiplied by r to produce the physical r–θ component:

$$C_{(12)} = U_r\frac{\partial}{\partial r}\overline{u_r u_\theta} + \frac{U_\theta}{r}\frac{\partial}{\partial \theta}\overline{u_r u_\theta} + U_z\frac{\partial}{\partial z}\overline{u_r u_\theta} + \frac{U_\theta}{r}(\overline{u_r^2} - \overline{u_\theta^2}). \tag{7.10}$$

(b) Transformation of diffusion term The diffusion term involves the lengthiest calculations. We explicitly evaluate each of the three terms in the diffusion expression for the cases $i = 1, j = 1$ and $i = 1, j = 2$, beginning with the term $vg^{kl}\overline{u^i u^j}_{,kl}$. Performing the summations over k and l gives

$$v\left(\overline{u^i u^j}_{,11} + \frac{1}{r^2}\overline{u^i u^j}_{,22} + \overline{u^i u^j}_{,33}\right). \tag{7.11}$$

Now

$$\overline{u^i u^j}_{,1} = \frac{\partial}{\partial x^1}(\overline{u^i u^j}) + \Gamma^i_{l1}\overline{u^l u^j} + \Gamma^j_{l1}\overline{u^i u^l}, \tag{7.12}$$

by (4.40). Moreover, by the general rule in (4.44) for the covariant derivatives of higher-order tensors.

$$(\overline{u^i u^j})_{,11} = (\overline{u^i u^j}_{,1})_{,1}$$

$$= \frac{\partial}{\partial x^1}(\overline{u^i u^j}_{,1}) + \Gamma^i_{l1}\overline{u^l u^j}_{,1} + \Gamma^j_{l1}\overline{u^i u^l}_{,1} - \Gamma^l_{11}\overline{u^i u^j}_{,l}, \tag{7.13}$$

since $u^i u^j{}_{,k}$ is a third-order tensor.

With $i = j = 1$, summing over non-zero components gives

$$\frac{\partial}{\partial x^1}[\overline{(u^1)^2}_{,1}] = \frac{\partial}{\partial r}[\overline{(u^1)^2}_{,1}] = \frac{\partial}{\partial r}\left(\frac{\partial}{\partial r}[\overline{(u^1)^2}]\right) \quad \text{by (7.12)}$$

$$= \frac{\partial^2}{\partial r^2}(\overline{u_r^2}).$$

Similarly, using (4.44),

$$(\overline{u^i u^j})_{,22} = \frac{\partial}{\partial x^2}(\overline{u^i u^j}_{,2}) + \Gamma^i_{l2}\overline{u^l u^j}_{,2} + \Gamma^j_{l2}\overline{u^i u^l}_{,2} - \Gamma^l_{22}\overline{u^i u^j}_{,l}$$

$$= \frac{\partial}{\partial x^2}(\overline{u^i u^j}_{,2}) + \Gamma^i_{12}\overline{u^1 u^j}_{,2} + \Gamma^i_{22}\overline{u^2 u^j}_{,2} + \Gamma^j_{12}\overline{u^i u^1}_{,2} + \Gamma^j_{22}\overline{u^i u^2}_{,2} - \Gamma^1_{22}\overline{u^i u^j}_{,1} \tag{7.14}$$

carrying out the l-summation over possibly non-zero terms, while, by (4.40),

$$\overline{u^i u^j}_{,2} = \frac{\partial}{\partial x^2}(\overline{u^i u^j}) + \Gamma^i_{l2}\overline{u^l u^j} + \Gamma^j_{l2}\overline{u^i u^l}.$$

For the case $i = j = 1$, we obtain

$$\overline{(u^1)^2}_{,22} = \frac{\partial}{\partial x^2}[\overline{(u^1)^2}_{,2}] + \Gamma^1_{22}\overline{u^2 u^1}_{,2} + \Gamma^1_{22}\overline{u^1 u^2}_{,2} - \Gamma^1_{22}\overline{(u^1)^2}_{,1}$$

$$= \frac{\partial}{\partial \theta}\left(\frac{\partial}{\partial \theta}(\overline{u_r^2}) - r\overline{u^2 u^1} - r\overline{u^1 u^2}\right) - 2r\left(\frac{\partial}{\partial \theta}(\overline{u^1 u^2}) - r\overline{(u^2)^2} + \frac{1}{r}\overline{(u^1)^2}\right) + r\frac{\partial}{\partial r}(\overline{u_r^2})$$

$$= \frac{\partial^2}{\partial \theta^2}(\overline{u_r^2}) - 4\frac{\partial}{\partial \theta}(\overline{u_r u_\theta}) + 2\overline{u_\theta^2} - 2\overline{u_r^2} + r\frac{\partial}{\partial r}(\overline{u_r^2}).$$

The final term in (7.11) is easily evaluated, and the resulting diffusion component for $i = j = 1$ is

$$\nu\left(\frac{\partial^2}{\partial^2}\overline{u_r^2} + \frac{1}{r^2}\frac{\partial^2}{\partial \theta^2}\overline{u_r^2} - \frac{4}{r^2}\frac{\partial}{\partial \theta}\overline{u_r u_\theta} + \frac{2}{r^2}\overline{u_\theta^2} - \frac{2}{r^2}\overline{u_r^2} + \frac{1}{r}\frac{\partial}{\partial r}\overline{u_r^2} + \frac{\partial^2}{\partial z^2}\overline{u_r^2}\right),$$

which may be expressed as

$$\nu\left(\nabla^2 \overline{u_r^2} - \frac{4}{r^2}\frac{\partial}{\partial \theta}\overline{u_r u_\theta} - \frac{2}{r^2}(\overline{u_r^2} - \overline{u_\theta^2})\right), \tag{7.15}$$

making use of the laplacian operator ∇^2 in cylindrical polars,

$$\nabla^2(\ ) = \frac{1}{r}\frac{\partial}{\partial r}\left(r\frac{\partial}{\partial r}(\ )\right) + \frac{1}{r^2}\frac{\partial^2}{\partial \theta^2}(\ ) + \frac{\partial^2}{\partial z^2}(\ ).$$

Sec. 7.2]	**Examples**	187

For $i = 1$, $j = 2$, (7.13) becomes

$$\overline{(u^1 u^2)}_{,11} = \frac{\partial}{\partial x^1}(\overline{u^1 u^2}_{,1}) + \Gamma^1_{l1}\overline{u^l u^2}_{,1} + \Gamma^2_{l1}\overline{u^1 u^l}_{,1} - \Gamma^l_{11}\overline{u^1 u^2}_{,l}$$

$$= \frac{\partial}{\partial x^1}(\overline{u^1 u^2}_{,1}) + \frac{1}{r}\overline{u^1 u^2}_{,1}$$

$$= \frac{\partial}{\partial r}\left(\frac{\partial}{\partial r}(\overline{u^1 u^2}) + \Gamma^2_{21}(\overline{u^1 u^2})\right) + \frac{1}{r}\left(\frac{\partial}{\partial r}(\overline{u^1 u^2}) + \Gamma^2_{21}(\overline{u^1 u^2})\right),$$

using (7.12) to evaluate $\overline{u^1 u^2}_{,1}$. Thus

$$\overline{u^1 u^2}_{,11} = \frac{\partial}{\partial r}\left[\frac{\partial}{\partial r}\left(\frac{1}{r}\overline{u_r u_\theta}\right) + \frac{1}{r^2}\overline{u_r u_\theta}\right] + \frac{1}{r}\left[\frac{\partial}{\partial r}\left(\frac{1}{r}\overline{u_r u_\theta}\right) + \frac{1}{r^2}\overline{u_r u_\theta}\right]$$

$$= \frac{1}{r}\frac{\partial^2}{\partial r^2}(\overline{u_r u_\theta}).$$

Moreover, (7.14) gives

$$\overline{u^1 u^2}_{,22} = \frac{\partial}{\partial x^2}(\overline{u^1 u^2}_{,2}) + \Gamma^1_{l2}\overline{u^l u^2}_{,2} + \Gamma^2_{l2}\overline{u^1 u^l}_{,2} - \Gamma^l_{22}\overline{u^1 u^2}_{,l}$$

$$= \frac{\partial}{\partial x^2}\left(\frac{\partial}{\partial x^2}(\overline{u^1 u^2}) + \Gamma^1_{l2}\overline{u^l u^2} + \Gamma^2_{l2}\overline{u^1 u^l}\right) - r\overline{(u^2)^2}_{,2} + \frac{1}{r}\overline{(u^1)^2}_{,2} + r\overline{u^1 u^2}_{,1}$$

$$= \frac{\partial}{\partial \theta}\left[\frac{\partial}{\partial \theta}\left(\frac{1}{r}\overline{u_r u_\theta}\right) - r\frac{\overline{u_\theta^2}}{r^2} + \frac{1}{r}\overline{u_r^2}\right] - r\left(\frac{\partial}{\partial \theta}[\overline{(u^2)^2}] + \Gamma^1_{l2}\overline{u^l u^2} + \Gamma^2_{l2}\overline{u^2 u^l}\right)$$

$$+ \frac{1}{r}\left(\frac{\partial}{\partial \theta}[\overline{(u^1)^2}] + \Gamma^1_{l2}\overline{u^l u^1} + \Gamma^1_{l2}\overline{u^1 u^l}\right) + r\left(\frac{\partial}{\partial r}(\overline{u^1 u^2}) + \Gamma^1_{l1}\overline{u^l u^2} + \Gamma^2_{l1}\overline{u^1 u^l}\right)$$

$$= \left(\frac{1}{r}\frac{\partial^2}{\partial \theta^2}(\overline{u_r u_\theta}) - \frac{1}{r}\frac{\partial}{\partial \theta}(\overline{u_\theta^2}) + \frac{1}{r}\frac{\partial}{\partial \theta}(\overline{u_r^2})\right) - r\left[\frac{\partial}{\partial \theta}\left(\frac{1}{r^2}\overline{u_\theta^2}\right) + \frac{2}{r^2}\overline{u_r u_\theta}\right]$$

$$+ \frac{1}{r}\left(\frac{\partial}{\partial \theta}(\overline{u_r^2}) - 2r\frac{\overline{u_r u_\theta}}{r}\right) + r\left[\frac{\partial}{\partial r}\left(\frac{1}{r}\overline{u_r u_\theta}\right) + \frac{1}{r^2}\overline{u_r u_\theta}\right]$$

$$= \frac{1}{r}\frac{\partial^2}{\partial \theta^2}(\overline{u_r u_\theta}) - \frac{2}{r}\frac{\partial}{\partial \theta}\overline{u_\theta^2} + \frac{2}{r}\frac{\partial}{\partial \theta}\overline{u_r^2} - \frac{4}{r}\overline{u_r u_\theta} + \frac{\partial}{\partial r}\overline{u_r u_\theta}.$$

Finally,

$$\overline{u^1 u^2}_{,33} = \frac{1}{r}\frac{\partial^2}{\partial z^2}(\overline{u_r u_\theta}),$$

giving the diffusion term for $i = 1$, $j = 2$, as

$$\nu\left(\overline{u^1 u^2}_{,11} + \frac{1}{r^2}\overline{u^1 u^2}_{,22} + \overline{u^1 u^2}_{,33}\right)$$

$$= \frac{\nu}{r}\left(\frac{\partial^2}{\partial r^2}\overline{u_r u_\theta} + \frac{1}{r}\frac{\partial}{\partial r}\overline{u_r u_\theta} + \frac{1}{r^2}\frac{\partial^2}{\partial \theta^2}\overline{u_r u_\theta} + \frac{\partial^2}{\partial z^2}\overline{u_r u_\theta} + \frac{2}{r^2}\frac{\partial}{\partial \theta}\overline{u_r^2} - \frac{2}{r^2}\frac{\partial}{\partial \theta}\overline{u_\theta^2} - \frac{4}{r^2}\overline{u_r u_\theta}\right)$$

$$= \frac{\nu}{r}\left[\nabla^2\overline{u_r u_\theta} + \frac{2}{r^2}\frac{\partial}{\partial\theta}(\overline{u_r^2} - \overline{u_\theta^2}) - \frac{4}{r^2}\overline{u_r u_\theta}\right]. \tag{7.16}$$

The corresponding physical r–θ component is obtained by multiplying (7.16) by r.

The next term in the diffusion expression is $\overline{u^k u^i u^j}_{,k}$ and is, fortunately, easier to evaluate. By (4.44) and (4.23) we have

$$\overline{u^k u^i u^j}_{,k} = \frac{\partial}{\partial x^k}(\overline{u^k u^i u^j}) + \Gamma^k_{mk}\overline{u^m u^i u^j} + \Gamma^i_{mk}\overline{u^k u^m u^j} + \Gamma^j_{mk}\overline{u^k u^i u^m}$$

$$= \frac{\partial}{\partial x^k}(\overline{u^k u^i u^j}) + \frac{1}{r}\frac{\partial}{\partial x^m}(r)\overline{u^m u^i u^j} + \Gamma^i_{mk}\overline{u^k u^m u^j} + \Gamma^j_{mk}\overline{u^k u^i u^m}.$$

Hence, for $i = j = 1$,

$$\overline{u^k(u^1)^2}_{,k} = \frac{\partial}{\partial x^k}(\overline{u^k(u^1)^2}) + \frac{1}{r}\overline{(u^1)^3} + 2\Gamma^1_{mk}\overline{u^k u^m u^1}$$

$$= \frac{\partial}{\partial r}(\overline{u_r^3}) + \frac{1}{r}\frac{\partial}{\partial\theta}(\overline{u_r^2 u_\theta}) + \frac{\partial}{\partial z}(\overline{u_r^2 u_z}) + \frac{1}{r}\overline{u_r^3} - \frac{2}{r}\overline{u_r u_\theta^2}. \tag{7.17}$$

The incompressibility condition may be used to express this as

$$\overline{u_r \frac{\partial}{\partial r}u_r^2} + \frac{1}{r}\overline{u_\theta \frac{\partial u_r^2}{\partial\theta}} + \overline{u_z \frac{\partial u_r^2}{\partial z}} - \frac{2}{r}\overline{u_r u_\theta^2}. \tag{7.18}$$

For $i = 1$, $j = 2$, we have

$$\overline{u^k u^1 u^2}_{,k} = \frac{\partial}{\partial x^k}(\overline{u^k u^1 u^2}) + \frac{1}{r}\overline{(u^1)^2 u^2} + \Gamma^1_{mk}\overline{u^k u^m u^2} + \Gamma^2_{mk}\overline{u^k u^m u^1}$$

$$= \frac{\partial}{\partial r}\left(\frac{1}{r}\overline{u_r^2 u_\theta}\right) + \frac{1}{r^2}\frac{\partial}{\partial\theta}(\overline{u_r u_\theta^2}) + \frac{1}{r}\frac{\partial}{\partial z}(\overline{u_r u_\theta u_z}) + \frac{1}{r^2}\overline{u_r^2 u_\theta} - \frac{1}{r^2}\overline{u_\theta^3} + \frac{2}{r^2}\overline{u_r^2 u_\theta}$$

$$= \frac{1}{r}\left(\frac{\partial}{\partial r}(\overline{u_r^2 u_\theta}) + \frac{1}{r}\frac{\partial}{\partial\theta}(\overline{u_r u_\theta^2}) + \frac{\partial}{\partial z}(\overline{u_r u_\theta u_z}) + \frac{2}{r}\overline{u_r^2 u_\theta} - \frac{1}{r}\overline{u_\theta^3}\right). \tag{7.19}$$

Again, the incompressibility condition may be used to reduce this expression to

$$\frac{1}{r}\left(\overline{u_r\frac{\partial}{\partial r}(u_r u_\theta)} + \frac{1}{r}\overline{u_\theta\frac{\partial}{\partial\theta}(u_r u_\theta)} + \overline{u_z\frac{\partial}{\partial z}(u_r u_\theta)} + \frac{\overline{u_r^2 u_\theta}}{r} - \frac{\overline{u_\theta^3}}{r}\right). \tag{7.20}$$

Multiplication by r gives the physical component.

The final term in the diffusion expression is also straightforward to evaluate. For $i = j = 1$, we obtain

$$\frac{1}{\rho}[\overline{(p'u^1)}_{,k}g^{1k} + \overline{(p'u^1)}_{,k}g^{1k}] = \frac{2}{\rho}\overline{(p'u^1)}_{,1} = \frac{2}{\rho}(\frac{\partial}{\partial x^1}\overline{(p'u^1)} + \Gamma^1_{k1}\overline{p'u^k}) = \frac{2}{\rho}\frac{\partial}{\partial r}\overline{(p'u_r)} \tag{7.21}$$

while, for $i = 1$, $j = 2$,

$$\frac{1}{\rho}[\overline{(p'u^1)}_{,k}g^{2k} + \overline{(p'u^2)}_{,k}g^{1k}$$

Sec. 7.2] **Examples** 189

$$= \frac{1}{\rho}\left(\frac{1}{r^2}\overline{p'u^1}_{,2} + \overline{p'u^2}_{,1}\right)$$

$$= \frac{1}{\rho}\left[\frac{1}{r^2}\left(\frac{\partial}{\partial x^2}(\overline{p'u^1}) + \Gamma^1_{k2}\overline{p'u^k}\right) + \left(\frac{\partial}{\partial x^1}(\overline{p'u^2}) + \Gamma^2_{k1}\overline{p'u^k}\right)\right]$$

$$= \frac{1}{\rho}\left[\frac{1}{r^2}\frac{\partial}{\partial\theta}\overline{p'u_r} - \frac{1}{r^2}\overline{p'u_\theta} + \frac{\partial}{\partial r}\left(\frac{1}{r}\overline{p'u_\theta}\right) + \frac{1}{r^2}\overline{p'u_\theta}\right]$$

$$= \frac{1}{\rho}\left(\frac{1}{r}\frac{\partial}{\partial r}\overline{p'u_\theta} + \frac{1}{r^2}\frac{\partial}{\partial\theta}\overline{p'u_r} - \frac{1}{r^2}\overline{p'u_\theta}\right). \tag{7.22}$$

Collecting together the expressions (7.15), (7.18) and (7.21), we obtain the contravariant component D^{11} of the diffusion term, which is equal to the physical component $D_{(11)}$,

$$\nu\left(\nabla^2\overline{u_r^2} - \frac{4}{r}\frac{\partial}{\partial\theta}\overline{u_r u_\theta} - \frac{2}{r^2}(\overline{u_r^2} - \overline{u_\theta^2})\right)$$

$$-\left(\overline{u_r\frac{\partial}{\partial r}u_r^2} + \frac{1}{r}\overline{u_\theta\frac{\partial}{\partial\theta}u_r^2} + \overline{u_z\frac{\partial}{\partial z}u_r^2} - \frac{2}{r}\overline{u_r u_\theta^2}\right) - \frac{2}{\rho}\frac{\partial}{\partial r}(\overline{p'u_r}). \tag{7.23}$$

Similarly, the physical component $D_{(12)}$ of the diffusion term comes from (7.16), (7.20) and (7.22) and may be written as

$$\nu\left(\nabla^2\overline{u_r u_\theta} + \frac{2}{r^2}\frac{\partial}{\partial\theta}(\overline{u_r^2} - \overline{u_\theta^2}) - \frac{4}{r^2}\overline{u_r^2 u_\theta}\right)$$

$$-\left(\overline{u_r\frac{\partial}{\partial r}(u_r u_\theta)} + \frac{1}{r}\overline{u_\theta\frac{\partial}{\partial\theta}(u_r u_\theta)} + \overline{u_z\frac{\partial}{\partial z}(u_r u_\theta)} + \frac{\overline{u_r^2 u_\theta}}{r} - \frac{\overline{u_\theta^3}}{r}\right)$$

$$-\frac{1}{\rho}\left(\frac{\partial}{\partial r}\overline{p'u_\theta} + \frac{1}{r}\frac{\partial}{\partial\theta}\overline{p'u_r} - \frac{1}{r}\overline{p'u_\theta}\right). \tag{7.24}$$

(c) Transformation of production term If we define 3×3 matrices A and B with i–j entries $\overline{u^i u^j}$ and $U^i_{,j}$, respectively, then the production term

$$P^{ij} = -\overline{u^i u^k}\,U^j_{,k} - \overline{u^j u^k}\,U^i_{,k}$$

corresponds to the i–j entry of the 3×3 matrix given by

$$-AB^\mathrm{T} - BA^\mathrm{T}.$$

Since B is given by the array (6.13) with u replaced by U, it is simple to evaluate all the components of the production term.

For $i = j = 1$, the component is

$$P^{11} = -\overline{u^1 u^k}\,U^1_{,k} - \overline{u^1 u^k}\,U^1_{,k}$$

$$= -2\overline{u^1 u^k}\,U^1_{,k}$$

$$= -2\overline{(u^1)^2}\,U^1_{,1} - 2\overline{u^1 u^2}\,U^1_{,2} - 2\overline{u^1 u^3}\,U^1_{,3}$$

$$= -2\left[\overline{u_r^2}\frac{\partial U_r}{\partial r} + \frac{1}{r}\overline{u_r u_\theta}\left(\frac{\partial U_r}{\partial \theta} - U_\theta\right) + \overline{u_r u_z}\frac{\partial U_r}{\partial z}\right] \quad (7.25)$$

while, for $i = 1, j = 2$,

$$P^{12} = -\overline{u^1 u^k}U^2{}_{,k} - \overline{u^2 u^k}U^1{}_{,k}$$

$$= -[\overline{(u^1)^2}U^2{}_{,1} + \overline{u^1 u^2}U^2{}_{,2} + \overline{u^1 u^3}U^2{}_{,3}] - [\overline{u^2 u^1}U^1{}_{,1} + \overline{(u^2)^2}U^1{}_{,2} + \overline{u^2 u^3}U^1{}_{,3}]$$

$$= -\left[\overline{u_r^2}\frac{1}{r}\frac{\partial U_\theta}{\partial r} + \frac{\overline{u_r u_\theta}}{r^2}\left(\frac{\partial U_\theta}{\partial \theta} + U_r\right) + \frac{\overline{u_r u_z}}{r}\frac{\partial U_\theta}{\partial z}\right]$$

$$-\left[\frac{\overline{u_r u_\theta}}{r}\frac{\partial U_r}{\partial r} + \frac{\overline{u_\theta^2}}{r^2}\left(\frac{\partial U_r}{\partial \theta} - U_\theta\right) + \frac{\overline{u_\theta u_z}}{r}\frac{\partial U_r}{\partial z}\right].$$

This contravariant component must be multiplied by r to give the physical component

$$-\left[\overline{u_r^2}\frac{\partial U_\theta}{\partial r} + \overline{u_r u_\theta}\left(\frac{\partial U_r}{\partial r} + \frac{1}{r}\frac{\partial U_\theta}{\partial \theta} + \frac{U_r}{r}\right) + \frac{\overline{u_\theta^2}}{r}\left(\frac{\partial U_r}{\partial \theta} - U_\theta\right) + \overline{u_r u_z}\frac{\partial U_\theta}{\partial z} + \overline{u_\theta u_z}\frac{\partial U_r}{\partial z}\right]. \quad (7.26)$$

(d) Transformation of redistribution term This term is also evaluated easily using (6.13). We have

$$\Phi^{11} = \frac{1}{\rho}(\overline{p'u^1{}_{,k}}g^{1k} + \overline{p'u^1{}_{,k}}g^{1k}) = \frac{2}{\rho}\overline{p'u^1{}_{,k}}g^{1k} = \frac{2}{\rho}\overline{p'u^1{}_{,1}}g^{11}$$

because of the orthogonality of the co-ordinate system. Here we have, by (6.13),

$$\Phi^{11} = \Phi_{(11)} = \frac{2}{\rho}\overline{p'\frac{\partial u_r}{\partial r}}. \quad (7.27)$$

Next,

$$\Phi^{12} = \frac{1}{\rho}(\overline{p'u^1{}_{,k}}g^{2k} + \overline{p'u^2{}_{,k}}g^{1k}) = \frac{1}{\rho}(\overline{p'u^1{}_{,2}}g^{22} + \overline{p'u^2{}_{,1}}g^{11})$$

$$= \frac{1}{\rho}\left[\frac{1}{r^2}\overline{p'\left(\frac{\partial u_r}{\partial \theta} - u_\theta\right)} + \frac{1}{r}\overline{p'\frac{\partial u_\theta}{\partial r}}\right].$$

Multiplying by r to obtain the physical component gives

$$\Phi_{(12)} = \frac{1}{\rho}\left(\frac{1}{r}\overline{p'\frac{\partial u_r}{\partial \theta}} + \overline{p'\frac{\partial u_\theta}{\partial r}} - \frac{1}{r}\overline{p'u_\theta}\right). \quad (7.28)$$

(e) Transformation of dissipation term The array (6.13) of covariant derivatives of contravariant vector components in cylindrical polars is again useful.

For $i = j = 1$, the dissipation term is

$$-\mathscr{E}^{11} = -2\nu\overline{u^1{}_{,l}u^1{}_{,k}}g^{kl}$$

$$= -2\nu\left(\overline{(u^1{}_{,1})^2} + \frac{1}{r^2}\overline{(u^1{}_{,2})^2} + \overline{(u^1{}_{,3})^2}\right)$$

$$= -2\nu\left[\overline{\left(\frac{\partial u_r}{\partial r}\right)^2} + \frac{1}{r^2}\overline{\left(\frac{\partial u_r}{\partial \theta} - u_\theta\right)^2} + \overline{\left(\frac{\partial u_r}{\partial z}\right)^2}\right]$$

$$= -2\nu\left[\overline{\left(\frac{\partial u_r}{\partial r}\right)^2} + \frac{1}{r^2}\overline{\left(\frac{\partial u_r}{\partial \theta}\right)^2} + \overline{\left(\frac{\partial u_r}{\partial z}\right)^2} - \frac{2}{r^2}\overline{u_\theta \frac{\partial u_r}{\partial \theta}} + \frac{1}{r^2}\overline{u_\theta^2}\right] \quad (7.29)$$

and, for $i = 1, j = 2$,

$$-\mathscr{E}^{12} = -2\nu \overline{u^1_{,k} u^2_{,k}} g^{kl} = -2\nu\left(\overline{u^1_{,1} u^2_{,1}} + \frac{1}{r^2}\overline{u^1_{,2} u^2_{,2}} + \overline{u^1_{,3} u^2_{,3}}\right)$$

$$= -2\nu\left[\frac{1}{r}\overline{\frac{\partial u_r}{\partial r}\frac{\partial u_\theta}{\partial r}} + \frac{1}{r^3}\overline{\left(\frac{\partial u_r}{\partial \theta} - u_\theta\right)\left(\frac{\partial u_\theta}{\partial \theta} + u_r\right)} + \frac{1}{r}\overline{\frac{\partial u_r}{\partial z}\frac{\partial u_\theta}{\partial z}}\right].$$

Multiplying by r gives the physical component

$$-2\nu\left[\overline{\frac{\partial u_r}{\partial r}\frac{\partial u_\theta}{\partial r}} + \frac{1}{r}\overline{\frac{\partial u_r}{\partial \theta}\frac{\partial u_\theta}{\partial \theta}} + \overline{\frac{\partial u_r}{\partial z}\frac{\partial u_\theta}{\partial z}} - \frac{1}{r^2}\left(\overline{u_r u_\theta} - \overline{u_r \frac{\partial u_r}{\partial \theta}} + \overline{u_\theta \frac{\partial u_\theta}{\partial \theta}}\right)\right]. \quad (7.30)$$

Finally, adding together the expressions (7.23), (7.25), (7.27) and (7.28) and equating the result to (7.8) gives the physical r–r component of the Reynolds stress transport equations, while adding (7.24), (7.26), (7.28) and (7.30) and equating to (7.10) gives the physical r–θ component.

The remaining four independent components (r–z, θ–θ, θ–z and z–z) may be obtained similarly.

Example 3

Transform the Reynolds stress transport equation for high Reynolds number flows from rectangular cartesians to a streamline co-ordinate system.

Method

In high Reynolds number turbulent flow calculations the 'molecular diffusion' term $\nu \partial(\overline{u_i u_j})/\partial y_k$ in the general diffusion expression in (7.6) is neglected, and the dissipation term may be 'modelled' according to

$$2\nu \overline{\frac{\partial u_i}{\partial y_k}\frac{\partial u_j}{\partial y_k}} = \tfrac{2}{3}\delta_{ij}\varepsilon, \quad (7.31)$$

where ε can be regarded as the total dissipation of the 'turbulent kinetic energy' k defined as $\tfrac{1}{2}\overline{u_i u_i}$ in cartesians. This modelling is based on the assumption of isotropy for small-scale motion. Thus, instead of (7.6), we have

$$U_k \frac{\partial}{\partial y_k}\overline{u_i u_j} = -\frac{\partial}{\partial y_k}\left(\overline{u_i u_j u_k} + \frac{1}{\rho}\overline{p'(\delta_{ik}u_j + \delta_{jk}u_i)}\right)$$

$$-\overline{u_i u_k}\frac{\partial U_j}{\partial y_k} - \overline{u_j u_k}\frac{\partial U_i}{\partial y_k} + \frac{1}{\rho}\overline{p'\left(\frac{\partial u_i}{\partial y_j} + \frac{\partial u_j}{\partial y_i}\right)} - \tfrac{2}{3}\delta_{ij}\varepsilon, \quad (7.32)$$

which generalizes to

$$\underbrace{U^k\overline{u^i u^j}_{,k}}_{\text{convection}} = -\underbrace{\left(\overline{u^i u^j u^k} + \frac{1}{\rho}\overline{p'(g^{ik}u^j + g^{jk}u^i)}\right)_{,k}}_{\text{diffusion}}$$

$$\underbrace{-\overline{u^i u^k}U^j_{,k} - \overline{u^j u^k}U^i_{,k}}_{\text{production}} + \underbrace{\frac{1}{\rho}\overline{p'(g^{kj}u^i_{,k} + g^{ki}u^j_{,k})}}_{\text{redistribution}} - \underbrace{\tfrac{2}{3}g^{ij}\varepsilon}_{\text{dissipation}}. \qquad (7.33)$$

In the streamline co-ordinate system illustrated in Fig. 5.9 we introduce a third co-ordinate z measured perpendicular to the s–n plane shown. The physical components of the mean velocity and the fluctuating velocity corresponding to the s, n, z directions will be denoted by (U, V, W) and (u, v, w), respectively. No new non-zero Christoffel symbols besides those given in (5.59) are generated by considering this third co-ordinate, and the third scale factor is $h_3 = 1$.

It is convenient to have on hand the array of covariant derivatives $U^i_{,j}$ of contravariant vector components. Evaluating (4.26) with (5.59) gives

$$\begin{bmatrix} U^1_{,1} & U^1_{,2} & U^1_{,3} \\ U^2_{,1} & U^2_{,2} & U^2_{,3} \\ U^3_{,1} & U^3_{,2} & U^3_{,3} \end{bmatrix} = \begin{bmatrix} \dfrac{1}{h_1}\dfrac{\partial U}{\partial s} + \dfrac{V}{Rh_1} & \dfrac{1}{h_1}\dfrac{\partial U}{\partial n} & \dfrac{1}{h_1}\dfrac{\partial U}{\partial z} \\ \dfrac{\partial V}{\partial s} - \dfrac{U}{R} & \dfrac{\partial V}{\partial n} & \dfrac{\partial V}{\partial z} \\ \dfrac{\partial W}{\partial s} & \dfrac{\partial W}{\partial n} & \dfrac{\partial W}{\partial z} \end{bmatrix}. \qquad (7.34)$$

We carry out the analysis only for the cases $i = j = 1$ and $i = 1, j = 2$. Since $h_2 = 1$, we obtain the physical components corresponding to the contravariant 1–1 components by multiplying by h_1^2 and those corresponding to the 1–2 components by multiplying by h_1, according to (5.15).

(a) Transformation of convection term   As in Example 2,

$$C^{11} = U^k\left(\frac{\partial}{\partial x^k}\overline{[(u^1)^2]} + \Gamma^1_{lk}\overline{u^l u^1} + \Gamma^1_{lk}\overline{u^1 u^l}\right)$$

$$= U^1\frac{\partial}{\partial x^1}\left(\frac{\overline{(u)^2}}{h_1^2}\right) + U^2\frac{\partial}{\partial x^2}\left(\frac{\overline{(u)^2}}{h_1^2}\right) + U^3\frac{\partial}{\partial x^3}\left(\frac{\overline{(u)^2}}{h_1^2}\right) + 2U^k\Gamma^1_{lk}\overline{u^1 u^l}$$

$$= \frac{U}{h_1}\frac{\partial}{\partial s}\left(\frac{\overline{u^2}}{h_1^2}\right) + V\frac{\partial}{\partial n}\left(\frac{\overline{u^2}}{h_1^2}\right) + W\frac{\partial}{\partial z}\left(\frac{\overline{u^2}}{h_1^2}\right) + 2\frac{U}{h_1^4}\frac{\partial h_1}{\partial s}\overline{u^2} + \frac{2V}{Rh_1^3}\overline{u^2} + \frac{2U}{Rh_1^3}\overline{uv},$$

carrying out the summation over l and k. Note carefully the potential source of confusion here in that our notation for the square of the physical component u is identical with that for the contravariant component u^2. However, since we are dealing with double-velocity correlations $\overline{u^i u^j}$ at this stage (and triple-velocity correlations later) it should be clear to the reader what is meant.

Examples

We obtain, finally,

$$C^{11} = \frac{U}{h_1{}^3} \frac{\partial}{\partial s} \overline{u^2} + \frac{V}{h_1{}^2} \frac{\partial}{\partial n} \overline{u^2} + \frac{W}{h_1{}^2} \frac{\partial}{\partial z} \overline{u^2} + \frac{2U}{Rh_1{}^3} \overline{uv} \qquad (7.35)$$

with associated physical component

$$C_{(11)} = \frac{U}{h_1} \frac{\partial}{\partial s} \overline{u^2} + V \frac{\partial}{\partial n} \overline{u^2} + W \frac{\partial}{\partial z} \overline{u^2} + \frac{2U}{Rh_1} \overline{uv}. \qquad (7.36)$$

Similarly, we find

$$C^{12} = U^k \left(\frac{\partial}{\partial x^k} (\overline{u^1 u^2}) + \Gamma^1_{lk} \overline{u^l u^2} + \Gamma^2_{lk} \overline{u^1 u^l} \right)$$

$$= \frac{U}{h_1{}^2} \frac{\partial}{\partial s} \overline{uv} + \frac{V}{h_1} \frac{\partial}{\partial n} \overline{uv} + \frac{W}{h_1} \frac{\partial}{\partial z} \overline{uv} + \frac{U}{Rh_1{}^2} (\overline{v^2} - \overline{u^2}) \qquad (7.37)$$

with physical components

$$C_{(12)} = \frac{U}{h_1} \frac{\partial}{\partial s} \overline{uv} + V \frac{\partial}{\partial n} \overline{uv} + W \frac{\partial}{\partial z} \overline{uv} + \frac{U}{Rh_1} (\overline{v^2} - \overline{u^2}). \qquad (7.38)$$

(b) Transformation of diffusion term There are now two terms to evaluate, the first being the covariant derivative of the triple-velocity correlation $\overline{u^i u^j u^k}$. As in Example 2,

$$\overline{u^i u^j u^k}_{,k} = \frac{\partial}{\partial x^k} (\overline{u^i u^j u^k}) + \Gamma^i_{mk} \overline{u^m u^j u^k} + \Gamma^j_{mk} \overline{u^i u^m u^k} + \Gamma^k_{mk} \overline{u^i u^j u^m},$$

and, by (4.23),

$$\Gamma^k_{mk} = \frac{1}{\sqrt{g}} \frac{\partial}{\partial x^m} (\sqrt{g}) = \frac{1}{h_1} \frac{\partial}{\partial x^m} (h_1).$$

Hence, with $i = j = 1$,

$$\overline{u^1 u^1 u^k}_{,k} = \frac{\partial}{\partial x^k} [\overline{(u^1)^2 u^k}] + 2\Gamma^1_{mk} \overline{u^1 u^m u^k} + \frac{1}{h_1} \frac{\partial h_1}{\partial x^m} \overline{(u^1)^2 u^m}$$

$$= \frac{\partial}{\partial s} [\overline{(u^1)^3}] + \frac{\partial}{\partial n} [\overline{(u^1)^2 u^2}] + \frac{\partial}{\partial z} [\overline{(u^1)^2 u^3}]$$

$$+ \frac{2}{h_1} \frac{\partial h_1}{\partial s} \overline{(u^1)^3} + \frac{4}{Rh_1} \overline{(u^1)^2 u^2} + \frac{1}{h_1} \frac{\partial h_1}{\partial s} \overline{(u^1)^3} + \frac{1}{h_1} \frac{\partial h_1}{\partial n} \overline{(u^1)^2 u^2}$$

$$= \frac{\partial}{\partial s} \left(\frac{\overline{u^3}}{h_1{}^3} \right) + \frac{\partial}{\partial n} \left(\frac{\overline{u^2 v}}{h_1{}^2} \right) + \frac{\partial}{\partial z} \left(\frac{\overline{u^2 w}}{h_1{}^2} \right) + \frac{3}{h_1{}^4} \frac{\partial h_1}{\partial s} \overline{u^3} + \frac{5}{Rh_1{}^3} \overline{u^2 v}$$

$$= \frac{1}{h_1{}^3} \frac{\partial}{\partial s} \overline{u^3} + \frac{1}{h_1{}^2} \frac{\partial}{\partial n} \overline{u^2 v} + \frac{1}{h_1{}^2} \frac{\partial}{\partial z} \overline{u^2 w} + \frac{3}{Rh_1{}^3} \overline{u^2 v}. \qquad (7.39)$$

Similarly, with $i = 1, j = 2$,

$$\overline{u^1 u^2 u^k}_{,k} = \frac{\partial}{\partial x^k} (\overline{u^1 u^2 u^k}) + \Gamma^1_{mk} \overline{u^m u^2 u^k} + \Gamma^2_{mk} \overline{u^m u^1 u^k} + \frac{1}{h_1} \frac{\partial h_1}{\partial x^m} \overline{u^1 u^2 u^m}$$

$$= \frac{\partial}{\partial s}\left(\frac{\overline{u^2 v}}{h_1{}^2}\right) + \frac{\partial}{\partial n}\left(\frac{\overline{uv^2}}{h_1}\right) + \frac{\partial}{\partial z}\left(\frac{\overline{uvw}}{h_1}\right)$$

$$+ \frac{1}{h_1{}^3}\frac{\partial h_1}{\partial s}\overline{u^2 v} + \frac{2}{Rh_1{}^2}\overline{uv^2} - \frac{1}{Rh_1{}^2}\overline{u^3} + \left(\frac{1}{h_1{}^3}\frac{\partial h_1}{\partial s}\overline{u^2 v} + \frac{1}{Rh_1{}^2}\overline{uv^2}\right)$$

$$= \frac{1}{h_1{}^2}\frac{\partial}{\partial s}\overline{u^2 v} + \frac{1}{h_1}\frac{\partial}{\partial n}\overline{uv^2} + \frac{1}{h_1}\frac{\partial}{\partial z}\overline{uvw} + \frac{2}{Rh_1{}^2}\overline{uv^2} - \frac{1}{Rh_1{}^2}\overline{u^3}. \quad (7.40)$$

Multiplication of (7.39) by $h_1{}^2$ and (7.40) by h_1 gives the corresponding physical components.

The second term to evaluate is the 'pressure–diffusion' term $\rho^{-1}\overline{p'(g^{ik}u^j + g^{jk}u^i)}_{,k}$. For $i = j = 1$, this becomes

$$\frac{1}{\rho}\overline{p'(g^{1k}u^1 + g^{1k}u^1)}_{,k} = \frac{2}{\rho h_1{}^2}\overline{p'u^1}_{,1},$$

since $g^{11} = h_1{}^{-2}, g^{12} = 0$, etc. Evaluating the covariant derivative $\overline{p'u^1}_{,1}$ gives an expression similar to the 1–1 entry of (7.34); thus

$$\frac{2}{\rho h_1{}^2}\overline{p'u^1}_{,1} = \frac{2}{\rho h_1{}^2}\left(\frac{1}{h_1}\frac{\partial}{\partial s}(\overline{p'u}) + \frac{1}{Rh_1}\overline{p'v}\right). \quad (7.41)$$

Multiplying by $h_1{}^2$ gives the physical component

$$\frac{2}{\rho h_1}\left(\frac{\partial}{\partial s}\overline{p'u} + \frac{1}{R}\overline{p'v}\right). \quad (7.42)$$

For $i = 1, j = 2$, we have

$$\frac{1}{\rho}\overline{p'(g^{1k}u^2 + g^{2k}u^1)}_{,k} = \frac{1}{\rho}\left[\overline{p'\left(\frac{1}{h_1{}^2}u^2\right)}_{,1} + \overline{p'u^1}_{,2}\right]$$

$$= \frac{1}{\rho}\left[\frac{1}{h_1{}^2}\left(\frac{\partial}{\partial s}(\overline{p'v}) - \frac{1}{R}\overline{p'u}\right) + \frac{1}{h_1}\frac{\partial}{\partial n}(\overline{p'u})\right], \quad (7.43)$$

by (7.34), and multiplication by h_1 gives the physical component

$$\frac{1}{\rho}\left(\frac{1}{h_1}\frac{\partial}{\partial s}\overline{p'v} + \frac{\partial}{\partial n}\overline{p'u} - \frac{1}{Rh_1}\overline{p'u}\right). \quad (7.44)$$

The physical components of the complete diffusion term for the cases $i = j = 1$ and $i = 1, j = 2$, are thus

$$D_{(11)} = -\left[\frac{1}{h_1}\frac{\partial}{\partial s}\overline{u^3} + \frac{\partial}{\partial n}\overline{u^2 v} + \frac{\partial}{\partial z}\overline{u^2 w} + \frac{3}{Rh_1}\overline{u^2 v} + \frac{2}{\rho h_1}\left(\frac{\partial}{\partial s}\overline{p'u} + \frac{1}{R}\overline{p'v}\right)\right], \quad (7.45)$$

and

$$D_{(12)} = -\left[\frac{1}{h_1}\frac{\partial}{\partial s}\overline{u^2 v} + \frac{\partial}{\partial n}\overline{uv^2} + \frac{\partial}{\partial z}\overline{uvw} + \frac{2}{Rh_1}\overline{uv^2} - \frac{1}{Rh_1}\overline{u^3}\right.$$

$$\left. + \frac{1}{\rho}\left(\frac{1}{h_1}\frac{\partial}{\partial s}\overline{p'v} + \frac{\partial}{\partial n}\overline{p'u} - \frac{1}{Rh_1}\overline{p'u}\right)\right]. \quad (7.46)$$

Sec. 7.2] Examples

(c) Transformation of production term The components of this term are easily obtained, using (7.34). For $i = j = 1$, we have

$$P^{11} = -\overline{u^1 u^k} U^1{}_{,k} - \overline{u^1 u^k} U^1{}_{,k}$$

$$= -2\overline{u^1 u^k} U^1{}_{,k}$$

$$= -2\overline{(u^1)^2} U^1{}_{,1} - 2\overline{u^1 u^2} U^1{}_{,2} - 2\overline{u^1 u^3} U^1{}_{,3}$$

$$= -\frac{2}{h_1{}^2}\overline{u^2}\left(\frac{1}{h_1}\frac{\partial U}{\partial s} + \frac{V}{Rh_1}\right) - \frac{2}{h_1{}^2}\overline{uv}\frac{\partial U}{\partial n} - \frac{2}{h_1{}^2}\overline{uw}\frac{\partial U}{\partial z},$$

giving the physical component

$$P_{(11)} = -2\overline{u^2}\left(\frac{1}{h_1}\frac{\partial U}{\partial s} + \frac{V}{Rh_1}\right) - 2\overline{uv}\frac{\partial U}{\partial n} - 2\overline{uw}\frac{\partial U}{\partial z}. \quad (7.47)$$

Moreover,

$$P^{12} = -\overline{u^1 u^k} U^2{}_{,k} - \overline{u^2 u^k} U^1{}_{,k}$$

$$= -\overline{(u^1)^2} U^2{}_{,1} - \overline{u^1 u^2} U^2{}_{,2} - \overline{u^1 u^3} U^2{}_{,3} - \overline{u^2 u^1} U^1{}_{,1} - \overline{(u^2)^2} U^1{}_{,2} - \overline{u^2 u^3} U^1{}_{,3}$$

$$= -\frac{1}{h_1{}^2}\overline{u^2}\left(\frac{\partial V}{\partial s} - \frac{U}{R}\right) - \frac{1}{h_1}\overline{uv}\left(\frac{\partial V}{\partial n} + \frac{1}{h_1}\frac{\partial U}{\partial s} + \frac{V}{Rh_1}\right)$$

$$- \frac{1}{h_1}\overline{uw}\frac{\partial V}{\partial z} - \frac{1}{h_1}\overline{v^2}\frac{\partial U}{\partial n} - \frac{1}{h_1}\overline{vw}\frac{\partial U}{\partial z},$$

with corresponding physical component

$$P_{(12)} = -\frac{1}{h_1}\overline{u^2}\left(\frac{\partial V}{\partial s} - \frac{U}{R}\right) - \overline{uv}\left(\frac{1}{h_1}\frac{\partial U}{\partial s} + \frac{\partial V}{\partial n} + \frac{V}{Rh_1}\right) - \overline{uw}\frac{\partial V}{\partial z} - \overline{v^2}\frac{\partial U}{\partial n} - \overline{vw}\frac{\partial U}{\partial z}. \quad (7.48)$$

(d) Transformation of redistribution term We have

$$\Phi^{11} = \frac{2}{\rho}\overline{p'\left(\frac{1}{h_1{}^2}u^1{}_{,1}\right)} = \frac{2}{\rho h_1{}^2}\overline{p'\left(\frac{1}{h_1}\frac{\partial u}{\partial s} + \frac{v}{Rh_1}\right)},$$

by (7.34), i.e.

$$\Phi^{11} = \frac{2}{\rho h_1{}^3}\left(\overline{p'\frac{\partial u}{\partial s}} + \frac{1}{R}\overline{p'v}\right),$$

with physical component

$$\Phi_{(11)} = \frac{2}{\rho h_1}\left(\overline{p'\frac{\partial u}{\partial s}} + \frac{1}{R}\overline{p'v}\right). \quad (7.49)$$

Also

$$\Phi^{12} = \frac{1}{\rho}\left(\overline{p'u^1{}_{,2}} + \frac{1}{h_1{}^2}\overline{p'u^2{}_{,1}}\right) = \frac{1}{\rho}\left[\frac{1}{h_1}\overline{p'\frac{\partial u}{\partial n}} + \frac{1}{h_1{}^2}\overline{p'\left(\frac{\partial v}{\partial s} - \frac{u}{R}\right)}\right],$$

with physical component

$$\Phi_{(12)} = \frac{1}{\rho}\left(\frac{1}{h_1}\overline{p'\frac{\partial v}{\partial s}} + \overline{p'\frac{\partial u}{\partial n}} - \frac{1}{Rh_1}\overline{p'u}\right). \tag{7.50}$$

(e) Transformation of dissipation term   Here we simply obtain

$$\mathscr{E}^{11} = \tfrac{2}{3}g^{11}\varepsilon = \tfrac{2}{3}\frac{1}{h_1^2}\varepsilon, \quad \mathscr{E}^{12} = \tfrac{2}{3}g^{12}\varepsilon = 0,$$

with physical components

$$\mathscr{E}_{(11)} = \tfrac{2}{3}\varepsilon, \quad \mathscr{E}_{(12)} = 0. \tag{7.51}$$

Collecting together the physical components (7.36), (7.45), (7.47), (7.49) and (7.51) gives us the physical *s–s* component of (7.33), which may be expressed after some rearrangement as

$$\left(\frac{U}{h_1}\frac{\partial}{\partial s} + V\frac{\partial}{\partial n} + W\frac{\partial}{\partial z}\right)\overline{u^2}$$

$$= -2\frac{\overline{u^2}}{h_1}\left(\frac{\partial U}{\partial s} + \frac{V}{R}\right) - 2\overline{uv}\left(\frac{\partial U}{\partial n} + \frac{U}{Rh_1}\right) - \frac{2}{h_1}\frac{\partial}{\partial s}\left(\frac{\overline{p'u}}{\rho} + \tfrac{1}{2}\overline{u^3}\right)$$

$$- \frac{1}{h_1}\frac{\partial}{\partial n}(h_1\overline{u^2 v}) - \frac{\partial}{\partial z}(\overline{u^2 w}) - \frac{2\overline{u^2 v}}{Rh_1} - 2\overline{uw}\frac{\partial U}{\partial z} + \frac{2}{\rho h_1}\overline{p'\frac{\partial u}{\partial s}} - \tfrac{2}{3}\varepsilon. \tag{7.52}$$

Similarly, the physical *s–n* component of (7.33) may now be written, after rearrangement,

$$\left(\frac{U}{h_1}\frac{\partial}{\partial s} + V\frac{\partial}{\partial n} + W\frac{\partial}{\partial z}\right)\overline{uv}$$

$$= -\frac{\overline{u^2}}{h_1}\frac{\partial V}{\partial s} - \overline{v^2}\left(\frac{\partial U}{\partial n} - \frac{U}{Rh_1}\right) + \frac{2U}{Rh_1}(\overline{u^2} - \overline{v^2}) - \frac{1}{h_1}\frac{\partial}{\partial s}\left(\frac{\overline{p'v}}{\rho} + \overline{u^2 v}\right)$$

$$- \frac{\partial}{\partial n}\left(\frac{\overline{p'u}}{\rho} + \overline{uv^2}\right) - \frac{\partial}{\partial z}\overline{uvw} - \frac{2\overline{uv^2} - \overline{u^3}}{Rh_1} + \frac{1}{\rho}\left(\frac{1}{h_1}\overline{p'\frac{\partial v}{\partial s}} + \overline{p'\frac{\partial u}{\partial n}}\right) - \overline{uw}\frac{\partial V}{\partial z}$$

$$- \overline{vw}\frac{\partial U}{\partial z} - \overline{uv}\left(\frac{1}{h_1}\frac{\partial U}{\partial s} + \frac{\partial V}{\partial n} + \frac{V}{Rh_1}\right). \tag{7.53}$$

Note that for two-dimensional flows the last term disappears because of the incompressibility condition (6.80) (see Gibson and Rodi 1981). The remaining four equations for the physical *n–n*, *s–z*, *n–z* and *z–z* components of (7.33) may be obtained similarly.

Example 4

Transform the equation for the transport of turbulent kinetic energy (according to the k–ε model) from rectangular cartesians to a toroidal co-ordinate system (2).

Method

Contraction of the Reynold's stress transport equation (7.6) and use of incompressibility produce the equation for k transport:

$$\underbrace{U_j\frac{\partial k}{\partial y_j}}_{\text{convection}} = \underbrace{-\frac{\partial}{\partial y_j}\left(-v\frac{\partial k}{\partial y_j} + \overline{u_j k} + \frac{1}{\rho}\overline{p' u_j}\right)}_{\text{diffusion}} \underbrace{- \overline{u_i u_j}\frac{\partial U_i}{\partial y_j}}_{\text{production}} \underbrace{- v\frac{\overline{\partial u_i}}{\partial y_j}\frac{\partial u_i}{\partial y_j}}_{\text{dissipation}}. \quad (7.54)$$

Representing the named terms symbolically by C, D, P and ε, this equation may be expressed as

$$C = D + P - \varepsilon.$$

Modelling of the diffusion term (for high-Reynolds-number flows) according to what is usually referred to as the k–ε model, and modelling of the double-velocity correlation by the Boussinesq eddy viscosity law

$$-\overline{u_i u_j} = v_t\left(\frac{\partial U_i}{\partial y_j} + \frac{\partial U_j}{\partial y_i}\right) - \tfrac{2}{3}\delta_{ij} k, \quad (7.55)$$

produce the representations

$$D = \frac{\partial}{\partial y_j}\left[\left(v + \frac{v_t}{\sigma_k}\right)\frac{\partial k}{\partial y_j}\right]$$

and

$$P = \left[v_t\left(\frac{\partial U_i}{\partial y_j} + \frac{\partial U_j}{\partial y_i}\right) - \tfrac{2}{3}\delta_{ij} k\right]\frac{\partial U_i}{\partial y_j}$$

$$= v_t\left(\frac{\partial U_i}{\partial y_j}\frac{\partial U_i}{\partial y_j} + \frac{\partial U_j}{\partial y_i}\frac{\partial U_i}{\partial y_j}\right),$$

since

$$\delta_{ij}\frac{\partial U_i}{\partial y_j} = \frac{\partial U_i}{\partial y_i} = 0$$

by incompressibility. Here v_t is the eddy viscosity and σ_k is the turbulent kinetic energy Prandtl number.

Then (7.54) becomes

$$U_j\frac{\partial k}{\partial y_j} = \frac{\partial}{\partial y_j}\left[\left(v + \frac{v_t}{\sigma_k}\right)\frac{\partial k}{\partial y_j}\right] + v_t\left(\frac{\partial U_i}{\partial y_j}\frac{\partial U_i}{\partial y_j} + \frac{\partial U_i}{\partial y_j}\frac{\partial U_j}{\partial y_i}\right) - \varepsilon. \quad (7.56)$$

Written in generalized co-ordinates, this becomes

$$U^j k_{,j} = g^{lj}\left[\left(v + \frac{v_t}{\sigma_k}\right)k_{,j}\right]_{,l} + v_t(U^i{}_{,j} U_{i,l} g^{jl} + U^i{}_{,j} U^j{}_{,i}) - \varepsilon. \quad (7.57)$$

Since k is a scalar (and, indeed, equation (7.57) is a scalar equation), its covariant derivatives are identical with partial derivatives.

Thus, in toroidal co-ordinates,

$$C = U^j k_{,j} = U^1 k_{,1} + U^2 k_{,2} + U^3 k_{,3} = \frac{U_{(1)}}{h_1} k_{,1} + \frac{U_{(2)}}{h_2} k_{,2} + \frac{U_{(3)}}{h_3} k_{,3}$$

$$= \frac{U_\phi}{r} \frac{\partial k}{\partial \phi} + U_r \frac{\partial k}{\partial r} + \frac{U_\theta}{R + r\cos\phi} \frac{\partial k}{\partial \theta}, \qquad (7.58)$$

while, since $(v + v_t/\sigma_k) k_{,j}$ is a covariant vector,

$$D = g^{jl} \left(v + \frac{v_t}{\sigma_k} k_{,j} \right)_{,l} = g^{jl} \left\{ \frac{\partial}{\partial x^l} \left[\left(v + \frac{v_t}{\sigma_k} \right) \frac{\partial k}{\partial x^j} \right] - \Gamma^i_{jl} \left[\left(v + \frac{v_t}{\sigma_k} \right) \frac{\partial k}{\partial x^i} \right] \right\}.$$

Performing the summation over i, j and l with the non-zero components of the Christoffel symbol (5.56) gives

$$\frac{1}{r^2} \frac{\partial}{\partial \phi} \left[\left(v + \frac{v_t}{\sigma_k} \right) \frac{\partial k}{\partial \phi} \right] + \frac{\partial}{\partial r} \left[\left(v + \frac{v_t}{\sigma_k} \right) \frac{\partial k}{\partial r} \right] + \frac{1}{(R + r\cos\phi)^2} \frac{\partial}{\partial \theta} \left[\left(v + \frac{v_t}{\sigma_k} \right) \frac{\partial k}{\partial \theta} \right]$$

$$- \left(v + \frac{v_t}{\sigma_k} \right) \left(\frac{1}{h_3^2} \Gamma^1_{33} \frac{\partial k}{\partial x^1} + \frac{1}{h_3^2} \Gamma^2_{33} \frac{\partial k}{\partial x^2} + \frac{1}{h_1^2} \Gamma^2_{11} \frac{\partial k}{\partial x^2} \right)$$

$$= \frac{1}{r^2} \frac{\partial}{\partial \phi} \left[\left(v + \frac{v_t}{\sigma_k} \right) \frac{\partial k}{\partial \phi} \right] + \frac{\partial}{\partial r} \left[\left(v + \frac{v_t}{\sigma_k} \right) \frac{\partial k}{\partial r} \right] + \frac{1}{(R + r\cos\phi)^2} \frac{\partial}{\partial \theta} \left[\left(v + \frac{v_t}{\sigma_k} \right) \frac{\partial k}{\partial \theta} \right]$$

$$- \left(v + \frac{v_t}{\sigma_k} \right) \left[\frac{1}{r(R + r\cos\phi)} \left(\sin\phi \frac{\partial k}{\partial \phi} - (R + 2r\cos\phi) \frac{\partial k}{\partial r} \right) \right]. \qquad (7.59)$$

To evaluate P it is convenient to have the arrays of covariant derivatives of covariant and contravariant components. For toroidal co-ordinates these are

$$\begin{bmatrix} U_{1,1} & U_{1,2} & U_{1,3} \\ U_{2,1} & U_{2,2} & U_{2,3} \\ U_{3,1} & U_{3,2} & U_{3,3} \end{bmatrix} = \begin{bmatrix} r\left(\frac{\partial U_\phi}{\partial \phi} + U_r\right) & r\frac{\partial U_\phi}{\partial r} & r\left(\frac{\partial U_\phi}{\partial \theta} + U_\theta \sin\phi\right) \\ \frac{\partial U_r}{\partial \phi} - U_\phi & \frac{\partial U_r}{\partial r} & \frac{\partial U_r}{\partial \theta} - U_\theta \cos\phi \\ h_3 \frac{\partial U_\theta}{\partial \phi} & h_3 \frac{\partial U_\theta}{\partial r} & h_3 \left(\frac{\partial U_\theta}{\partial \theta} + U_r \cos\phi - U_\phi \sin\phi\right) \end{bmatrix}$$

(7.60)

and

$$\begin{bmatrix} U^1_{,1} & U^1_{,2} & U^1_{,3} \\ U^2_{,1} & U^2_{,2} & U^2_{,3} \\ U^3_{,1} & U^3_{,2} & U^3_{,3} \end{bmatrix} = \begin{bmatrix} \frac{1}{r}\left(\frac{\partial U_\phi}{\partial \phi} + U_r\right) & \frac{1}{r}\frac{\partial U_\phi}{\partial r} & \frac{1}{r}\left(\frac{\partial U_\phi}{\partial \theta} + U_\theta \sin\phi\right) \\ \frac{\partial U_r}{\partial \phi} - U_\phi & \frac{\partial U_r}{\partial r} & \frac{\partial U_r}{\partial \theta} - U_\theta \cos\phi \\ \frac{1}{h_3} \frac{\partial U_\theta}{\partial \phi} & \frac{1}{h_3} \frac{\partial U_\theta}{\partial r} & \frac{1}{h_3}\left(\frac{\partial U_\theta}{\partial \theta} + U_r \cos\phi - U_\phi \sin\phi\right) \end{bmatrix}$$

(7.61)

where $h_3 = R + r\cos\phi$.

Performing the summation over i, j and k then gives

$$P = v_t(U^i{}_{,j} U_{i,k} g^{jk} + U^i{}_{,j} U^j{}_{,i})$$

$$= v_t \left[\frac{2}{r^2} \left(\frac{\partial U_\phi}{\partial \phi} + U_r \right)^2 + 2 \left(\frac{\partial U_r}{\partial r} \right)^2 + \frac{2}{h_3{}^2} \left(\frac{\partial U_\theta}{\partial \theta} + U_r \cos\phi - U_\phi \sin\phi \right)^2 \right.$$

$$+ \left(\frac{\partial U_\phi}{\partial r} + \frac{1}{r} \frac{\partial U_r}{\partial \phi} - \frac{U_\phi}{r} \right)^2 + \left(\frac{\partial U_\theta}{\partial r} + \frac{1}{h_3} \frac{\partial U_r}{\partial \theta} - \frac{U_\theta}{r} \cos\phi \right)^2$$

$$\left. + \left(\frac{1}{r} \frac{\partial U_\theta}{\partial \phi} + \frac{1}{h_3} \frac{\partial U_\phi}{\partial \theta} + \frac{U_\theta}{h_3} \sin\phi \right)^2 \right]. \tag{7.62}$$

Substituting (7.58), (7.59) and (7.62) into (7.57) gives the required result.

Note the similarity between (7.55) and the constitutive equation (6.15). The transformation of (7.55) to generalized co-ordinates is, in covariant form,

$$-\overline{u_i u_j} = v_t(U_{i,j} + U_{j,i}) - \tfrac{2}{3} g_{ij} k. \tag{7.63}$$

In the k–ε model the turbulent viscosity v_t is expressible, using dimensional analysis, as

$$v_t = \frac{C_\mu k^2}{\varepsilon}, \tag{7.64}$$

where C_μ is a constant of proportionality, and a separate equation for the flux of ε must be formulated to close the system of equations.

Example 5

Transform the time-averaged transport equation for a passive scalar from rectangular cartesians to a generalized co-ordinate system.

Method

We shall consider the transport equation for temperature T:

$$\frac{\partial T}{\partial t} + U_j \frac{\partial T}{\partial y_j} = \frac{\partial}{\partial y_j} \left(\kappa \frac{\partial T}{\partial y_j} \right), \tag{7.65}$$

neglecting source terms, as a typical example. The transport equation for any other 'passive scalar' will have the same form except that the thermal conductivity κ will be replaced by the appropriate coefficient of diffusion.

Employing Reynold's decomposition of variables into mean and fluctuating components followed by time averaging, we obtain the time-averaged transport equation

$$\frac{\partial}{\partial y_j}(U_j T) = \frac{\partial}{\partial y_j} \left(\kappa \frac{\partial T}{\partial y_j} - \overline{u_j T'} \right), \tag{7.66}$$

where T is now the mean and T' the fluctuating temperature. The correlation $\overline{u_j T'}$

represents the rate of transport of the scalar T' in the y_j direction due to turbulent velocity fluctuations.

In transforming (7.66) to generalized co-ordinates we must ensure that its scalar character is preserved. This is achieved by the equation

$$(U^j T)_{,j} = (\kappa T_{,k} g^{jk} - \overline{u^j T'})_{,j}. \tag{7.67}$$

The procedure of time averaging introduces the heat-flux correlation $\overline{u_j T'}$, and, in order to produce a closed set of equations, further relations between the various correlations must be assumed. One proposal of Daley and Harlow (1970), for example, is that the heat-flux correlation is related to the Reynolds stress by the equation

$$\overline{u_j T'} = - C_T \frac{k}{\varepsilon} \overline{u_j u_l} \frac{\partial T}{\partial y_l} \tag{7.68}$$

in rectangular cartesians, where C_T is a constant. This 'generalized gradient diffusion hypothesis' may be expressed as

$$\overline{u^j T'} = - C_T \frac{k}{\varepsilon} \overline{u^j u^l} T_{,l} \tag{7.69}$$

in generalized co-ordinates.

Example 6

Transform the transport equation for the heat-flux correlation for high Reynolds number turbulent flow from rectangular cartesians to toroidal co-ordinates (2).

Method

In cartesian co-ordinates, this transport equation may be written as

$$\underbrace{U_j \frac{\partial}{\partial y_j}(\overline{u_i T'})}_{\text{convection}} = -\underbrace{\left(\overline{u_i u_j} \frac{\partial T}{\partial y_j} + \overline{u_j T'} \frac{\partial U_i}{\partial y_j} \right)}_{\text{production}} + \underbrace{\frac{\overline{\rho' T'}}{\rho} g_i}_{\text{buoyant generation}}$$

$$\underbrace{- (\kappa + \nu) \overline{\frac{\partial T'}{\partial y_j} \frac{\partial u_i}{\partial y_j}}}_{\text{pressure scrambling}} + \underbrace{\overline{\frac{p'}{\rho} \frac{\partial T'}{\partial y_i}}}_{\text{dissipation}}$$

$$\underbrace{- \frac{\partial}{\partial y_j} \left(\overline{u_i u_j T'} - \kappa U_i \frac{\partial T'}{\partial y_j} - \nu \overline{T' \frac{\partial u_i}{\partial y_j}} + \frac{\overline{p' T'}}{\rho} \delta_{ij} \right)}_{\text{diffusion}} \tag{7.70}$$

where ρ' is density fluctuation. This equation can be derived by multiplying (7.65) by u_i, the Navier–Stokes equations (6.53) by T', adding and then time averaging. The Boussinesq approximation for density is also incorporated in (7.70), according to which variations in ρ are neglected except for their effect on body-force terms. Thus buoyancy effects are retained.

Equation (7.70) is a *vector* equation and generalizes to

$$U^j(\overline{u^iT'})_{,j} = -(\overline{u^iu^j}T_{,j} + \overline{u^jT'}U^i_{,j}) + \frac{\overline{\rho'T'}}{\rho}g^i$$

$$-(\kappa+\nu)\overline{T'_{,k}u^i_{,j}}g^{kj} + \frac{\overline{p'T'_{,j}}}{\rho}g^{ij}$$

$$-\left(\overline{u^iu^jT'} - \kappa\overline{u^iT'_{,k}}g^{kj} - \nu\overline{T'u^i_{,k}}g^{kj} + \frac{\overline{p'T'}}{\rho}g^{ij}\right)_{,j} \tag{7.71}$$

in contravariant form. Note, for example, that the expression

$$\overline{\frac{\partial T'}{\partial y_j}\frac{\partial u_i}{\partial y_j}}$$

in (7.70) does not become $\overline{T'_{,j}u^i_{,j}}$ in generalized form since the repeated j do not here appear at different levels. Thus $\overline{T'_{,j}u^i_{,j}}$ does not represent a (first-order) tensor. We must write the valid tensor expression $\overline{T'_{,k}u^i_{,j}}g^{jk}$, which reduces to the original term in rectangular cartesians.

In high-Reynolds-number flows the molecular and thermal diffusion terms in ν and κ can be neglected, and density and temperature fluctuations are related by

$$\frac{\rho'}{\rho} = -\alpha\frac{T'}{T},$$

where α is the coefficient of volumetric expansion of the fluid; (7.71) becomes (see, for example, Launder 1975)

$$\underbrace{U^j(\overline{u^iT'})_{,j}}_{\text{convection}} = \underbrace{-(\overline{u^iu^j}T_{,j} + \overline{u^jT'}U^i_{,j})}_{\text{production}} \underbrace{- \alpha\frac{\overline{T'^2}}{T}g^i}_{\text{buoyant generation}}$$

$$\underbrace{-(\kappa+\nu)\overline{T'_{,k}u^i_{,j}}g^{kj}}_{\text{dissipation}} + \underbrace{\frac{\overline{p'}}{\rho}\overline{T'_{,j}}g^{ij}}_{\text{pressure scrambling}} - \underbrace{\left(\overline{u^iu^jT'} + \frac{\overline{p'T'}}{\rho}g^{ij}\right)_{,j}}_{\text{diffusion}}. \tag{7.72}$$

We shall evaluate here only the term corresponding to $i = 1$ in a toroidal co-ordinate system and leave the other two components ($i = 2$ and 3) as an exercise for the reader.

(a) Convection Using (7.61) with $\overline{u^iT'}$ in place of U^i, we obtain

$$U^1\overline{u^1T'}_{,1} + U^2\overline{u^1T'}_{,2} + U^3\overline{u^1T'}_{,3}$$

$$= \frac{U_\phi}{r^2}\left(\frac{\partial}{\partial\phi}\overline{u_\phi T'} + \overline{u_r T'}\right) + \frac{U_r}{r}\frac{\partial}{\partial r}\overline{u_\phi T'} + \frac{U_\theta}{h_3 r}\left(\frac{\partial}{\partial\theta}\overline{u_\phi T'} + \overline{u_\theta T'}\sin\phi\right). \tag{7.73}$$

(b) Production Again making use of (7.61),

$$-(\overline{u^i u^j} T_{,j} + \overline{u^j T'} U^1_{,j})$$
$$= -(\overline{u^1 u^1} T_{,1} + \overline{u^1 u^2} T_{,2} + \overline{u^1 u^3} T_{,3} + \overline{u^1 T'} U^1_{,1} + \overline{u^2 T'} U^1_{,2} + \overline{u^3 T'} U^1_{,3})$$
$$= -\left[\frac{\overline{u_\phi^2}}{r^2}\frac{\partial T}{\partial \phi} + \frac{\overline{u_\phi u_r}}{r}\frac{\partial T}{\partial r} + \frac{\overline{u_\phi u_\theta}}{rh_3}\frac{\partial T}{\partial \theta}\right.$$
$$\left.+ \frac{\overline{u_\phi T'}}{r^2}\left(\frac{\partial U_\phi}{\partial \phi} + U_r\right) + \frac{\overline{u_r T'}}{r}\frac{\partial U_\phi}{\partial r} + \frac{\overline{u_\theta T'}}{h_3 r}\left(\frac{\partial U_\phi}{\partial \theta} + U_\theta \sin\phi\right)\right]. \quad (7.74)$$

(c) Buoyant generation

$$-\alpha\frac{\overline{T'^2}}{T}g^1 = -\alpha\frac{\overline{T'^2}}{T}\frac{g_\phi}{r}, \quad (7.75)$$

where g_ϕ is the physical component of gravitational acceleration in the ϕ direction.

(d) Dissipation

$$-(\kappa + \nu)\overline{T'_{,k} u^1_{,j}} g^{kj}$$
$$= -(\kappa + \nu)(\overline{T'_{,1} u^1_{,1}} g^{11} + \overline{T'_{,2} u^1_{,2}} g^{22} + \overline{T'_{,3} u^1_{,3}} g^{33})$$
$$= -(\kappa + \nu)\left[\frac{1}{r}\frac{\overline{\partial T'}}{\partial \phi}\left(\frac{\partial u_\phi}{\partial \phi} + u_r\right)\frac{1}{r^2} + \frac{1}{r}\frac{\overline{\partial T'}}{\partial r}\frac{\partial u_\phi}{\partial r} + \frac{1}{r}\frac{\overline{\partial T'}}{\partial r}\left(\frac{\partial u_\phi}{\partial \theta} + u_\theta \sin\phi\right)\frac{1}{h_3^2}\right]$$
$$= -(\kappa + \nu)\left(\frac{1}{r^3}\frac{\overline{\partial T'}}{\partial \phi}\frac{\partial u_\phi}{\partial \phi} + \frac{1}{r}\frac{\overline{\partial T'}}{\partial r}\frac{\partial u_\phi}{\partial r} + \frac{1}{rh_3^2}\frac{\overline{\partial T'}}{\partial \theta}\frac{\partial u_\phi}{\partial \theta} + \frac{1}{r^3}\overline{u_r \frac{\partial T'}{\partial \phi}} + \frac{\sin\phi}{rh_3^2}\overline{u_\theta \frac{\partial T'}{\partial \theta}}\right). \quad (7.76)$$

(e) Pressure scrambling

$$\frac{\overline{p' T'_{,j}}}{\rho}g^{1j} = \frac{\overline{p' T'_{,1}}}{\rho}g^{11} \quad \text{(since } g^{12} = g^{13} = 0\text{)}$$
$$= \frac{1}{\rho r^2}\overline{p' \frac{\partial T'}{\partial \phi}}. \quad (7.77)$$

(f) Diffusion

$$-\left(\overline{u^1 u^j T'} + \frac{\overline{p' T'}}{\rho}g^{1j}\right)_{,j}$$
$$= -(\overline{u^1 u^j T'})_{,j} - \left(\frac{\overline{p' T'}}{\rho}g^{11}\right)_{,1}$$
$$= -\left(\frac{1}{rh_3}\frac{\partial}{\partial x^j}(rh_3 \overline{u^1 u^j T'}) + \Gamma^1_{lj}\overline{u^l u^j T'}\right) - \frac{\partial}{\partial \phi}\left(\frac{1}{r^2\rho}\overline{p' T'}\right)$$

using (6.32)

Sec. 7.2] **Examples** 203

$$= -\frac{1}{rh_3}\left[\frac{\partial}{\partial\phi}\left(\frac{h_3}{r}\overline{u_\phi^2 T'}\right) + \frac{\partial}{\partial r}(h_3\overline{u_\phi u_r T'}) + \frac{\partial}{\partial\theta}(\overline{u_\phi u_\theta T'})\right]$$

$$-\frac{2}{r^2}\overline{u_\phi u_r T'} + \frac{h_3 \sin\phi}{r}\frac{\overline{u_\theta^2 T'}}{h_3^2} - \frac{1}{\rho r^2}\frac{\partial}{\partial\phi}\overline{p' T'}$$

$$= -\left(\frac{1}{r^2}\frac{\partial}{\partial\phi}\overline{u_\phi^2 T'} + \frac{1}{r}\frac{\partial}{\partial r}\overline{u_\phi u_r T'} + \frac{1}{rh_3}\frac{\partial}{\partial\theta}\overline{u_\phi u_\theta T'}\right) - \frac{1}{\rho r^2}\frac{\partial}{\partial\phi}\overline{p' T'}$$

$$+\frac{1}{rh_3}\sin\phi\,\overline{u_\phi^2 T'} - \left(\frac{\cos\phi}{rh_3} + \frac{2}{r^2}\right)\overline{u_\phi u_r T'} + \frac{\sin\phi}{rh_3}\overline{u_\theta^2 T'}. \quad (7.78)$$

Collecting together terms (7.73)–(7.78) gives the $i = 1$ contravariant component of the heat-flux equation. We would have to multiply through by $h_1 = r$ to obtain the corresponding physical component.

Appendix: Chain rules

Chain rules relating partial derivatives are a natural extension of the rule for differentiating a composite function of a single variable. The 'function of a function' rule for differentiating the dependent variable y with respect to the 'independent' variable x, when y is a function of a variable u and u is a function of x, is

$$\frac{dy}{dx} = \frac{dy}{du}\frac{du}{dx}.$$

A useful starting point is the increment formula

$$\delta y \approx \frac{dy}{du}\delta u \qquad (A.1)$$

to first order in the increment δu, the approximation involving neglect of second- and higher-order terms such as $(\delta u)^2$ and $(\delta u)^3$. The exact form

$$dy = \frac{dy}{du}du,$$

relating the 'differentials' dy and du, may also be used.

For a function $\psi(x, y, z)$ of three variables, the increment $\delta\psi$ corresponding to increments $\delta x, \delta y, \delta z$ is given by

$$\delta\psi = \psi(x + \delta x, y + \delta y, z + \delta z) - \psi(x, y, z).$$

The identity

$$\begin{aligned}\delta\psi &= [\psi(x + \delta x, y + \delta y, z + \delta z) - \psi(x, y + \delta y, z + \delta z)] \\ &+ [\psi(x, y + \delta y, z + \delta z) - \psi(x, y, z + \delta z)] \\ &+ [\psi(x, y, z + \delta z) - \psi(x, y, z)]\end{aligned} \qquad (A.2)$$

Appendix: Chain rules

leads immediately to the extension

$$\delta\psi \approx \frac{\partial\psi}{\partial x}\delta x + \frac{\partial\psi}{\partial y}\delta y + \frac{\partial\psi}{\partial z}\delta z \tag{A.3}$$

of (A.1), since, for example, the first set of brackets in (A.2) involves fixed values of the variables y and z but varying x values, giving the approximation $\delta x\, \partial\psi/\partial x$. Relabelling $\{x, y, z\}$ as $\{y_1, y_2, y_3\}$, we can express (A.3) by the summation convention as

$$\delta\psi \approx \frac{\partial\psi}{\partial y_j}\delta y_j. \tag{A.4}$$

If the y_j are each functions of a single variable t, say, dividing (A.4) throughout by δt and taking the limit as $\delta t \to 0$ gives the 'total derivative' formula

$$\frac{d\psi}{dt} = \frac{\partial\psi}{\partial y_j}\frac{\partial y_j}{dt}. \tag{A.5}$$

However, if each y_j is a function of the three variables y_1', y_2', y_3', each increment δy_j is given by a formula similar to (A.4) in terms of increments $\delta y_1'$, $\delta y_2'$, $\delta y_3'$, i.e.

$$\delta y_j \approx \frac{\partial y_j}{\partial y_i'}\delta y_i',$$

where, since j is already in use, i has been used for summation. Hence, by (A.4),

$$\delta\psi \approx \frac{\partial\psi}{\partial y_j}\left(\frac{\partial y_j}{\partial y_i'}\delta y_i'\right) = \left(\frac{\partial\psi}{\partial y_j}\frac{\partial y_j}{\partial y_i'}\right)\delta y_i'. \tag{A.6}$$

However,

$$\delta\psi \approx \frac{\partial\psi}{\partial y_i'}\delta y_i', \tag{A.7}$$

similarly to (A.4), when ψ is expressed as a function of y_1', y_2', y_3'. Identifying coefficients of $\delta y_i'$ in (A.6) and (A.7) (since the y_i' are independent variables) gives the following chain rule:

$$\frac{\partial\psi}{\partial y_i'} = \frac{\partial\psi}{\partial y_j}\frac{\partial y_j}{\partial y_i'}, \qquad j = 1, 2, 3. \tag{A.8}$$

or, in terms of differential operators, we may write

$$\frac{\partial}{\partial y_i'}(\ ) = \frac{\partial y_j}{\partial y_i'}\frac{\partial}{\partial y_j}(\ ). \tag{A.9}$$

A similar argument gives

$$\frac{\partial}{\partial y_i}(\ ) = \frac{\partial y_j'}{\partial y_i}\frac{\partial}{\partial y_j'}(\ ). \tag{A.10}$$

Expressions such as

$$\frac{\partial \bar{x}^i}{\partial x^l}\frac{\partial x^s}{\partial \bar{x}^i}$$

regularly appear in this book. We note the appearance of the variables $\bar{x}^i$, with repeated index i, on the top and bottom of this expression. By (A.10) we can immediately equate it, effectively cancelling the $\bar{x}^i$ terms, to

$$\frac{\partial}{\partial x^l}(x^s).$$

Since $\{x^1, x^2, x^3\}$ are an independent set of variables, we have by the nature of partial differentiation,

$$\frac{\partial x^s}{\partial x^l} = \begin{cases} 1, & \text{if } s = l, \\ 0, & \text{if } s \neq l. \end{cases}$$

Hence

$$\frac{\partial x^s}{\partial x^l} = \delta_l^s,$$

and we have

$$\frac{\partial \bar{x}^i}{\partial x^l}\frac{\partial x^s}{\partial \bar{x}^i} = \delta_l^s.$$

Answers to problems

2.1.3 $AA^T = \begin{bmatrix} 62 & 103 \\ 103 & 185 \end{bmatrix}$; $A^TA = \begin{bmatrix} 106 & 120 & 13 \\ 120 & 136 & 14 \\ 13 & 14 & 5 \end{bmatrix}$.

2.1.4 $A = \begin{bmatrix} 1 & 2 & 0 \\ 0 & 3 & -4 \\ 0 & -4 & -1 \end{bmatrix}$.

2.3.3 $A^{-1} = \begin{bmatrix} 1 & \frac{3}{7} & \frac{2}{7} \\ 1 & \frac{4}{7} & \frac{5}{7} \\ 1 & \frac{1}{7} & \frac{3}{7} \end{bmatrix}$, $\begin{bmatrix} x_1 \\ x_2 \\ x_3 \end{bmatrix} = \begin{bmatrix} 1 \\ -2 \\ 1 \end{bmatrix}$.

2.3.4 $\begin{bmatrix} x_1 \\ x_2 \\ x_3 \end{bmatrix} = \begin{bmatrix} 2\alpha \\ \alpha \\ -3\alpha \end{bmatrix}$, with arbitrary α.

3.2.4 $\begin{bmatrix} 1 & 1 & -1 \\ 0 & 1 & -1 \\ 0 & 0 & 1 \end{bmatrix}$.

3.2.5 (i) $\mathbf{i}_1' = -\mathbf{i}_1$, $\mathbf{i}_2' = -\mathbf{i}_2$, $\mathbf{i}_3' = \mathbf{i}_3$.
(ii) $\mathbf{i}_1' = \mathbf{i}_2$, $\mathbf{i}_2' = -\mathbf{i}_1$, $\mathbf{i}_3' = \mathbf{i}_3$.

3.2.6 $S_{11} = 2\mu E_{11} + \lambda(E_{11} + E_{22} + E_{33})$,
$S_{12} = 2\mu E_{12}$, etc.

3.2.15 $\alpha = \frac{\pi}{3}$, $n_1 = n_2 = n_3 = \frac{1}{\sqrt{3}}$.

3.2.17 (a) $\{6, 2, 0\}$; $\left\{\mathbf{i}_1, \dfrac{1}{\sqrt{2}}\mathbf{i}_2 - \dfrac{1}{\sqrt{2}}\mathbf{i}_3, \dfrac{1}{\sqrt{2}}\mathbf{i}_2 + \dfrac{1}{\sqrt{2}}\mathbf{i}_3\right\}$;

$$\begin{bmatrix} 1 & 0 & 0 \\ 0 & \dfrac{1}{\sqrt{2}} & \dfrac{-1}{\sqrt{2}} \\ 0 & \dfrac{1}{\sqrt{2}} & \dfrac{1}{\sqrt{2}} \end{bmatrix}.$$

(b) $\{-1, 1, 4\}$;

$$\left\{\dfrac{1}{\sqrt{2}}\mathbf{i}_2 - \dfrac{1}{\sqrt{2}}\mathbf{i}_3, \dfrac{-2}{\sqrt{6}}\mathbf{i}_1 + \dfrac{1}{\sqrt{6}}\mathbf{i}_2 + \dfrac{1}{\sqrt{6}}\mathbf{i}_3, \dfrac{1}{\sqrt{3}}\mathbf{i}_1 + \dfrac{1}{\sqrt{3}}\mathbf{i}_2 + \dfrac{1}{\sqrt{3}}\mathbf{i}_3\right\};$$

$$\begin{bmatrix} 0 & \dfrac{1}{\sqrt{2}} & \dfrac{-1}{\sqrt{2}} \\ \dfrac{-2}{\sqrt{6}} & \dfrac{1}{\sqrt{6}} & \dfrac{1}{\sqrt{6}} \\ \dfrac{1}{\sqrt{3}} & \dfrac{1}{\sqrt{3}} & \dfrac{1}{\sqrt{3}} \end{bmatrix}.$$

3.2.18 Principal moments of inertia: $4ma^2, 10ma^2, 10ma^2$. Principal axes of inertia: the long diagonal of the cube ($4ma^2$), and any axis through O perpendicular to the long diagonal ($10ma^2$).

3.2.23 (i) (a) $\begin{bmatrix} 1 & 0 & 0 \\ 0 & 0 & 0 \\ 0 & 0 & 0 \end{bmatrix}$;

(b) $\begin{bmatrix} 0 & 1 & 0 \\ 0 & 0 & 0 \\ 0 & 0 & 0 \end{bmatrix}$;

(c) $\begin{bmatrix} 0 & 0 & 0 \\ 1 & 0 & 0 \\ 0 & 0 & 0 \end{bmatrix}$;

(d) $\begin{bmatrix} 1 & 0 & 0 \\ 0 & 2 & 0 \\ 0 & 0 & 0 \end{bmatrix}.$

(a) $(u_1, 0, 0)$;
(b) $(u_2, 0, 0)$;
(c) $(0, u_1, 0)$
(d) $(u_1, 2u_2, 0)$.

(ii) $\begin{bmatrix} 0 & 0 & 3 \\ 8 & 0 & 0 \\ 0 & 21 & 0 \end{bmatrix}.$

3.3.2 (b) $\dfrac{\partial \bar{v}^i}{\partial \bar{x}^j} = \dfrac{\partial \bar{x}^i}{\partial x^k} \dfrac{\partial x^l}{\partial \bar{x}^j} \dfrac{\partial v^k}{\partial x^l} + \dfrac{\partial^2 \bar{x}^i}{\partial x^k \partial x^l} \dfrac{\partial x^l}{\partial \bar{x}^j} v^k.$

References

Anderson, D., Tannehill, J. C. and Pletcher, R. H. (1984) *Computational fluid mechanics and heat transfer*. McGraw-Hill, New York.
Aris, R. (1962) *Vectors, tensors, and the basic equations of fluid mechanics*. Prentice-Hall, Englewood Cliffs, NJ.
Bradshaw, P. *et al.* (eds.) (1976) *Turbulence*. Springer, Berlin.
Daley, B. J. and Harlow, F. H. (1970) Transport equations in turbulence. *Phys. Fluids* **13**, 2634–2649.
Gibson, M. M. and Rodi, W. (1981) A Reynolds-stress closure model of turbulence applied to the calculation of a highly curved mixing layer. *J. Fluid Mech.* **103**, 161–182.
Hinze, J. O. (1959) *Turbulence*. McGraw-Hill, New York.
Hunter, S. C. (1983) *Mechanics of continuous media*, 2nd edn. Ellis Horwood, Chichester.
Jones, C. W. and Watson, E. J. (1963) Two-dimensional boundary layers. In: Rosenhead, L. (ed.) *Laminar boundary layers*. Oxford University Press, Oxford.
Launder, B. E. (1975) On the effects of a gravitational field on the turbulent transport of heat and momentum. *J. Fluid Mech.* **67**, 569–581.
Rodi, W. (1982). *Turbulent buoyant jets and plumes*. Pergamon, Oxford.
Schlichting, H. (1968) *Boundary layer theory*, 6th edn. McGraw-Hill, New York.
Spiegel, M. R. (1959) *Vector analysis*, Schaum's outline series. McGraw-Hill, New York.

FURTHER READING

Goodbody, A. M. (1982) *Cartesian tensors*. Ellis Horwood, Chichester.
Jeffreys, H. (1963) *Cartesian tensors*. Cambridge University Press, London.
Renton, J. D. (1987) *Applied elasticity*. Ellis Horwood, Chichester.
Simmonds, J. G. (1982) *A brief on tensor analysis*. Springer, New York.
Sokolnikoff, I. S. (1964) *Tensor analysis*, 2nd edn. John Wiley, New York.
Spain, B. (1956) *Tensor calculus*. Oliver & Boyd, Edinburgh.
Synge, J. L. and Schild, A. (1949) *Tensor calculus*. University of Toronto Press, Toronto.

Index

acceleration vector in particle kinematics, 145–146
adjoint matrix, 54
    adjoint operator, 100
alternating symbol, 13
    alternating tensor, 83, 102, 125
angular momentum, 76
anti-symmetric, 13
associated components, 93, 97, 98, 99

basis, 7–8
    change of, 19–21
    reciprocal, 23
bipolar co-ordinates, 136–138
Boussinesq approximation, 200
Boussinesq eddy viscosity law, 197

cartesian tensors, 64–85
    alternating, 83
    contraction, 74, 84
    diagonalization, 73, 81
    first-order, 65
    higher-order, 83
    invariants, 73–75, 82
    isotropic, 70,72,83
    principal axes, 81
    second-order, 67
    tensor fields,84
chain rule, 205
characteristic equation, 57
Christoffel symbols
    of the first kind, 113
    of the second kind, 112
    transformation equations, 118
closure problem of turbulence, 180
cofactors, 50
column vectors, 42
components of tensors, 63
    contravariant, 94
    covariant, 94
    mixed, 94
    physical, 129–130,
    rectangular cartesian, 67
constitutive equation for linear elastick solid in
    generalized form, 155
    rectangular cartesians, 155
    spherical polar co-ordinates, 156
constitutive equation for newtonian fluid in
    generalized form, 154
    rectangular cartesians, 154
    spherical polar co-ordinates, 155
continuity equation in
    generalized form, 157
    parabolic cylindrical co-ordinates, 158
    rectangular cartesians, 157
    spherical polar co-ordinates, 158
    toroidal co-ordinates (2), 157
contraction, 74, 84, 103–104
contravariant
    base vectors, 38
    components of second-order tensors, 94

Index

components of vectors, 25, 37, 93
covariant
    base vector, 35
    components of second-order tensors, 94
    components of vectors, 25, 37, 93
covariant derivatives of
    contravariant vectors, 118
    covariant vectors, 119
    second-order and higher-order tensors, 122–123
curl
    in rectangular cartesians, 31, 84
    in generalized form, 121–122
    in orthogonal curvilinear co-ordinates, 130
curvilinear co-ordinate systems, 35–40, 86–89
cylindrical polar co-ordinates, 34–36, 114, 131, 146, 149, 153, 160, 183, 184–191

determinants, 49–52
    differentiation of, 117
diagonalization
    of cartesian tensors, 73, 81–82
    of generalized tensors, 108–110
    of symmetric matrices, 56–62
direction cosines, 11
directional derivative, 29
dissipation function, 170–171
divergence of a vector
    in rectangular cartesians, 31, 74
    in generalized form, 120–121
    in orthogonal curvilinear co-ordinates, 130
dummy suffix, 14
dyadic product, 68–69, 76, 85, 97–98

eddy viscosity, 197, 199
eigenvalues, 56, 109
eigenvectors, 56, 109
    orthogonality, 57–58
elliptic cylindrical co-ordinates, 134–136
energy equations, 169–171
enthalpy transport equation, 171
equations of motion of a continuum in
    generalized form, 163
    rectangular cartesians, 163
    spherical polar co-ordinates, 163
equilibrium equations for a continuum in
    cylindrical polar co-ordinates, 160
    rectangular cartesians, 158
    spherical polar co-ordinates, 160
euclidean spaces, 126

fields
    scalar, 27
    vector, 29, 31
    tensor, 84
finite strain tensor, 151–154

free suffix, 11, 15

general cartesian systems, 22–27
generalized gradient diffusion
    hyphothesis, 200
generalized tensors, 93–110
    alternating, 102
    associated components, 93, 97–99
    contraction, 103–104
    contravariant first-order, 93
    contravariant second-order, 94
    covariant first-order, 93
    covariant second-order, 94
    diagonalization, 108–110
    dyadic products, 97–98
    fields, 111
    higher order, 101
    metric tensor, 39, 95
    mixed second-order, 94
    zero order, 104
gradient vector
    in rectangular cartesians, 28–31, 65–66
    in generalized form, 92
    in orthogonal curvilinear co-ordinates, 130
Green strain tensor, 151–154

heat–flux correlation, 200

identity matrix, 44
    identity operator, 85, 99
index
    raising, 39, 98
    lowering, 39, 98
inertia tensor, 77
infinitesimal strain tensor, 148
internal energy transport equation, 170
invariants, 73–75, 82, 104
inverse matrix, 53–55
isotropic tensors, 70, 72, 83

Jacobian, 86–89

k–ε model, 197
Kronector delta, 11, 25, 70, 95

linear operator, 63, 75

material derivative of velocity in
    cylindrical polar co-ordinates, 162
    generalized form, 161
    orthogonal curvilinear co-ordinates, 162
    rectangular cartesians, 161

Index

spherical polar co-ordinates, 162
matrices, 41–47, 53–62
    addition, 43
    adjoint, 54
    conformable, 44
    diagonalization, 56–62
    eigenvalues and eigenvectors, 56
    identity, 44
    inverse, 53–55
    minors, 49
    multiplication, 44–47
    non-singular, 54
    orthogonal, 56, 58, 78
    proper orthogonal, 56
    singular, 54
    skew-symmetric, 42
    square, 41
    symmetric, 42
    trace, 74
    transpose, 42
metric tensor, 39, 95
mixed components of second-order tensors, 94
momentum (transport) equation in
    generalized form, 172
    rectangular cartesians, 171
    streamline co-ordinates, 172–175

Navier's equation in
    rectangular cartesians, 164
    spherical polar co-ordinates, 165–166
Navier–Stokes equations in
    column vector form, 175–179
    generalized form, 166
    rectangular cartesians, 166
    spherical polar co-ordinates, 168
Navier–Stokes equations, time-averaged (Reynolds equations) in
    cylindrical polar co-ordinates, 183
    generalized form, 181
    rectangular cartesians, 181
non-euclidean spaces, 126
non-singular matrices, 54

oblate spheroidal co-ordinates, 139–140, 146
orthogonal curvilinear co-ordinate systems, 34 127–144
    bipolar co-ordinates, 136–138
    cylindrical polar co-ordinates, 34–36, 114, 131, 146, 149, 153, 160, 183, 184
    elliptic cylindrical co-ordinates, 134–136
    oblate spheroidal co-ordinates, 139–140, 146
    parabolic co-ordinates, 138–139
    parabolic cylindrical co-ordinates, 132–134
    prolate spheroidal co-ordinates, 140
    spherical polar co-ordinates, 34, 36, 115, 131, 146, 150, 155, 156, 157, 160, 163, 168
    streamline co-ordinates, 143–144, 192–196
    toroidal co-ordinates (1), 140–142
    toroidal co-ordinates (2), 142–143, 150
orthogonal matrices, 56, 58, 78
orthonormal, 8
Paplacian, 121
    in generalized form, 121
    in cylindrical polar co-ordinates, 122
    in orthogonal curvilinear co-ordinates, 130
    in spherical polar co-ordinates, 122
parabolic co-ordinates, 138–139
parabolic cylindrical co-ordinates, 132–134
parallelogram law of addition, 2
particle kinematics, 145
permutation identity, 13–14, 103
physical components
    of vectors, 129
    of tensors, 129–130
pressure–strain correlation, 184
principal axes of a symmetric tensor, 81
principal moments of inertia, 81
prolate spheroidal co-ordinates, 140
proper orthogonal matrices, 56

quotient theorem, 75–76, 84, 105–107

rectangular cartesian systems, 7–8
repeated suffix, 14
Reynolds equations, see Navier–Stokes equations, time-averaged
Reynolds stress, 181
Reynolds stress transport equation in
    cylindrical polar co-ordinates, 184–191
    generalized form, 184
    rectangular cartesians, 184
    streamline co-ordinates (for high Reynolds number flows), 191–196
Riemann–Christoffel tensor, 126
rigid body rotation, 76–78
rotation tensor, 78
row vector, 42

scalar, 1
    fields, 27
    product of vectors, 4, 10, 15, 104
    triple product of vectors,, 17–18
scale factors, 127
second-order modelling, 184
singular matrix, 54
skew-symmetric, 13, 17, 69, 108
spectral representation, 110

Index

spherical polar co-ordinates, 34, 36, 115, 131, 146, 150, 155, 156, 157, 160, 163, 165, 168
strain-rate tensor, 148–151
   in cylindrical polars, 149
   in orthogonal curvilinear co-ordinates, 151
   in spherical polar co-ordinates, 150
strain tensor (finite), 151–154
strain tensor (infinitesimal), 148
streamline co-ordinates, 143–144, 192–196
stress
   matrix, 79
   tensor, 154
strong conservation form, 170
substitution operator, 16
suffix
   free, 11
   repeated (dummy), 14
summation convention, 14–16, 26
   symmetric, 17, 69, 107

tensors, *see* cartesian tensors, generalized tensors,
thin layer approximation, 178
toroidal co-ordinates (1), 140–142
toroidal co-ordinates (2), 142–143, 150, 198–199, 201–203
trace, 74
traction, 79
transformation law for
   base vectors, 90–91
   Christoffel symbols, 118
   first-order cartesian tensors, 22, 65
   second-order cartesians tensors, 67–68
   second-order cartesian tensors in two dimensions, 72–73
   higher order cartesian tensors, 83
   contravariant vectors, 91, 92
   contravariant second-order tensors, 94, 96
   covariant vectors, 92, 93
   covariant second-order tensors, 94, 96
   mixed second-order tensors, 94, 96
   higher order tensors, 101, 103
transformations
   between rectangular cartesian co-ordinates, 19–21
   between general co-ordinate systems, 86–89
transport equations for heat–flux correlation (high Reynolds number) in
   generalized form, 201
   rectangular cartesians, 200
   toroidal co-ordinates (2), 201–203
transport equations for a passive scalar, 199
   time-averaged, 199–200
transport equation for turbulent kinetic energy in
   generalized form, 197
   rectangular cartesians, 197
   toroidal co-ordinates (2), 198–199

transpose
   matrix, 42
   of tensor, 99
turbulence modelling, 180
turbluence properties, 180
turbulent kinetic energy dissipation, 191
turbulent kinetic energy Prandtl number, 197

unit matrix, 44

vector, 1
   addition, 2
   components, 9
   field, 29, 31
   magnitude, 3
   position, 6
   product, 5, 15, 105
   standard identities, 33
   triple product, 18
   zero, 3

zero vector, 3

Mathematics and its Applications

Series Editor: G. M. BELL,
Professor of Mathematics, King's College London, University of London

Ball, M.A.	Mathematics in the Social and Life Sciences: Theories, Models and Methods
de Barra, G.	Measure Theory and Integration
Bartak, J., Herrmann, L., Lovicar, V. & Vejvoda, D.	Partial Differential Equations of Evolution
Bejancu, A.	Finsler Geometry and Applications
Bell, G.M. and Lavis, D.A.	Statistical Mechanics of Lattice Models, Vols. 1 & 2
Berry, J.S., Burghes, D.N., Huntley, I.D., James, D.J.G. & Moscardini, A.O.	Mathematical Modelling Courses
Berry, J.S., Burghes, D.N., Huntley, I.D., James, D.J.G. & Moscardini, A.O.	Mathematical Modelling Methodology, Models and Micros
Berry, J.S., Burghes, D.N., Huntley, I.D., James, D.J.G. & Moscardini, A.O.	Teaching and Applying Mathematical Modelling
Blum, W.	Applications and Modelling in Learning and Teaching Mathematics
Brown, R.	Topology: A Geometric Account of General Topology, Homotopy Types and the Fundamental Groupoid
Burghes, D.N. & Borrie, M.	Modelling with Differential Equations
Burghes, D.N. & Downs, A.M.	Modern Introduction to Classical Mechanics and Control
Burghes, D.N. & Graham, A.	Introduction to Control Theory, including Optimal Control
Burghes, D.N., Huntley, I. & McDonald, J.	Applying Mathematics
Burghes, D.N. & Wood, A.D.	Mathematical Models in the Social, Management and Life Sciences
Butkovskiy, A.G.	Green's Functions and Transfer Functions Handbook
Cartwright, M.	Fourier Methods: for Mathematicians, Scientists and Engineers
Cerny, I.	Complex Domain Analysis
Chorlton, F.	Textbook of Dynamics, 2nd Edition
Chorlton, F.	Vector and Tensor Methods
Cohen, D.E.	Computability and Logic
Cordier, J.-M. & Porter, T.	Shape Theory: Categorical Methods of Approximation
Crapper, G.D.	Introduction to Water Waves
Cross, M. & Moscardini, A.O.	Learning the Art of Mathematical Modelling
Cullen, M.R.	Linear Models in Biology
Dunning-Davies, J.	Mathematical Methods for Mathematicians, Physical Scientists and Engineers
Eason, G., Coles, C.W. & Gettinby, G.	Mathematics and Statistics for the Biosciences
El Jai, A. & Pritchard, A.J.	Sensors and Controls in the Analysis of Distributed Systems
Exton, H.	Multiple Hypergeometric Functions and Applications
Exton, H.	Handbook of Hypergeometric Integrals
Exton, H.	q-Hypergeometric Functions and Applications
Faux, I.D. & Pratt, M.J.	Computational Geometry for Design and Manufacture
Firby, P.A. & Gardiner, C.F.	Surface Topology
Gardiner, C.F.	Modern Algebra
Gardiner, C.F.	Algebraic Structures
Gasson, P.C.	Geometry of Spatial Forms
Goodbody, A.M.	Cartesian Tensors
Goult, R.J.	Applied Linear Algebra
Graham, A.	Kronecker Products and Matrix Calculus: with Applications
Graham, A.	Matrix Theory and Applications for Engineers and Mathematicians
Graham, A.	Nonnegative Matrices and Applicable Topics in Linear Algebra
Griffel, D.H.	Applied Functional Analysis
Griffel, D.H.	Linear Algebra and its Applications: Vol. 1, A First Course; Vol. 2, More Advanced
Guest, P. B.	The Laplace Transform and Applications
Hanyga, A.	Mathematical Theory of Non-linear Elasticity
Harris, D.J.	Mathematics for Business, Management and Economics
Hart, D. & Croft, A.	Modelling with Projectiles
Hoskins, R.F.	Generalised Functions
Hoskins, R.F.	Standard and Nonstandard Analysis
Hunter, S.C.	Mechanics of Continuous Media, 2nd (Revised) Edition
Huntley, I. & Johnson, R.M.	Linear and Nonlinear Differential Equations
Irons, B. M. & Shrive, N. G.	Numerical Methods in Engineering and Applied Science
Ivanov, L. L.	Algebraic Recursion Theory
Johnson, R.M.	Theory and Applications of Linear Differential and Difference Equations
Johnson, R.M.	Calculus: Theory and Applications in Technology and the Physical and Life Sciences
Jones, R.H. & Steele, N.C.	Mathematics in Communication Theory
Jordan, D.	Geometric Topology
Kelly, J.C.	Abstract Algebra
Kim, K.H. & Roush, F.W.	Applied Abstract Algebra
Kim, K.H. & Roush, F.W.	Team Theory
Kosinski, W.	Field Singularities and Wave Analysis in Continuum Mechanics
Krishnamurthy, V.	Combinatorics: Theory and Applications
Lindfield, G. & Penny, J.E.T.	Microcomputers in Numerical Analysis
Livesley, K.	Mathematical Methods for Engineers
Lord, E.A. & Wilson, C.B.	The Mathematical Description of Shape and Form
Malik, M., Riznichenko, G.Y. & Rubin, A.B.	Biological Electron Transport Processes and their Computer Simulation
Martin, D.	Manifold Theory: An Introduction for Mathematical Physicists
Massey, B.S.	Measures in Science and Engineering
Meek, B.L. & Fairthorne, S.	Using Computers

Mathematics and its Applications
Series Editor: G. M. BELL,
Professor of Mathematics, King's College London, University of London

Menell, A. & Bazin, M.	Mathematics for the Biosciences
Mikolas, M.	Real Functions and Orthogonal Series
Moore, R.	Computational Functional Analysis
Moshier, S.L.B.	Methods and Programs for Mathematical Functions
Murphy, J.A., Ridout, D. & McShane, B.	Numerical Analysis, Algorithms and Computation
Nonweiler, T.R.F.	Computational Mathematics: An Introduction to Numerical Approximation
Norcliffe, A. & Slater, G.	Mathematics of Software Construction
Ogden, R.W.	Non-linear Elastic Deformations
Oldknow, A.	Microcomputers in Geometry
Oldknow, A. & Smith, D.	Learning Mathematics with Micros
O'Neill, M.E. & Chorlton, F.	Ideal and Incompressible Fluid Dynamics
O'Neill, M.E. & Chorlton, F.	Viscous and Compressible Fluid Dynamics
Page, S. G.	Mathematics: A Second Start
Prior, D. & Moscardini, A.O.	Model Formulation Analysis
Rankin, R.A.	Modular Forms
Sewell, G.	Computational Methods of Linear Algebra
Scorer, R.S.	Environmental Aerodynamics
Sharma, O.P.	Markovian Queues
Shivamoggi, B.K.	Stability of Parallel Gas Flows
Smith, D.K.	Network Optimisation Practice: A Computational Guide
Srivastava, H.M. & Manocha, L.	A Treatise on Generating Functions
Stirling, D.S.G.	Mathematical Analysis
Sweet, M.V.	Algebra, Geometry and Trigonometry in Science, Engineering and Mathematics
Temperley, H.N.V.	Graph Theory and Applications
Temperley, H.N.V.	Liquids and Their Properties
Thom, R.	Mathematical Models of Morphogenesis
Toth, G.	Harmonic and Minimal Maps and Applications in Geometry and Physics
Townend, M. S.	Mathematics in Sport
Townend, M.S. & Pountney, D.C.	Computer-aided Engineering Mathematics
Trinajstic, N.	Mathematical and Computational Concepts in Chemistry
Twizell, E.H.	Computational Methods for Partial Differential Equations
Twizell, E.H.	Numerical Methods, with Applications in the Biomedical Sciences
Vince, A. and Morris, C.	Discrete Mathematics for Computing
Walton, K., Marshall, J., Gorecki, H. & Korytowski, A.	Control Theory for Time Delay Systems
Warren, M.D.	Flow Modelling in Industrial Processes
Wheeler, R.F.	Rethinking Mathematical Concepts
Willmore, T.J.	Total Curvature in Riemannian Geometry
Willmore, T.J. & Hitchin, N.	Global Riemannian Geometry

Statistics, Operational Research and Computational Mathematics
Editor: B. W. CONOLLY,
Emeritus Professor of Mathematics (Operational Research), Queen Mary College, University of London

Abaffy, J. & Spedicato, E.	ABS Projection Algorithms: Mathematical Techniques for Linear and Nonlinear Equations
Beaumont, G.P.	Introductory Applied Probability
Beaumont, G.P.	Probability and Random Variables
Conolly, B.W.	Techniques in Operational Research: Vol. 1, Queueing Systems
Conolly, B.W.	Techniques in Operational Research: Vol. 2, Models, Search, Randomization
Conolly, B.W.	Lecture Notes in Queueing Systems
Conolly, B.W. & Pierce, J.G.	Information Mechanics: Transformation of Information in Management, Command, Control and Communication
French, S.	Sequencing and Scheduling: Mathematics of the Job Shop
French, S.	Decision Theory: An Introduction to the Mathematics of Rationality
Griffiths, P. & Hill, I.D.	Applied Statistics Algorithms
Hartley, R.	Linear and Non-linear Programming
Jolliffe, F.R.	Survey Design and Analysis
Jones, A.J.	Game Theory
Kapadia, R. & Andersson, G.	Statistics Explained: Basic Concepts and Methods
Lootsma, F.	Operational Research in Long Term Planning
Moscardini, A.O. & Robson, E.H.	Mathematical Modelling for Information Technology
Moshier, S.L.B.	Mathematical Functions for Computers
Oliveira-Pinto, F.	Simulation Concepts in Mathematical Modelling
Ratschek, J. & Rokne, J.	New Computer Methods for Global Optimization
Schendel, U.	Introduction to Numerical Methods for Parallel Computers
Schendel, U.	Sparse Matrices
Sehmi, N.S.	Large Order Structural Eigenanalysis Techniques: Algorithms for Finite Element Systems
Späth, H.	Mathematical Software for Linear Regression
Stoodley, K.D.C.	Applied and Computational Statistics: A First Course
Stoodley, K.D.C., Lewis, T. & Stainton, C.L.S.	Applied Statistical Techniques
Thomas, L.C.	Games, Theory and Applications
Whitehead, J.R.	The Design and Analysis of Sequential Clinical Trials